Advances in Geological Science

Studies in the twentieth century uncovered groundbreaking facts in geophysics and produced a radically new picture of the Earth's history. However, in some respects it also created more puzzles for the research community of the twenty-first century to tackle. This book series aims to present the state of the art of contemporary geological studies and offers the opportunity to discuss major open problems in geosciences and their phenomena. The main focus is on physical geological features such as geomorphology, petrology, sedimentology, geotectonics, volcanology, seismology, glaciology, and their environmental impacts. The monographs in the series, including multi-authored volumes, will examine prominent features of past events up to their current status, and possibly forecast some aspects of the foreseeable future. The guiding principle is that understanding the fundamentals and applied methodology of overlapping fields will be key to paving the way for the next generation.

More information about this series at http://www.springer.com/series/11723

Yukio Himiyama · Kenji Satake ·
Taikan Oki
Editors

Human Geoscience

Springer

Editors
Yukio Himiyama
Hokkaido University of Education
Sapporo, Japan

Taikan Oki
Institute for Future Initiatives
The University of Tokyo
Tokyo, Japan

Kenji Satake
Earthquake Research Institute
The University of Tokyo
Tokyo, Japan

ISSN 2524-3829 ISSN 2524-3837 (electronic)
Advances in Geological Science
ISBN 978-981-32-9223-9 ISBN 978-981-32-9224-6 (eBook)
https://doi.org/10.1007/978-981-32-9224-6

This Springer imprint is published by the registered company Springer Nature Singapore Pte Ltd.
The registered company address is: 152 Beach Road, #21-01/04 Gateway East, Singapore 189721, Singapore

Preface

Human geoscience is a new genre of geoscience, or earth science, which is committed to sustainability of human existence on the earth. It is hence concerned not only with the natural phenomena that occur on the surface of the earth, but also with their relations with human activities. It is related with humanities, social sciences, environmental sciences, agricultural sciences, and engineering, particularly, in relation to disaster prevention and mitigation, as well as with more traditional fields of geoscience including physical geography, geomorphology, geology, soil science, climatology, meteorology, hydrology, seismology, and volcanology. They are all important fields for resource use and management, geo disaster prevention and mitigation, and tackling global environmental problems for enhancement of sustainability.

Human geoscience started to take shape in Japan in the early 2000s and evolved while the world science community was advocating "science for society". In fact, "earth science for society" was the slogan of the United Nation's International Year of Planet Earth (IYPE) initiated by UNESCO and the International Union of Geological Sciences (IUGS) in order to help building safer, healthier, and wealthier societies by ensuring greater and more effective use of the knowledge held by earth scientists. It was hence timely for the geoscience community in Japan to establish Human Geoscience Committee in Science Council of Japan (SCJ) in 2006, and Human Geosciences Section in Japan Geoscience Union (JpGU) soon afterwards. In fact, they played important roles in the gigantic disasters that followed in Japan, such as the Great East Japan Disaster in March 2011.

This book represents the joint efforts of the members of the Human Geoscience Committee of SCJ. The Committee developed and promoted human geoscience through various activities, such as organizing SCJ open symposia, compiling official SCJ proposals, making the "road map" of human geoscience, which all contributed to intra-disciplinary understanding within human geoscience, interdisciplinary cooperation with broader geoscience and other sciences, and to trans-disciplinary challenges. Human geoscience thus developed has turned out to be uniquely relevant for supporting, promoting, and relating with each other the on-going global sustainability programmes such as Future Earth, United Nation's

Sustainable Development Goals (SDGs), and Education for Sustainable Development (ESD), as the reader will find in this book.

The publication of this book coincides with the birth of the International Science Council (ISC), the new scientific body created by the merger of the International Council for Science (ICSU) and the International Social Science Council (ISSC), in July 2018. The birth of the ISC is creating a new scope for sustainability sciences, such as human geoscience. It is hoped that the readers of this book join the task of constructing human geoscience, which is evolving, and has to evolve further for sustainable development of the world. This book is designed for those who may not necessarily have geoscience background but have broader scientific interests in understanding the causes, mechanisms, and consequences of geo-disasters and global environmental problems, and wish to make the world more sustainable based on them.

Sapporo, Japan
March 2019

Yukio Himiyama
Emeritus Professor

Contents

Contributors

Keiji Asano CTI Engineering Co., Ltd., Chuo-ku, Japan

Joel Bandibas Geological Survey of Japan, National Institute of Advanced Industrial Science and Technology, Tsukuba, Japan

Masahiro Chigira Disaster Prevention Research Institute, Kyoto University, Gokasho, Uji, Japan

Ronald C. Estoque National Institute for Environmental Studies, Tsukuba City, Ibaraki, Japan

Toshitaka Gamo Atmosphere and Ocean Research Institute, The University of Tokyo, Kashiwa, Japan

Akira Hama Graduate Course of Sciences, Chiba University, Chiba, Japan

Shigeko Haruyama Mie University, Tsu, Mie, Japan

Yukio Himiyama Hokkaido University of Education, Sapporo, Japan

Yoshiyasu Ida University of Tsukuba, Tsukuba, Japan

Tomokazu Ishihara Yamaguchi, Japan

Koki Iwao Geological Survey of Japan, National Institute of Advanced Industrial Science and Technology, Tsukuba, Japan

Natsuko Kasirajima PKUTECH Co., Ltd., Tokyo, Japan

Chihiro Kato Hirosaki University, Hirosaki, Japan

Akio Kitoh Japan Meteorological Business Support Center, Tsukuba, Japan

Dai Komatsubara Konoike Construction Co., Ltd., Ishinomaki, Japan

Akihiko Kondoh Center for Environmental Remote Sensing, Chiba University, Chiba, Japan

Tsuyoshi Miyazaki The University of Tokyo, Tokyo, Japan

Yuji Murayama Faculty of Life and Environmental Sciences, University of Tsukuba, Tsukuba City, Ibaraki, Japan

Setsuya Nakada National Research Institute for Earth Science and Disaster Resilience, Tsukuba, Japan

Takashi Oguchi Center for Spatial Information Science, The University of Tokyo, Tokyo, Japan

Taikan Oki Institute for Future Initiatives, The University of Tokyo, Tokyo, Japan

Kenji Satake Earthquake Research Institute, The University of Tokyo, Tokyo, Japan

Toshikazu Shinya Dia Consultant Co., Ltd., Fukuoka, Japan

Koichi Suzuki Hokkaido University, Sapporo, Japan

Yasuhiro Suzuki Disaster Mitigation Research Center, Nagoya University, Nagoya, Japan

Shinji Takarada Geological Survey of Japan, National Institute of Advanced Industrial Science and Technology, Tsukuba, Japan

Kazuhiro Tanaka Yamaguchi University, Yamaguchi, Japan

Shingo Tokuyasu NC Geophysical Survey Co., Ltd., Chiyoda-ku, Japan

Eikichi Tsukuda Geological Survey of Japan, National Institute of Advanced Industrial Science and Technology, Tsukuba, Japan

Yoshiki Wakabayashi Department of Geography, Tokyo Metropolitan University, Tokyo, Japan

Masatoshi Watanabe Soja, Japan

Koshun Yamaoka Graduate School of Environmental Studies, Nagoya University, Nagoya, Japan

Part I
Introduction

Chapter 1
Emergence of Human Geoscience

Yukio Himiyama

Abstract Restructuring of Science Council of Japan in 2005 gave an opportunity to a broad range of geosciences to work together in Japan, and one of its outcomes was the birth of a new discipline of geoscience, which was later called the *human geoscience*. It is the science of natural phenomena that occur on the surface of the earth and their relations with human activities. The persistent threat of global environmental problems and geo disasters have made geoscientists in Japan aware of the necessity of developing this new discipline of science together in order to tackle these problems. The newly established Human Geoscience Committee of Science Council of Japan, with support from the Human Geosciences Section of Japan Geoscience Union (JpGU), has played major role in the creation and development of this new science through its various activities, such as compilation of official proposals, organization of open symposia, presentation of the *human geoscience* roadmap, publications, and committee meetings themselves, that all contributed greatly to intra-disciplinary, interdisciplinary, and trans-disciplinary understanding and cooperation. *Human geoscience* is thus growing not only for itself as a strongly interdisciplinary science for sustainability, but also as a "science for society".

Keywords Human geoscience · Earth science · Disaster · Global environmental problem · Sustainable development

1.1 Introduction

The world has been facing various serious global environmental problems and geo disasters threatening its sustainability, and the International Council for Science (ICSU), together with the International Social Science Council (ISSC) and the other related organizations, has been leading the world academic community in organizing and re-organizing international research programmes for sustainability, such as IGBP (International Geosphere-Biosphere Programme 1986–2015), IHDP (International Human Dimensions Programme 1990–2015) and Future Earth (2015-) (Future Earth

Y. Himiyama (✉)
Hokkaido University of Education, Sapporo, Japan
e-mail: himiyamay@kkd.biglobe.ne.jp

Y. Himiyama et al. (eds.), *Human Geoscience*, Advances in Geological Science,
https://doi.org/10.1007/978-981-32-9224-6_1

2014). When Science Council of Japan (SCJ) was restructured in 2005, Japan's contribution to such international programmes was seriously interrupted because the new SCJ initially lacked proper structure to support them. It was the Earth and Planetary Science Committee that was desired to fill the gap. Earth and planetary science, which was one of the thirty disciplines officially acknowledged by SCJ, was broadly divided into two sub-disciplines, namely "earth and planetary sphere science" and "earth and human sphere science". The latter, which was considered as the science of the natural phenomena that occur on the surface of the earth and their relations with human activities was later-named as a more straight-forward English term "*human geoscience*". It was used publicly for the first time in the English version of the Road Map of this discipline proposed according to the recommendation of SCJ in 2011 (SCJ 2011).

Japan Geoscience Union (JpGU), which was created in 2005 as the representative body of the broad geoscience community in Japan in order to support this discipline in SCJ, established five academic sections, namely Space and Planetary Sciences Section, Atmospheric and Hydrospheric Sciences Section, Solid Earth Sciences Section, Biogeosciences Section, and a section which was later called Human Geosciences Section. In this way, the Human Geoscience Committee of SCJ and the Human Geosciences Section of JpGU were born almost simultaneously and worked closely together for the progress of this new discipline. The two bodies are concerned about the academic development of this science, and about practical means to counter geo-disasters and to alleviate environmental problems such as those related to global warming.

Human geoscience thus started may be characterized by its holistic views and diverse research methods originated in various disciplines of natural sciences, social sciences, and engineering. The roots of *human geoscience* are found in more traditional geo-sciences, such as geography, geology, geomorphology, geophysics, hydrology, oceanology, and soil sciences. Consequently, the research often involves field surveys with observation, measurement, interviews, recording, and mapping, as well as data compilation, storage, management, analyses, and modeling in order to enhance understanding and make predictions, and to offer knowledge, ideas, and proposals to the society.

Among the leading scholars of them, Arthur Newell Strahler (1918–2002), an American geomorphologist, geologist and physical geographer, is noted as an outstanding pioneer of *human geoscience*, although he didn't use the term. He realized and showed the importance of multi-disciplinary, multi-scale and multi-regional comprehensive views, understanding in physical processes on the earth's surface and human action, starting from the earth's planetary setting, gravity, spherical form, rotation and tilt of its axis, and reaching even cultural and philosophical perspective, in his numerous books of physical geography, such as *Elements of Physical Geography* (Strahler and Strahler 1976).

More recently, the United Nations General Assembly declared 2008 as the UN International Year of Planet Earth (IYPE) in order to increase awareness of the importance of earth sciences for sustainable development (UN 2006). It was initiated by the International Union of Geological Sciences (IUGS) in 2011, and was developed with the Earth Science Division of UNESCO. The basic concept of IYPE, namely, "earth sciences for society" and the whole plan of IYPE indicated the necessity for geosciences to cooperate with each other, to widen the scope of geoscience, and to work with and for the society. For these purposes, Japan National Committee for IYPE was established, and a large number of geo-scientists with diverse backgrounds were involved in its activities. IYPE in no doubt had a great impact on the progress of *human geoscience*.

With such backgrounds, *human geoscience* has been evolving fast in Japan in the present century under the leadership of the Human Geoscience Committee of SCJ. It played major role in the creation and development of this new science through its various activities, such as presentation of *human geoscience* roadmap, compilation of official proposals, organization of open symposia, publications, and committee meetings themselves that all contributed greatly to intra-disciplinary understanding within *human geoscience*, interdisciplinary cooperation with broader geoscience and other sciences, and to trans-disciplinary challenges. The progress of this new science can be traced by reviewing some of these activities, as discussed below.

1.2 Roadmap of *Human Geoscience*

The ICSU proposed a plan of earth system science for global sustainability called the "Grand Challenges" in 2010 (ICSU 2010). It was a timely initiative for those who were seeking a break-through for their sometimes stagnant global sustainability research. At about the same time, SCJ asked each of its thirty disciplinary committees to present its roadmap designed for the coming thirty years in 2010. In response to this call, the Human Geoscience Committee, with assistance from the JpGU Human Geosciences Section, issued a *human geoscience* road map as part of more inclusive roadmaps of geosciences, in 2011 (SCJ 2011). Figure 1.1 is its English version produced by the present writer. The 2011 roadmap of *human geoscience* was made so that it incorporates the ICSU/ISSC Grand Challenges within it, with its five "Grand Challenges" listed in Table 1.1 as the driving wheel for sustainability during the anticipated ten-year period of the Grand Challenges activated by circulation of human, information and knowledge supported by research-education network and partnership desired to be developed into the second and the third ten-year phases.

The 2011 *Human Geoscience* Road Map was updated in 2014 with the following amendments (SCJ 2014a):

(1) "Grand Challenges" was changed to "Future Earth", following the creation of the latter out of the former. As a result, problem finding/solving aspect was made more explicit.
(2) "Risk management study" was refined in order to strengthen the research related with disaster.

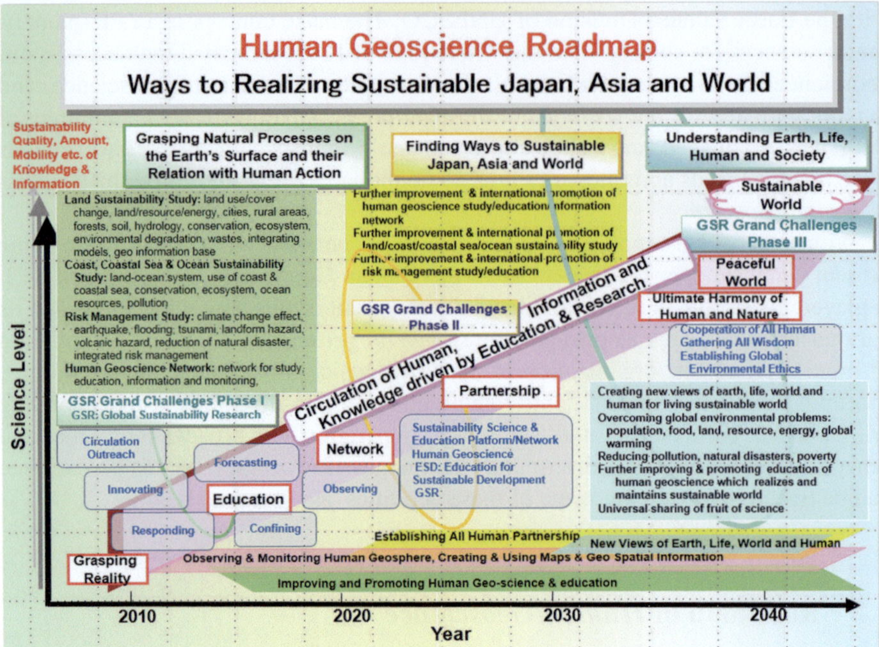

Fig. 1.1 2014 human geoscience road map

Table 1.1 Grand challenges

Challenge 1	Forecasting—improve the usefulness of forecasts of future environmental conditions and their consequences for people
Challenge 2	Observing— develop, enhance, and integrate the observation systems needed to manage global and regional environmental change
Challenge 3	Confining— determine how to anticipate, recognize, avoid and manage disruptive global environmental change
Challenge 4	Responding— determine what institutional economic, and behavioral changes can enable effective steps toward global sustainability
Challenge 5	Innovating— encourage innovation (coupled with sound evaluation mechanisms) in developing technological, policy, and social responses to achieve global sustainability

After ICSU (2010)

The revised contents of the three phases of the 2014 roadmap are outlined as follows:

Phase I (ca. 2014–2023): To grasp the situation and problems of human geosphere and find ways to solve the problems, and to enhance research and education on *human geoscience*, particularly on the following themes, and ensure contribution to the world sustainability, by working closely with Future Earth:

(1) Land sustainability study: including land use/cover change, land/resource/energy, cities, rural areas, forests, soil, hydrology, environmental conservation, ecosystem protection, environmental degradation, wastes, integrative models, geo information bases, etc.
(2) Coast, coastal sea & ocean sustainability study: including land-coastal sea system, use of coast and coastal sea, environmental conservation, ecosystem protection, ocean resources, pollution, and its removal.
(3) Hazard study: including effect of climate change, earthquake, flooding, tsunami, landform hazard, volcanic hazard, etc.
(4) Integrated study of disaster risk: including establishment of the domestic research core for such international programme as IRDR (Integrated Research on Disaster Risk), scientific investigation and evaluation of the causes of disasters, integration of related data and information, risk governance, resilience study built on regional characteristics, investigation of nature-society-human interrelationships in relation to disaster risk and its contribution to disaster prevention or reduction, interdisciplinary cooperation, consolidation of science-society relation, etc.
(5) Study of mitigation of crustal disasters: including prediction of earthquake and volcanic eruption hazard as disaster study, such as governmental "monitoring research program on earthquake and volcanic eruption for mitigation of disaster".
(6) *Human geoscience* network of research, education, and information: including monitoring network of researchers, schools and citizens, ESD (Education for Sustainable Development), *human geoscience* education, identification and promotion of good practices.

Phase II (ca. 2024–2033): To build on the achievements of Phase I and develop them further, and to seek ways to sustainable Japan, Asia, and the World by establishing universal partnership of the whole humanity for sustainability, with particular emphases on the following themes:

(1) Further improvement and international promotion of *human geoscience* study/education/information network.
(2) Further improvement and international promotion of land/coast/coastal sea/ocean sustainability study of risk management.
(3) Further improvement and international promotion of risk management study and education.
(4) Consolidation and promotion of good practices.

Phase III (ca. 2034–2043): To build on the achievements of Phase I & II, consolidating *human geoscience* research/education and developing them further worldwide, and to contribute to the achievement of the following targets by realizing harmony of the whole humanity, gathering of all wisdom and establishment of viable global environmental ethics:

(1) Creating new views of earth, life, world and human for living sustainable world.
(2) Overcoming global environmental problems, such as those related with population, food, land, resource, energy, and global warming.
(3) Reducing pollution, natural disasters, and poverty.
(4) Further improving & promoting education of *human geoscience* which realizes and maintains sustainable world.
(5) Universal sharing of fruit of science.

In short, the target of *human geoscience* in Japan is, as demonstrated in the 2014 roadmap, to make maximum contribution to realization of peaceful, sustainable Japan, Asia, and World, where human and nature maintains harmony. The 2014 version of the *human geoscience* roadmap is expected to be revised in 2019.

1.3 Proposals Compiled by the Human Geoscience Committee

SCJ is a special governmental body belonging to the Cabinet Office with high-level of independence. There are about 2000 members and 210 council members representing some 840,000 scientists in Japan ranging over all fields of sciences including humanities, social sciences, life sciences, natural sciences, and engineering. There are about 300 committees entitled to issue official proposals to the central or local governments, to the society, or to the science or education communities.

SCJ issues a number of proposals of several different categories each year, usually directed to the central government ministries, local government departments, scientific or educational sectors, or to the society. SCJ regards compilation of official proposals as a high priority duty of its committees. The draft of a proposal has to go through the sophisticated referee system and has to be approved by the SCJ Executive Committee. Because of it, however, the SCJ proposals maintain high quality, credibility, and influence in Japan.

Compilation of a proposal, therefore, requires a substantial amount of work by the committee concerned. In the case of the Human Geoscience Committee, which consists of around 35 members representing a broad range of *human geosciences*, it offered an irreplaceable opportunity to cooperate, to exchange views and information, to enhance understanding each other, and to challenge broader range of issues in more comprehensive and integrative ways than otherwise. The committee meetings, which are held four to five times annually, are a very important place where the leading specialists with different backgrounds gather, exchange views and information, and

Table 1.2 Proposals compiled by the human geoscience committee of SCJ

(1) June 2008	Towards sustainable co-existence of nature and human on the land and in the coastal sea (SCJ 2008)
(2) January 2013	Towards sharing of geo-information—development of laws on geo-information for construction of safe and secure society (SCJ 2013a)
(3) September 2014	Towards formation of safe, secure and sustainable society based on the lessons of the Great East Japan Earthquake/Tsunami Disaster of March 2011 (SCJ 2014b)
(4) August 2017	Proposal—Promotion of cooperation and co-creation by science and society for reduction of disasters and formation of sustainable society (SCJ 2017a)

work together. A twenty minutes lecture by a member of the Committee at its every ordinary meeting also helped mutual understanding within the Committee.

The interdisciplinary character and membership of the Human Geoscience Committee made it uniquely well-prepared to cover a wide range of disasters and global environmental problems. The Committee started to function in 2006, and compiled four proposals in ten years, as listed in Table 1.2. The following section outlines the first proposal compiled by the Human Geoscience Committee, which marked the birth of *human geoscience*, in order to show how the new, more problem/solution oriented integrated science has been formed.

1.3.1 SCJ Proposal—Towards Sustainable Co-existence of Nature and Human on the Land and in the Coastal Sea

The above-titled SCJ Proposal was compiled by the Human Geoscience Committee of SCJ and was issued in September 2008 (SCJ 2008). Land and coastal sea are where most of the human activities, including production, consumption, and living, take place. They offer food, minerals, water, and energy, and are also used as dumping places for unwanted wastes and other materials. There are many natural disasters and serious global environmental problems and resource problems which threaten sustainable co-existence of nature and human. Within the land, the coastal areas, particularly coastal plains, are of particular concern because of high concentration of human activities, and consequently of disasters as well, there. In Japan coastal plains are often known as high-risk areas with many disasters caused by floods, earthquakes, and tsunamis.

The urgency of the increasing vulnerability to natural hazards, resource development and use, and environmental protection has been pointed in a number of reports. The UNEP's report in 2007 entitled GEO-4, for example, warned the limit of resources as well as the danger of global warming and stressed the importance of a joint effort of the government and the society to overcome the problems (UNEP

2007). It reflected the serious environmental constraints related with rapid increase of the world population, consumption, and waste, and economic development of developing countries such as China and India. It also warned the increase of disasters due to such global environmental changes as global warming, population increase, urbanization or globalization, and delay in grasping the evolving situation and taking countermeasures.

At present, we lack reasonable monitoring systems nor data bases to grasp the situation of the global environmental problems or large-scale disasters accurately and in time. The research on land and coastal sea as a continuous system based on the data bases compiled in the way mentioned as above is delayed, and it hinders the discussion of desirable use and development of land-coastal sea system from comprehensive viewpoints.

With these backgrounds, the proposal stated that in order to solve the global environmental problems and natural disasters problems deeply rooted in the limits of the land and the coastal sea, first we should provide geo-information bases of the land and the coastal sea needed for grasping the situation of them; second we should pursue research by using the above-mentioned geo-information bases for solving the problems of the land and the coastal sea; and third we should pursue policies and education based on the above research results, and drive this cycle based on the concept of "integration of information cycle driven by science" which was proposed by SCJ in its 2002 Report titled "Japan Perspective" (SCJ 2002).

The main part of the proposal is outlined below:

(1) **Provision of geo-information bases related with global environment and disasters on the land and in the coastal sea**: First, the governmental bodies in charge of geo-information, such as the Ministry of Land, Infrastructure and Transport (MLIT), should collect geo-information intensively in cooperation with such international organizations as GEO (Group on Earth Observations), and provide the geo-information bases for easy evaluation, selection and use without delay. Second, the governmental bodies concerned should facilitate legal system that respect protection of personal information, intellectual property right and secondary use of the data, so that many stake holders can contribute positively to collection, use and preservation of the geo-information, without delay. Third, universities and research institutions should be encouraged to do research on monitoring of global environmental change and the related information, and on the development of information gathering technology and systems, and funding bodies should support them financially.

(2) **Promotion of interdisciplinary studies for grasping the situation, and solving the problems, of the land and the coastal sea: In order to protect the land and the coastal sea and to realize their sustainable use, we should promote the following research**: First, comprehensive research on the land and the coastal sea that contributes to solution of the problems rooted in their limit as resource on the earth and their degradation should be carried out on the land and at the coastal sea, with the land and the coastal sea treated as a contiguous system, and with multiple factors taken into account integrally. Second, in

order to avoid or mitigate disasters that are increasing due to the rise of sea level and increase of abnormal weather generated by global warming or rapid socio-economic change, research for relevant land use, as well as more conventional disaster prevention research, should be promoted. Third, comprehensive research on land use and disaster prevention should be carried out in cooperation with such international global environmental programmes as IHDP, IGBP, LOICZ (Land–Ocean Interactions in the Coastal Zone) or GLP (Global Land Project).

(3) **Towards sustainable and safe use and development of the land and the coastal sea**: MLIT, Ministry of Education, Culture, Sports, Science and Technology (MEXT), Ministry of Agriculture, Forestry and Fisheries, Ministry of the Environment, and the other ministries concerned should make use of the findings of the research carried out as above, and should do the followings: First, they should pursue comprehensive scheme of land use and development based on the understanding of the limit of this sphere, where diverse uses are competing with each other, and it is desired that this concept is respected in the Government's "National Land Use Plan" or "Basic Plan for the Ocean"; Second, a new type of disaster prevention policy relevant in the era of global environmental crises, should be pursued under such comprehensive scheme, with close cooperation of land use/development scheme, regional disaster prevention scheme, and environmental conservation scheme; Third, they should enrich geography education, geology education, environmental education, and education for disaster prevention in school education and in social education.

1.3.2 SCJ Proposal—Promotion of Cooperation and Co-creation by Science and Society for Reduction of Disasters and Formation of Sustainable Society

If the 2008 SCJ Proposal shown above represents the state of *human geoscience* at its early stage, the above-titled Proposal of 2017 represents where *human geoscience* in Japan had reached by 2017. Japan is a country with many big disasters and limited natural resources, and many researchers are involved in the studies of reduction and mitigation of disaster risks and in those related with formation of sustainable society. However, a good research may not always contribute to the society, nor being appreciated by the society. There may be communication gap between science and society, or the society may be preoccupied by daily social or economic affairs. Therefore, the proposal first re-affirms the necessity of addressing disaster prevention and formation of sustainable society in close cooperation, and discusses the ways to promote them.

The proposal identified the following four types of cases the science-society cooperation and co-creation encountered in the effort of reducing disasters and achieving sustainability:

(1) The cases where short-term and medium-term risks are easy to perceive from the disaster perception/prevention view point, e.g., strong earthquakes, big tsunamis, typhoons, etc.
(2) The cases where medium-term and long-term risks are easily perceived from the global environment view point, e.g., global warming.
(3) The cases where risks tend to be overlooked in Japan, e.g., soil problems, population problems, water problems, land problems, etc.
(4) The cases related with construction of risk-sensitive society.

Based on these analyses, the proposal stated that reduction of disasters and formation of sustainable society should be aimed at with the broad view of the SDGs (Sustainable Development Goals) and attention to the kinds of risks that tended to be overlooked. The main part of the proposal is as follows:

(1) **Provision of the *places* (including opportunity) for cooperation and co-creation of science and society towards reduction of disasters and formation of sustainable society**: First, the Cabinet Office, local governments, research, and educational institutions should get rid of bureaucracy and cooperate at national and regional levels and expand the *places* for cooperation and co-creation of science and society. Second, focusing on the *places* of people's living, science and administration should send out information on what is likely to happen where at the time of future hazard from the local residents' perspective. Third, enhance the function of regional educational and cultural activities, such as UNESCO Schools and Geoparks, and museums as a *place* for cooperation and co-creation of science and society.
(2) **Promote compilation, disclosure, and visualization of regional information for reduction of disasters and formation of sustainable world**: First, MLIT and the other ministries concerned and local governments should collect, compile, disclose and use regional information such as geological information, historical records and tsunami deposits that are useful for reduction of disasters. Second, MLIT and the other ministries concerned and local governments should collect, compile, disclose and use regional information such as those of land form, geology, hydrology, biology, land use and pollution that are useful for formation of sustainable society. Third, MLIT and the other ministries concerned and local governments should develop technologies of high resolution three-dimensional regional geographic information systems that can be used effectively by the local administration and residents.
(3) **Provision of the opportunities of education and learning that form bases of cooperation and co-creation of science and society**: First, MEXT and the other ministries concerned and local governments should bring up science interpreters who can assist communication between science and society, systematically. Second, scientists involved in reduction of disasters or formation of sus-

tainable society should put themselves on the field or locality concerned, grasp the thoughts of local people and governments accurately, communicate research results to them in understandable ways, and make effort for mutual understanding. Such attitude should be desirably put into practice in university education and life-long education. Third, MEXT and the other ministries concerned should make effort to improve the contents of learning/education and education environment in order to enhance people's interest and basic knowledge in disasters and global environmental problems. In school education, cooperation of such subjects as geology, geography and history and enhancement of the "comprehensive geography" in the new senior high school curriculum is necessary.

1.4 Open Symposia Organized by the Human Geoscience Committee

Organization of open symposia is another important activity of the committees of SCJ. Every proposal of SCJ open symposium must be thoroughly checked and approved by the Executive Committee of SCJ, so that the SCJ open symposia are considered to be of very high academic quality and reliability, free from irrational bias, and open to the society, i.e., basically open to anyone with no entrance fee and photograph restriction. The Human Geoscience Committee organized seven open symposia during 2006–2017, once every year since 2012, as shown in Table 1.3. By comparing Table 1.3 with Table 1.2, one might notice that Proposal (1), (2) and (3) in Table 1.2 correspond to Open Symposium (1), (4) and (7) in Table 1.3, respectively. Table 1.3 indicate show the focus of *human geoscience* has been changing.

Table 1.3 Open symposia organized by the human geoscience committee of SCJ

(1) August 2009	Sea, land, and human (SCJ 2009)
(2) February 2012	The ways to reducing big disasters and creating sustainable society based on the lessons of the Great East Japan Disaster of March 2011 (SCJ 2012)
(3) December 2013	How can human geoscience tackle the increasing disasters and global environmental problems? (SCJ 2013b)
(4) September 2014	Towards formation of safe, secure and sustainable society based on the lessons of the Great East Japan Earthquake/Tsunami Disaster of March 2011 (SCJ 2014c)
(5) June 2015	Where can we live?—Let's make maps, read maps and protect ourselves from disasters (SCJ 2015) (Fig. 1.2)
(6) November 2016	Are scientists contributing to reduction of disasters and formation of sustainable society? (SCJ 2016)
(7) September 2017	Cooperation and co-creation by science and society for reduction of disasters and formation of sustainable society (SCJ 2017b)

Looking at the table again with focus on the intention of the members of the Human Geoscience Committee one may be able to say as follows:

(1) They do what they believe good for the society
(2) They learn from the Great East Japan Earthquake/Tsunami Disaster
(3) They re-think what to do
(4) They go forward with the experience of the Great East Japan Disaster
(5) They attempt to jump into the society with maps as a tool (Fig. 1.2)
(6) They ask if they are really needed by the society
(7) They try to work together with the society.

Table 1.4 is the programme of the SCJ Open Symposium—"Cooperation and co-creation by science and society for reduction of disasters and formation of sustainable society", the latest symposium organized by the Human Geoscience Committee in September 2017, i.e., a month after the issuance of the latest proposal of the Committee which is outlined in the previous section. There were ten lectures, and they effectively covered a wide range of geo disasters including earthquake disaster, volcanic disaster and tsunami disaster, and global environmental problems such as global warming, soil problem, water problem and food problem, and such new international frameworks for global sustainability as Future Earth and SDGs as well as more basic geoscientific issues.

Fig. 1.2 SCJ open symposium organized by the human geoscience committee

Table 1.4 Contents of SCJ open symposium on "cooperation and co-creation by science and society for reduction of disasters and formation of sustainable society"

(1)	How to solve global environmental problems?—the challenge of Future Earth
(2)	Security of water and food, and SDGs (Sustainable Development Goals)
(3)	Soil problems—healthy soil supports sustainability of the society
(4)	Meteorological and climatic disasters -risks under global warming
(5)	Public-private cooperation on compilation of strategic regional land information for risk cognition
(6)	Earthquake disasters—the scientific knowledge on the Kumamoto earthquake of 2016 un-transmitted to the society
(7)	A pilot study on clinical volcanic disaster prevention
(8)	Co-creation and co-promotion of the knowledge of the relationship between natural disasters and land condition
(9)	International joint research projects, capacity building and social implementation for disaster prevention
(10)	Civil engineering and construction—the gap between earthquake-proof of cities made by researchers and engineers and the citizens' expectation
(11)	Discussion

1.5 Scope of Human Geoscience

The persistent, if not increasing, threat of various global environmental problems and geo disasters, which the reader of the present book will find in detail in the following chapters, make geoscience an absolute necessity for a safer and more sustainable society. In Japan, rise of popularity of geoscience has been eminent for some years, partly because of the Great East Japan Disaster triggered by the M9.0 earth quake and the gigantic tsunami that followed on 11th March 2011 and the other major geo disasters before and after them, as well as global environmental problems such as climate change, pollution, water and land resource problems, that highlighted geoscientists with diverse disciplinary backgrounds at work on the field, on the TV, etc.

The present book as a book of geoscience covers many of the above issues. Moreover, it has additional significance as a *human geoscience* book, as seen in the titles of Part III to Part VI, namely,

Part III Natural Resources and Society
Part IV Natural Hazards and Society
Part V Global Environmental Problems
Part VI Global Sustainability Programmes and *Human Geoscience*.

The underlined terms strongly indicate the importance of human and social aspects in *human geoscience*. In fact, all the chapters of this book have human and social elements, as well as more natural geoscientific contents. One way of developing *human geoscience* is hence to promote each of its sub-disciplines, which is more or less represented by each chapter in this book, by incorporating more human and social elements.

A second way of developing this new science may be to encourage intra-disciplinary understanding and cooperation within *human geoscience*. The readers of this book might find it useful to read the chapters that are not familiar to them, or to compare different chapters, as there are good chances of finding new facts or ideas through these activities, or even creating new values. It is for this reason that the Human Geoscience Committee of SCJ has a 20 min lecture time at the beginning of its committee meeting, in which one of the members gives a lecture about his/her research.

A third way is to enhance understanding and cooperation between *human geoscience* and the other disciplines. Human geoscience is interdisciplinary in its nature, but it is far from enough. Science Council of Japan is proud of its interdisciplinarity, because it not only incorporates all scientific disciplines including human and social sciences, life sciences, natural sciences, and engineering but also encourages cross-disciplinary activities of its members. However, it is, in fact, easy to say, difficult to do. Even within a single committee of SCJ proclaiming interdisciplinarity, there are often invisible disciplinary barriers to be overcome. Japan Geoscience Union (JpGU), particularly its great annual meeting attended by nearly ten thousand people, offers a good opportunity to the *human geoscience* community to interact and cooperate with the other geoscience communities and other related communities.

A fourth way is to promote understanding and cooperation between *human geoscience* and the society. It was, in fact, the main theme of the 2017 SCJ Proposal compiled by the Human Geoscience Committee (SCJ 2017a). This challenge for trans-disciplinarity has just started, and it will be further explored. *Human geoscience* will thus grow not only for itself as a strongly interdisciplinary science, but also as a "science for society" that contributes to the formation of a sustainable human society.

The Human Geoscience Committee of SCJ will continue its effort to develop *human geoscience* further in coming years by organizing related meetings, symposia, and other activities, which contribute to intra-disciplinary understanding within *human geoscience*, interdisciplinary cooperation with broader geoscience and other sciences, and to trans-disciplinary challenges. Implementation of the 2017 SCJ proposal and submission of a new proposal, and formulation of the new roadmap of *human geoscience* will be among the priority issues of the Committee in the coming years.

The present book is also hoped to contribute to "internationalize" *human geoscience*. *Human Geoscience* will increase its involvement in the new international framework or programme for global sustainability, such as Future Earth, SDGs, and ESD, and programme for disaster reduction or mitigation, such as IRDR. It is intended to become an essential field of research for the modern world. *Human geoscience*, which is a new "science for society", will continue its growth with the society. The merger of ICSU and ISSC to form a new scientific body called International Science

Council (ISC) on 30th June 2018 is desired to have a positive effect on the promotion of intra-disciplinary, interdisciplinary and trans-disciplinary dialogue, understanding and cooperation of different disciplines. *Human geoscience* will welcome this move, and will be happy to go along with the ISC.

References

Future Earth (2014) Future Earth Strategic Research Agenda 2014. ICSU, Paris

ICSU (2010) Earth system science for global sustainability: The grand challenges. International Council for Science, Paris

Science Council of Japan (2002) Report—Japan perspective. SCJ, Tokyo, Japan (in Japanese). http://www.scj.go.jp/ja/info/kohyo/pdf/kohyo-18-t980-3.pdf

Science Council of Japan (2008) Proposal—Towards sustainable co-existence of nature and human on the land and in the coastal sea. SCJ, Tokyo, Japan (in Japanese). http://www.scj.go.jp/ja/info/kohyo/pdf/kohyo-20-t58-6.pdf

Science Council of Japan (2009) Open symposium—sea, land and human. SCJ, Tokyo, Japan (in Japanese). http://www.scj.go.jp/ja/event/pdf/78-s-3-3.pdf

Science Council of Japan (2011): Geoscience Road Map. SCJ, Tokyo, Japan (in Japanese). http://www.scj.go.jp/ja/info/kohyo/pdf/kohyo-21-h132-06.pdf

Science Council of Japan (2012) Open symposium—the ways to reduction of big disasters and formation of sustainable society based on the lessons of the Great East Japan Disaster of March 2011. SCJ, Tokyo, Japan (in Japanese). http://www.scj.go.jp/ja/event/pdf2/142-s-0211.pdf

Science Council of Japan (2013a) Proposal—Towards sharing of geo-information—development of laws on geo-information for construction of safe and secure society. SCJ, Tokyo, Japan (in Japanese). http://www.scj.go.jp/ja/info/kohyo/pdf/kohyo-22-t168-1.pdf

Science Council of Japan (2013b) Open symposium—how can human geoscience tackle the increasing disasters and global environmental problems? SCJ, Tokyo, Japan (in Japanese). http://www.scj.go.jp/ja/event/pdf2/179-s-3-3.pdf

Science Council of Japan (2014a) Geoscience Road Map. SCJ, Tokyo, Japan (in Japanese). http://www.scj.go.jp/ja/info/kohyo/pdf/kohyo-22-h201-3-4.pdf

Science Council of Japan (2014b) Proposal—Towards formation of a sustainable society based on what has been learned from the Great East Japan Earthquake-Tsunami Disaster. SCJ, Tokyo, Japan (in Japanese). http://www.scj.go.jp/ja/info/kohyo/pdf/kohyo-22-t198-3.pdf

Science Council of Japan (2014c) Open symposium—towards formation of safe, secure and sustainable society based on the lessons of the Great East Japan Earthquake/tsunami Disaster of March 2011.SCJ, Tokyo, Japan (in Japanese). http://www.scj.go.jp/ja/event/pdf2/197-s-3-2.pdf

Science Council of Japan (2015) Open symposium—where can we live?—let's make maps, read them and protect ourselves from disasters.SCJ, Tokyo, Japan (in Japanese). http://www.scj.go.jp/ja/event/pdf2/206-s-0620.pdf

Science Council of Japan (2016) Open symposium—are scientists really necessary for reduction of disasters and formation of sustainable society? Tokyo, Japan (in Japanese). http://www.scj.go.jp/ja/event/pdf2/230-s-1113.pdf

Science Council of Japan (2017a) Proposal—Promotion of cooperation and co-creation by science and society for reduction of disasters and formation of sustainable society.SCJ, Tokyo, Japan (in Japanese). http://www.scj.go.jp/ja/info/kohyo/pdf/kohyo-23-t247-7.pdf

Science Council of Japan (2017b) Open symposium—cooperation and co-creation by science and society for reduction of disasters and formation of sustainable society. Tokyo, Japan (in Japanese). http://www.scj.go.jp/ja/event/pdf2/243-s-3-2.pdf

Strahler AN, Strahler AH (1976) Elements of physical geography. Wiley, Canada
United Nations (2006) International Year of Planet Earth, 2008. Resolution adopted by the General Assemblyon 22 Dec. 2005 http://www.un.org/ga/search/view_doc.asp?symbol=A/RES/60/192
UNEP (2007) GEO 4—Environment for development. Progress Press Ltd., Valletta, Malta, p 540

Part II
Earth Surface Realms

Chapter 2
Atmosphere—The Gaseous Realm Which Supports Human Life

Akio Kitoh

Abstract Atmosphere is where we live. Atmosphere is a component of the climate system and comprises various weather systems. Water is the most variable component of the atmosphere and is central to the climate and its variability. Atmospheric greenhouse gases, carbon dioxide in particular due to its long lifetime, and aerosols are key components of the Earth's climate system, by regulating the temperature near the surface.

Keywords Atmosphere · Weather · Climate

2.1 Atmosphere Where We Live

Atmosphere is the air surrounding the earth. Although it is very thin, the atmosphere protects earth's life by various ways. Ozone layer absorbs solar ultraviolet radiation. Greenhouse gases keep the surface temperature sufficiently warm for water to exist in a liquid form that enables earth's life sustainable.

The atmosphere is a component of the climate system, together with the hydrosphere, the cryosphere (snow and ice), land surface and the biosphere. Atmosphere consists of troposphere, stratosphere, mesosphere, and thermosphere (Fig. 2.1). Troposphere occupies from the surface up to around 12 km altitude. The top of the troposphere (tropopause) largely varies with latitude, higher in the tropics and lower in the polar regions. Within the troposphere, temperature drops by around 6.5 K km^{-1}. Various weather phenomena such as cyclones, anticyclones, frontal systems, cloud, and precipitation (rain or snow) occur within the troposphere (Fig. 2.2). Stratosphere lies from 12 to 50 km altitude where temperature rises with altitude. From 50 to 80 km altitude (mesosphere), temperature drops with altitude. Temperature maximum around 50 km altitude is due to an ozone layer that absorbs solar radiation in the stratosphere. The ozone layer absorbs harmful ultraviolet radiation, thus enable plants and animals to live at the Earth's surface. Above 80 km altitude, temperature rises, so named the thermosphere. The upper limit of the atmosphere cannot

---

A. Kitoh (✉)
Japan Meteorological Business Support Center, Tsukuba, Japan
e-mail: akiokitoh@gmail.com

Y. Himiyama et al. (eds.), *Human Geoscience*, Advances in Geological Science,
https://doi.org/10.1007/978-981-32-9224-6_2

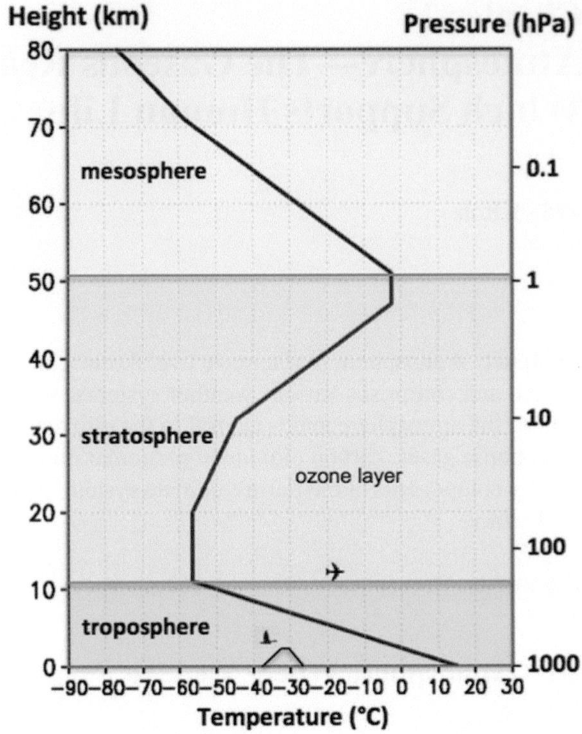

Fig. 2.1 Vertical profile of the atmospheric temperature

be explicitly defined as atmospheric pressure decreases exponentially. Atmospheric pressure is around 1000 hPa at the surface and rapidly decreases with the altitude. At the tropopause, atmospheric pressure is around 200 hPa. Atmospheric pressure at 50 km altitude is about 1 hPa, and that at 80 km altitude is about 0.01 hPa. If we practically assume that the vertical range of the Earth's atmosphere is within 100 km from the surface, it is very thin compared with the Earth's radius (6378 km). Annual mean surface pressure in Tokyo is 1011 hPa. At the top of Mt. Fuji (elevation at 3776 m), it becomes 639 hPa. Humans suffer altitude sickness generally below 800 hPa due to oxygen deficiency. Therefore, humans live within a very thin layer of the Earth's atmosphere, at the bottom of troposphere.

2.2 Weather and Climate

We experience various weather phenomena throughout season. Cyclones and anticyclones alternatively pass over our head, bringing us rain or snow. Monsoonal rain feeds tropical countries. Snowfall during winter becomes water resources after snowmelt. Extreme weather events (heavy rainfall, heavy snow, drought, heat wave, strong winds, storm surge) often affect human life.

Fig. 2.2 Infrared satellite imagery from Himawari-8 at 12:00 JST on 29 August 2016. Tropical cyclone Lionrock is seen south of Japan. (taken from the homepage of Japan Meteorological Agency)

Weather is a "state of the atmosphere, mainly with respect to its effect upon life and human activities at a particular time" and "consists of the short-term (minutes to weeks) variations of the atmosphere" (Geer 1996) such as temperature, precipitation (rain, snow) or winds. Climate is the "total of all statistical weather information that helps to describe the variation of weather at a given place for a specified interval of time" (Geer 1996). World Meteorological Organization (WMO) defines that period as 30 years.

2.3 Greenhouse Gas

Atmospheric components are almost uniform for N_2 (78.08%), O_2 (20.95%), Ar (0.93%), CO_2 (0.04%) between the troposphere and the mesosphere because they are not chemically active, but concentration ratio of other gases including water vapor varies largely with location.

Some trace gases (CO_2, O_2, CH_4, N_2O, chlorofluorocarbon) are called greenhouse gases because of the greenhouse effect. Water vapor also has a greenhouse effect, but as its amount cannot be controlled by human, water vapor is not categorized as an anthropogenic greenhouse gas.

Existence of greenhouse gases is vital for humans. Without greenhouse gases, surface air temperature of the Earth would be much colder than the present. Earth's temperature is determined by radiation balance of incoming solar radiation and outgoing terrestrial radiation (Fig. 2.3). Solar irradiance at the top of the atmosphere is 1360 W m^{-2}. As the Earth's surface area is four times of the area perpendicular to the sun, considering the ratio 1/4, the Earth's surface receives 340 W m^{-2}. About 30% of this incoming solar radiation is reflecting back by the atmosphere, clouds and the surface to the outer space. This ratio is called the planetary albedo. The value of 30% is not a universal constant, but depends on the characteristics of clouds, aerosols and surface conditions. Without greenhouse gases, the received energy (340 W m^{-2}) should be balanced with terrestrial radiation, which is determined by the surface temperature. A simple calculation yields 255 K (−18 °C) as the Earth's radiative equilibrium temperature.

Greenhouse gases trap the terrestrial radiation within the atmosphere, and re-emit it to the surface and the outer space at the value depending on the temperature of the atmosphere. The surface thus receives additional radiation and warms. This process is called the greenhouse effect, the resultant surface temperature being 288 K (+15 °C).

Due to the existence of greenhouse gases, the Earth's surface temperature is above the freezing point and the Earth is habitable. In this sense, we have to thank greenhouse gases. However, the amount of atmospheric concentration of greenhouse gases ultimately determines the global mean surface temperature. Rapid increase of anthro-

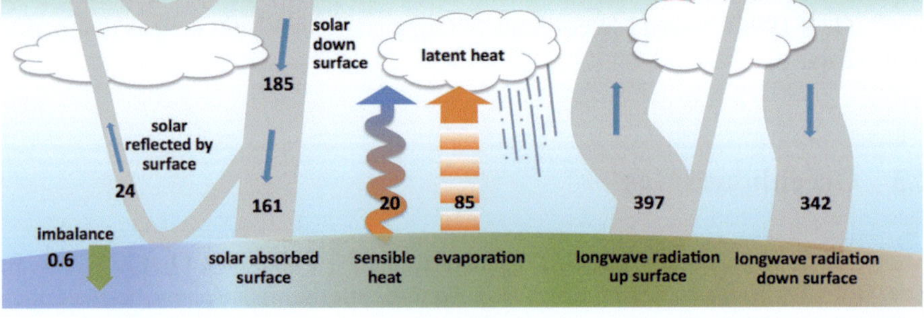

Fig. 2.3 Global mean energy balance of the Earth. Numbers are from Wild et al. (2013). Unit is W m^{-2}

pogenic greenhouse gas emissions leads to a sharp increase in atmospheric concentration of greenhouse gases and global surface temperature rise (IPCC 2013). Current observation-based estimate of the Earth's radiation balance shows an imbalance both at the top of the atmosphere and at the surface (Fig. 2.3). This net accumulation of energy can be considered due to rapidly increasing greenhouse effect by greenhouse gases. The global warming issue is discussed in Chap. 20.

2.4 Aerosols

Aerosols are small solid or liquid particles in the atmosphere. They are both natural and anthropogenic origin. Mineral dust and sea spray are of natural origin, while sulphates, biomass aerosols and soot (black carbon) are both of natural and anthropogenic origin. The size and shape of aerosols differ greatly from one aerosol to another.

Aerosols affect weather and climate in various ways by modifying solar radiation reaching the surface affecting global surface temperature by aerosol-radiation and aerosol-cloud interactions. First, aerosols scatter solar radiation, resulting less solar radiation reaching the surface (cooling effect). Some aerosols such as black carbon absorb solar radiation and heat local atmosphere. Overall these aerosol-radiation interactions have cooling effect.

In a very clean atmosphere without aerosols, clouds are hard to form. Clouds generally form around nuclei. Aerosols become these condensation nuclei. When there are more aerosols in the atmosphere, larger number of cloud droplets form but with smaller size. This causes to work as reflector of solar radiation (cooling effect). Distribution of aerosols greatly varies from region to region, cooling here and heating there, thus affecting atmospheric circulations (winds). Local effect of aerosols on clouds also affects precipitation. Anthropogenic emissions of aerosols thus influence global radiation balance and climate system but also affect regional weather and climate.

Aerosols drop to surface gravitationally and are washed out by precipitation in the troposphere. Therefore, lifetime of tropospheric aerosols is short and is from 1 day to 1 week. Once aerosols are entered into the stratosphere where no washed-out process works, aerosols affect climate for a year or two. For example, large volcanic eruptions inject aerosols into the stratosphere. A cooling effect of Mt. Pinatubo eruption in 1991 lasted for 2 years.

Since the industrial revolution, anthropogenic aerosol emissions have increased, but quantitative estimate is difficult. In the second half of the twentieth century, the cooling effect of anthropogenic aerosols counteracted the warming by increasing concentrations of greenhouse gases. Aerosols are the major source of local air quality problem. During the past few decades, anthropogenic aerosol emissions have decreased in developed countries due to air quality policies, but increased in developing countries because economic development was more important than health problem issues there. In the coming future, overall emissions of anthropogenic aerosols will decrease because all countries concern about air quality. This indicates that the cooling effect of aerosols will no more counteract the warming effect of greenhouse gases, implying acceleration of global warming.

References

Geer IW (ed) (1996) Glossary of weather and climate with related oceanic and hydrologic terms. American Meteorological Society, Boston

IPCC (2013) Climate change 2013: the physical science basis. In: Stocker TF, Qin D, Plattner GK, Tignor M, Allen SK, Boschung J, Nauels A, Xia Y, Bex V, Midgley PM (eds) Contribution of working group i to the fifth assessment report of the intergovernmental panel on climate change. Cambridge University Press, Cambridge, United Kingdom and New York, NY, USA

Wild M, Folini D, Schär C, Loeb N, Dutton EG, König-Langlo G (2013) The global energy balance from a surface perspective. Clim Dyn 40:3107–3134

Chapter 3
Lithosphere—The Solid Realm Which Supports Human Life

Takashi Oguchi

Abstract The lithosphere is geologically defined as the crust and upper mantle of the earth, and geographically the solid portion of the earth especially earth surfaces and shallow underground. This chapter deals with these two types of the lithosphere in relation to human activities. Plate tectonics represents a dynamic aspect of the lithosphere and is responsible for long-term shaping of the earth. It also causes hazardous earthquakes and volcanic eruptions. Rocks and minerals constituting the crust have been utilized for various human activities including architectural construction, agriculture, and industrial activities such as steel production. Landforms are an essential part of the lithosphere strongly related to human activities. Their quantitative representation and analysis using digital elevation models and GIS contribute to the appropriate social activities such as land-use planning. Geomorphological studies on Quaternary landform development are useful for archaeological understanding of ancient human activities. Knowledge about geomorphic processes is essential for mitigating natural hazards such as mass movements. These examples show that the lithosphere is a fundamental component of Human Geoscience.

Keywords Lithosphere · Geology · Geomorphology · Natural resources · Natural hazards

3.1 Introduction

The term lithosphere has two major definitions. One is a geological definition: the crust and upper mantle of the earth. The other is a geographical definition: the solid portion of the earth distinguished from atmosphere and hydrosphere. In the latter definition, "the solid portion" tends to mean earth surfaces and shallow underground, where human activities and their impacts are direct and strong. Deeper underground such as the lower crust is not the main scope in this case. The latter definition is also more directly related to Human Geoscience than the former definition. Therefore, landforms and their scientific research in the field of Geomorphology are particularly

T. Oguchi (✉)
Center for Spatial Information Science, The University of Tokyo, Tokyo, Japan
e-mail: oguchi@csis.u-tokyo.ac.jp

Y. Himiyama et al. (eds.), *Human Geoscience*, Advances in Geological Science,
https://doi.org/10.1007/978-981-32-9224-6_3

relevant to Human Geoscience. Sedimentology and Pedology (Soil Science) that deal with depositional materials on or near earth surfaces are also important. In some other scientific fields, however, the geological definition of the lithosphere seems to be more essential—Geology is one of them as a matter of course. For example, two textbooks entitled "The Lithosphere" (Gregory 2007; Artemieva 2012) were written mainly from the geological viewpoint. A scientific journal called "Lithosphere", published by the Geological Society of America since 2009, also focuses on geological issues.

This chapter deals with both geological and geographical characteristics of the lithosphere, with special emphasis on earth surfaces where the majority of human activities is ongoing. The geological aspects of the lithosphere are considered not for describing its physical characteristics but in relation to humans, for example, rocks and minerals used for industrial production, and natural hazards with geological backgrounds. The next section of this chapter (Sect. 3.2) focuses on such geological aspects of the lithosphere, and the subsequent section (Sect. 3.3) deals with geographical aspects.

3.2 Geological Aspects of the Lithosphere and Related Human Activities

3.2.1 Plate Tectonics and Related Natural Hazards

Although solid parts of the earth are much less mobile than the atmospheric and hydrological parts, they are not static. For example, the absolute location of each point on or within the upper crust is slowly moving due to plate tectonics. This crustal movement occurs with strong physical stress mainly along plate boundaries but also within a plate, and the resultant release of accumulated stress occasionally leads to earthquakes. Large earthquakes cause serious disasters that give strong impacts on human societies. Physical stress due to sudden ground motion may destruct houses and other buildings, as well as various artificial structures such as bridges. Ground motion may also lead to liquefaction of surficial sandy materials in lowlands, which may deform land and disturb objects built on it, including the tilting and distortion of houses. In steeper terrain, seismic ground motion may induce mass movements including slope failures and rock falls, which may become disastrous if they occur in and around residential, commercial, and industrial areas. For example, on May 12, 2008, Sichuan Province in China was struck by the $M = 7.9$ Wenchuan earthquake, resulting in more than 60,000 landslides that killed more than 20,000 people (e.g., Yin et al. 2009). Sudden deformation of sea floor due to a large earthquake at a plate subduction zone may cause a tsunami that can be highly disastrous, as evidenced by the Indian Ocean earthquake and tsunami on December 26, 2004, and the Tohoku earthquake and tsunami on March 11, 2011. Such catastrophic events give prolonged impacts on local societies, lasting more than several years.

Another outcome of plate tectonics is volcanic activities, which tend to be enhanced along convergent or divergent plate boundaries, although volcanoes associated with hot spots such as the Hawaiian Islands are more independent of plate movement. For example, the so-called Ring of Fire surrounding the Pacific Ocean is an extensive zone where volcanic activities are concentrated mainly due to plate convergence (Fig. 3.1). Volcanic eruptions often affect human societies through various types of geological phenomena including ash fall, cinder fall, gas emission, and pyroclastic flow. Lighter eruptions cause relatively limited impacts such as temporally decreased visibility and air pollution due to ash fall and smoke. In contrast, large-scale eruptions are much more dangerous and may cause a large number of fatalities. For instance, the Ultra-Plinian eruption of Mount Pinatubo on Luzon Island, the Philippines, on June 15, 1991, was the largest eruption in recent years, leading to the death of 847 people and the sudden collapse of local industries especially agriculture and forestry. The case of Mount Pinatubo also indicates prolonged impacts of huge eruption. The eruption produced broad devastated land covered with volcanic ash and pyroclastic deposits with little vegetation. Such areas are highly prone to erosion by rainfall and resultant surface flow. Therefore, lahars occurred repeatedly during several years after the eruption, leading to significant sedimentation in piedmont areas including residential areas (e.g., Newhall and Punongbayan 1996). Analysis of vegetation recovery in Mount Pinatubo since 1991 using remote sensing images suggests that more than 50 years are required for full recovery (De Rose et al. 2011). In addition, the abundant volcanic ash and gas emitted by the eruption altered global

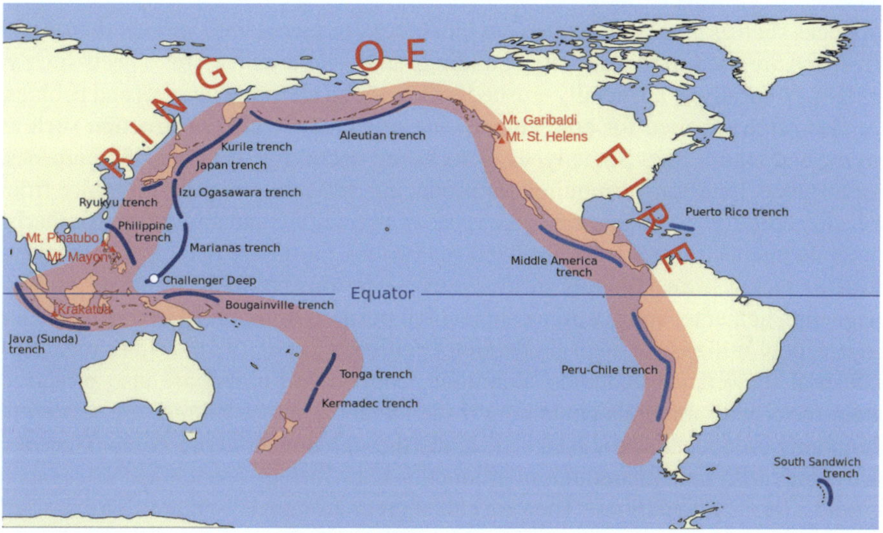

Fig. 3.1 Ring of Fire, where volcanic and seismic activities are concentrated. Submarine trenches occur at plate boundaries due to the subduction of oceanic plates. Public domain file from Wikimedia Commons

climate during the following months and even years with statistical significance (e.g., McCormick et al. 1995).

In spite of such hazards, plate tectonics and associated underground phenomena add some values to our lives. For example, heat in underground particularly that in volcanic areas is usable natural energy for power generation, and hot spring water available in such areas can be used for heating and leisure. In addition, scenery of volcanic areas is often beautiful and unique, attracting people and some of such areas have been developed for tourism (Sigurdsson and Lopes-Gautier 2000). A similar view is applicable to tectonics and resultant earthquakes. Mountains like the Himalayas are consequence of tectonic uplifts mainly originated from plate tectonics and uplift events often accompany earthquakes, but resultant spectacular natural landscapes can be resources of tourism and other human activities such as sports and arts.

3.2.2 Rocks and Minerals as Resources

The lithosphere in the geological sense consists of rocks and their fragments (sediments), and they consist of minerals. These rocks and minerals provide the earth surfaces where our daily lives are ongoing. Besides this basic importance, rocks and minerals extracted from the lithosphere are also materials supporting our lives in various ways. Thus, they are often regarded as important resources, not mere geological materials (e.g., Kogel 2006).

Rocks such as granites are often cut into blocks or pieces with various shapes and sizes, and are used for constructing buildings and other architectural objects such as bridges. Naturally or artificially crushed rocks, which are called gravels and pebbles, are also often required for civil engineering and architecture construction such as paving and land filling. Some types of rocks are useful for other specific purposes. For instance, limestone is required for producing cements, and porous pumice from volcanic eruption is suitable for some types of agriculture and gardening. Similarly, minerals included in rocks are utilized for various purposes. A typical example is irons extracted from minerals in iron ores; some rock-forming minerals such as magnetite, hematite, and goethite contain high portions of oxidized iron, which have been widely utilized for steel production (Yellishetty et al. 2010). Other metals of industrial importance including aluminum, copper, and nickel are also produced using rocks with minerals containing these metals. These examples are related to commonly used metals produced abundantly through industrial processes. There are also metals with smaller amounts of production but with high economical values such as gold, silver, and platinum. They are called precious metals, and are also extracted from rocks and minerals (Saviët̀skii 1989). Similarly, rare metals from rocks and minerals, i.e., rare earth resources, have been utilized to improve the quality of electronic products such as batteries, magnets, optical discs and apparatuses using laser (Zhanheng 2011). Moreover, fossil fuels including oil, coal, and gas in deep underground are strongly associated with rocks and minerals for their geological production and underground reservoirs or storage. Without fossil fuels, the rapid modernization and development of the whole world since the Industrial Revolution would have been impossible. In sum, rocks and minerals have been tremendously contributing to people and societies.

Rocks and minerals are useful not only for industrial purposes but also for fashion, decollation, and arts. Gemstones with beautiful reflection and/or color such as diamond, sapphire, and ruby are essential for producing jewelries. Beautiful rocks can also be used to decollate furniture and buildings (Price 2007). Rocks and their fragments are also important elements of gardening.

For the reasons above, rocks and minerals often become economic targets, and mining has been an important human activity for a long time. Because the distribution of rocks, minerals and fossil fuels is rather uneven, international trades of these materials are very common. They are not only economical but also political issues (e.g., Moody 2007). For example, the production of rare earth elements in China had markedly increased in the late twentieth century and the beginning of the twenty-first century, and the restriction of their export by the Chinese Government led to the sudden global rise of their prices in 2011. However, the share of China in the global market of rare earth products started to decrease since then, and their price also decreased (Fig. 3.2). To summarize, the relationships between the geological aspects of the lithosphere and human activities are diverse but very strong, and discussion on their relationships can be a core topic of Human Geoscience.

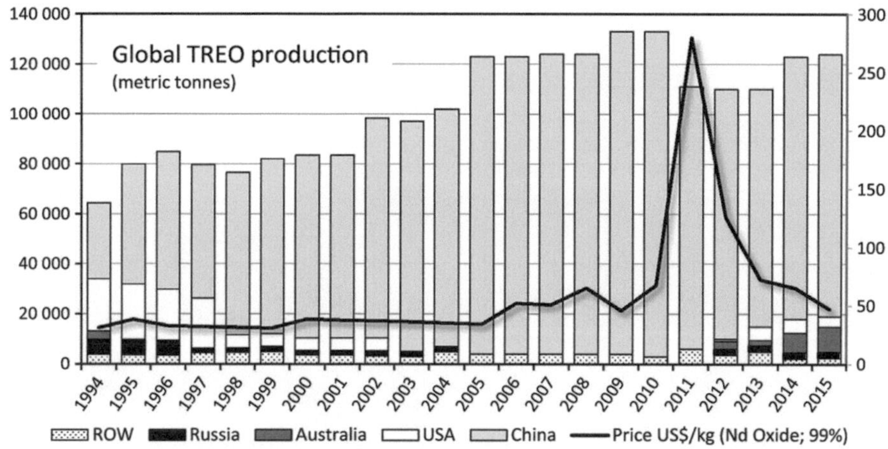

Fig. 3.2 Global annual production of Rare Earth Oxides (REO). Reprinted from Paulick and Machacek (2017), with permission from Elsevier

3.3 Geographical/Geomorphological Aspects of the Lithosphere

3.3.1 Perception of Landforms Using Topographic Maps and Digital Data

As noted, landforms on earth surfaces constitute the core of geographical aspects of the lithosphere. Landforms are components of terrain with various shapes, and their recognition and understanding are important for human lives. For instance, steep lands are generally unsuitable for settlement and other human activities such as agriculture, because lands are unstable, and activities there require more physical labors. The same applies to higher lands—they tend to be less accessible, less convenient, and with harsher environmental conditions than lowlands. However, for some particular purposes, inclined and/or higher lands may be more suitable than flat lowlands. For example, south-facing slopes receive more insolation than other slopes, and hence they are more suitable for growing some types of agricultural plants and constructing solar energy farms. Higher lands tend to have lower temperature, and thus facilitate the growth of some plants and animals that favor cool air.

Shapes of earth surfaces have components other than slope and height such as curvature describing concavity and convexity (Fig. 3.3). Curvature of lands significantly affects hydrological conditions, for example, surface water flow tends to be more concentrated in concave areas. Such hydrological conditions affect soil mois-

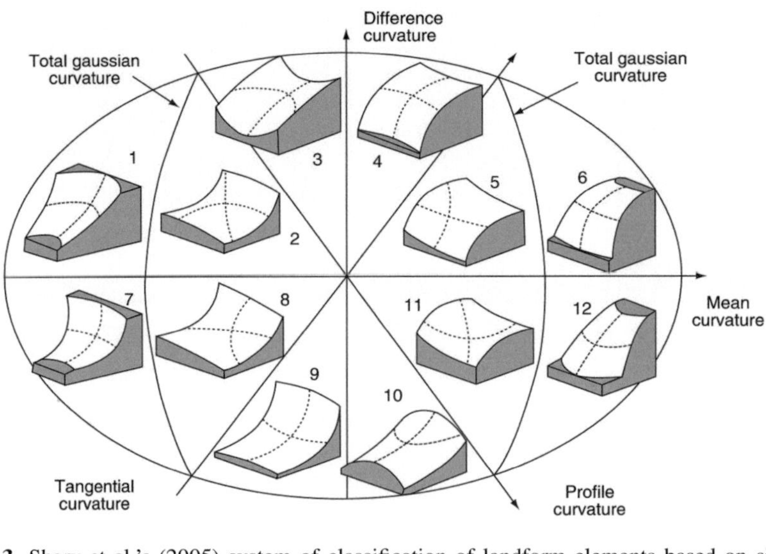

Fig. 3.3 Shary et al.'s (2005) system of classification of landform elements based on signs of tangential, profile, mean, difference, and total Gaussian curvatures. Reprinted from Wilson and Bishop (2013), with permission from Elsevier

ture as well as probability of surface flow or flooding, and hence exert influence on human activities including agriculture and settlement. Indeed, the global distribution of population is related to topographic relief and surface hydrological conditions (Meybeck et al. 2001).

Because landforms markedly affect human activities, accurate measurement of the shape of earth surfaces is important for societies. For this purpose, land surveying and mapping have been performed since long time ago. Ancient maps are generally not so accurate even if ground surveying was conducted, and many of them look more like paintings than scientific products. However, technological development such as the introduction of triangulation allowed cartographers and surveyors to produce realistic maps. Until the beginning of the twentieth century, however, adding height information over a map was difficult—most maps showed only horizontal locations of features such as roads and administrative boundaries. The introduction of aerial photogrammetry changed this situation, and permitted the publication of topographic maps with contour lines (Alspaugh 2004). By the middle of the twentieth century, many places of the world particularly already developed countries were covered with such topographic maps with scales typically 1:50,000–1:100,000. This shift facilitated appropriate usage of land resources based on the 3D characteristics of landforms.

In recent years, surveys on 3D shape of earth surfaces underwent drastic advancement. Even during the twentieth century, technology used for aerial photogrammetry saw significant development from analogue to digital (Alspaugh 2004). Nonetheless, it required professional instruments such as digital photogrammetric workstations (DPWs), specific metric cameras, and highly skilled personnel. Flights for taking high-quality aerial photographs suitable for detailed photogrammetry also costed a lot. In the last several years, however, new photogrammetric techniques including SfM (Structure from Motion) based on artificial intelligence have been applied to pictures taken by a commonly-used cheap digital camera attached to an unmanned aerial vehicle (UAV or drone). This allows the efficient production of high-quality surface 3D models with low cost (Remondino et al. 2011). UAVs can fly at heights close to the ground and thus suitable for producing very high-resolution topographic data. Aerial and terrestrial laser scanning, so-called LiDAR (light detection and ranging), is another surveying technique that has been widely used in recent years (e,g, Heritage and Large 2009). An advantage of laser over photogrammetry is that land surfaces below vegetation including forests can be measured, although the cost of laser scanning tends to be higher than the latest handy photogrammetry using UAVs.

Originally surveying on both horizontal location and height was meant for producing maps with contour lines. However, today's surveying using photogrammetry or laser scanning mostly aims to produce 3D point clouds and they are often converted into digital elevation models (DEMs). Typical DEMs have locational and altitudinal information for regularly spaced points, and their raster format is suitable for analyses using computers with geographical information systems (GIS). The combination of DEMs and GIS facilitates the quantitative and objective description and analysis of landform characteristics such as slope angle, aspect, and curvature. The scientific discipline of quantitative measurement and analysis of landforms is called Geomorphometry, which can be regarded as a subfield of Geomorphology, and saw a recent rapid progress thanks to the increased availability of DEMs and related analytical tools (e.g., Hengl and Reuter 2009). GIS also allows users to overlay other thematic layers on DEMs; for example, overlaying a current land use map on a DEM to investigate relationships between landforms and land use. Through such analyses, it is possible to examine whether current land use is appropriate from geomorphological viewpoints, and to propose a plan for better utilization of land resources (e.g., Bocco et al. 2001). Like this example, Geomorphometry and GIS facilitate interdisciplinary research in Human Geoscience.

3.3.2 Landform Development and Quaternary Environmental Changes

In addition to landforms as shape characteristics, their evolution in the past has been discussed extensively in the field of Geomorphology. This is a pure scientific issue in a sense, but also relevant to humans. Landforms in a historical period may be significantly different from those at present if they have been changing rapidly. For example, shift of river courses may produce new depositional landforms such as natural levees and alluvial fans, as well as erosional landforms such as incised channels and valleys. Environmental changes during the Quaternary including sea-level rise and fall, vegetation growth and decay, and climatological shifts including changes in rainfall regime may also lead to geomorphological changes. In such cases, perception of landforms and related environments by ancient people may have been different from that of today. Therefore, if Human Geoscience deals with time series from the past, it is important to understand landform development and its relation with Quaternary environmental changes. Extensive knowledge from the past Quaternary studies(e.g., Bell and Walker 2005) should be utilized to strengthen and expand Human Geoscience.

Archaeology is an academic field strongly related to Human Geoscience, because ancient people lived with higher dependence on geo-environmental conditions than modern people, for example, usable water was available from rivers, springs, and wells at that time, not from organized water supply systems as in cities today. Therefore, archaeological studies often discuss the effects of landforms and associated hydrological conditions on ancient people. Indeed, archaeologists often collaborate with geoscientists including physical geographers to examine relationships between their archeological findings and the natural environment, and such collaboration is considered as a scientific field called Geoarchaeology (Goldberg and Macphail 2008), which has common aspects to Human Geoscience. Geoarchaeological studies tend to be field-based, but in recent years, DEMs and GIS have also been utilized to effectively map relevant information (Fig. 3.4) and quantitatively analyze them in relation to archaeological findings (e.g., Clevis et al. 2006).

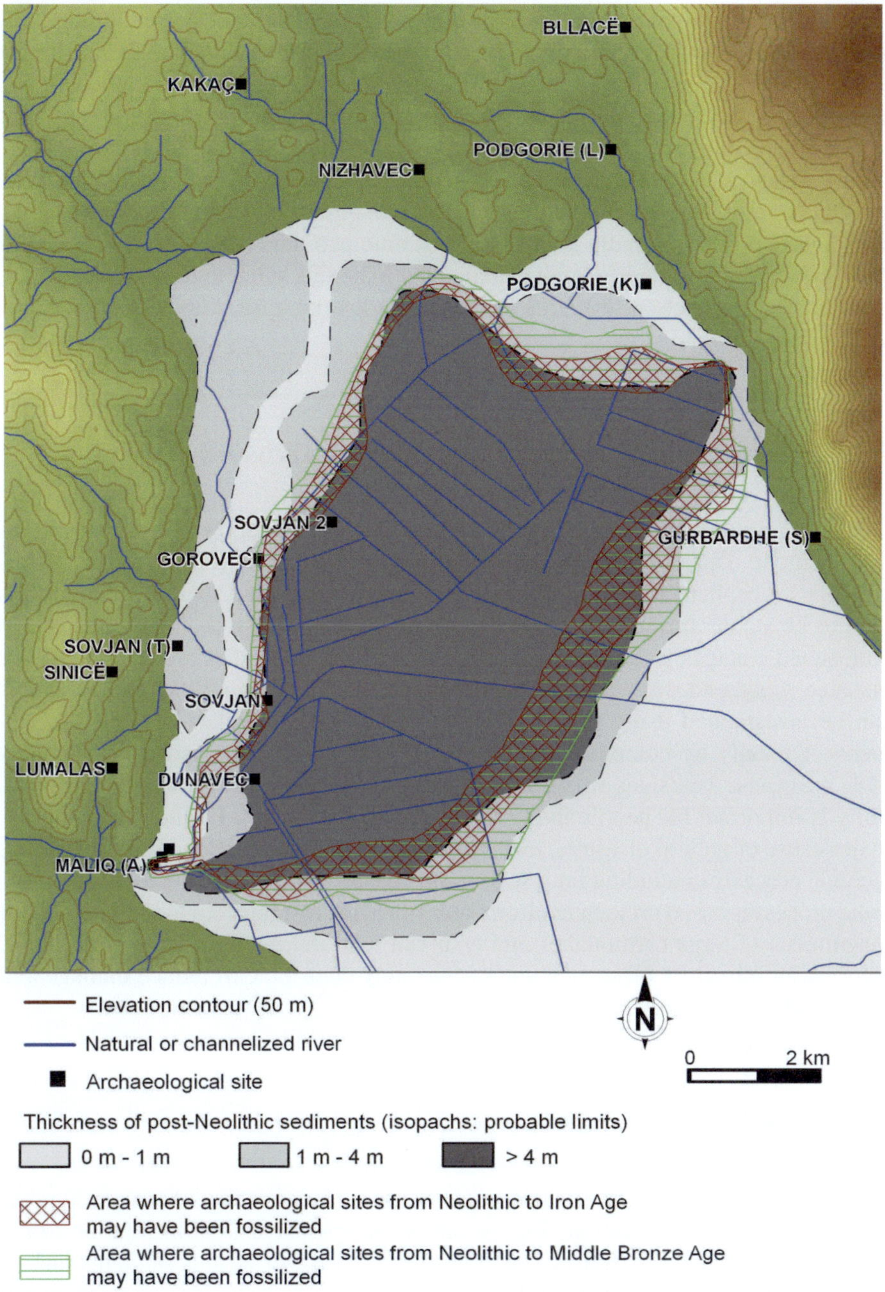

Fig. 3.4 Geoarchaeological map of the Korça Basin, Albania. From Ghilardi and Desruelles (2009), with permission CC BY 3.0

When development of landforms is analyzed, surficial deposits including soils are often investigated, because they were produced in the past, and thus may provide clues to the characteristics of ancient landforms and related environmental conditions. Soils and deposits significantly affect surface hydrological conditions and land fertility; therefore they play an important role in determining appropriate land use. As mentioned in Sect. 3.2.2 concerning pebbles and gravels as geological materials, soils and deposits are also sometimes regarded as important resources. For example, soils and deposits originally deposited long time ago, and subsequently subject to strong chemical weathering, may contain clay materials suitable for various usages, from ancient production of bricks to modern industrial applications (e.g., Bergaya and Lagaly 2006).

3.3.3 Geomorphic Processes and Human Activities

Geomorphological studies often investigate processes responsible for landform development. This is of pure scientific interest in a sense, but also useful to consider the lithosphere as earth surfaces in relation to people and societies. Landforms tend to be stable during most of time, but may occasionally become unstable. To understand changes in landform stability, process-based studies are important. For instance, whether a slope segment becomes unstable due to rainfall or an earthquake can be investigated using the concepts of soil mechanics related to mass movements, typically by computing the factor of safety (e.g., Anderson and Richards 1987). Because landsides may result in serious disasters, evaluation of slope stability is important for people living on or adjacent to slopes. Earthquakes, another major cause of natural disasters, lead to formation of particular landforms reflecting tectonic processes including fault scarps and coastal terraces. Reconstruction of tectonic processes based on such landforms may provide insights into the frequency and magnitudes of large earthquakes and resultant land deformation (e.g., Bull 2008). Information obtained from such approaches may contribute to reduce damage and casualty due to earthquakes.

3.4 Concluding Remarks

This chapter provides a brief summary concerning relationships between humans and the lithosphere in both geological and geographical senses, in relation to the concepts of Human Geoscience. The lithosphere consists of various elements such as landforms, rocks, soils, and minerals, and is dynamic with the tectonic movement and deformation of land and crust. All these elements are associated with humans, often support our activities such as agriculture, construction, and industrial manufacturing, but sometimes exert negative impacts thought natural hazards. Understanding the complex relationships between the lithosphere and humans is a key to sustainable development. In particular, issues related to landforms are important because most of human activities occur on them. People from various disciplines dealing with the lithosphere should collaborate under the umbrella of Human Geoscience for creating better environments and societies. This will be an important activity to support the Future Earth, a 10-year research Initiative launched in 2015 (http://www.futureearth.org). Geography includes both human and physical components since long time ago, and hence geographers should play a key role in coordinating such interdisciplinary research projects.

Currently humans live on the Earth, but some already landed on the Moon, and there are plans to send people to Mars. In future, some people will live on planets for a long period of time. There are also plans to utilize natural resources existing in space, for example, some planetary bodies such as asteroids contain a large amount of usable metals including rare and precious ones (Jakhu et al. 2017). Therefore, Human Geoscience in relation to the lithosphere is even becoming relevant to Planetary Science. This example indicates that research areas to be covered by Human Geoscience is very broad, and the lithosphere is a fundamental component in that field.

References

Alspaugh D (2004) A brief history of photogrammetry. In: McGlone JC (ed) Manual of photogrammetry, 5th edn. American Society for Photogrammetry and Remote Sensing, Bethesda, MD, pp 1–12

Anderson MG, Richards KS (eds) (1987) Slope stability: geotechnical engineering and geomorphology. Wiley, Chichester

Artemieva I (2012) The lithosphere: an interdisciplinary approach. Cambridge University Press, 794 p

Bell M, Walker MJ (2005) Late Quaternary environmental change: physical and human perspectives. Pearson Education

Bergaya F, Lagaly G (2006) General introduction: clays, clay minerals, and clay science. Develop Clay Sci 1:1–18

Bocco G, Mendoza M, Velázquez A (2001) Remote sensing and GIS-based regional geomorphological mapping—a tool for land use planning in developing countries. Geomorphology 39:211–219

Bull WB (2008) Tectonic geomorphology of mountains: a new approach to paleoseismology. Blackwell, Oxford

Chen Z (2011) Global rare earth resources and scenarios of future rare earth industry. J Rare Earths 29(1):1–6

Clevis Q, Tucker GE, Lock G, Lancaster ST, Gasparini N, Desitter A, Bras RL (2006) Geoarchaeological simulation of meandering river deposits and settlement distributions: a three dimensional approach. Geoarchaeology 21:843–874

De Rose R, Oguchi T, Morishima W, Collad M (2011) Land cover change on Mt Pinatubo, the Philippines, monitored using ASTER VNIR. Int J Remote Sens 32:9279–9305

Ghilardi M, Desruelles S (2009) Geoarchaeology: where human, social and earth sciences meet with technology. S.A.P.I.EN.S, 2.2

Goldberg P, Macphail RI (2008) Practical and theoretical geoarchaeology. Blackwell

Gregory V, (2007) The lithosphere: earth's crust. Twenty First Century Books, 80 p

Hengl T, Reuter H (eds) (2009) Geomorphometry: concepts, software, applications. Elsevier, Amsterdam

Heritage G, Large A (eds) (2009) Laser scanning for the environmental sciences. Wiley

Jakhu RS, Pelton JN, Nyampong YOM (2017) Space mining and its regulation. Praxis/Springer

Kogel JE (ed) (2006) Industrial minerals &rocks: commodities, markets, and uses. SME

McCormick MP, Thomason LW, Trepte CR (1995) Atmospheric effects of the Mt Pinatubo eruption. Nature 373(6513):399

Meybeck M, Green P, Vörösmarty C (2001) A new typology for mountains and other relief classes: an application to global continental water resources and population distribution. Mt Res Dev 21:34–45

Moody R (2007) Rocks and hard places: the globalization of mining (Global Issues Series). Zed Books, 224 p

Newhall CG, Punongbayan R (eds) (1996) Fire and mud: eruptions and lahars of Mount Pinatubo, Philippine. Philippine Institute of Volcanology and Seismology, Quezon City, p 1126

Paulick H, Machacek E (2017) The global rare earth element exploration boom: an analysis of resources outside of China and discussion of development perspectives. Res Policy 52:134–153

Price M (2007) The sourcebook of decorative stone: an illustrated identification guide. Firefly Books, 288 p

Remondino F, Barazzetti L, Nex F, Scaioni M, Sarazzi D (2011) UAV photogrammetry for mapping and 3d modeling–current status and future perspectives. Int Arch Photogrammetry Remote Sensing Spatial Info Sci 38(1):C22

Saviëtìskii EM (1989) Handbook of precious metals. Hemisphere Pub, Corp

Shary PA, Sharaya LS, Mitusov AV (2005) The problem of scale-specific and scale-free approaches to geomorphometry. GeografiaFisica e DinamicaQuaternaria 28:81–101

Sigurdsson H, Lopes-Gautier R (2000) Volcanoes and tourism. Encyclopedia of Volcanoes. Elsevier, Amsterdam, pp 1283–1299

Wilson JP, Bishop MP (2013) Geomorphometry. In: John F. Shroder (Editor-in-chief), Bishop, M.P. (Volume Editor). Treatise on geomorphology, vol 3. Remote Sensing and GIScience in Geomorphology. Academic Press, Elsevier

Yellishetty M, Ranjith PG, Tharumarajah A (2010) Iron ore and steel production trends and material flows in the world: is this really sustainable? Resour Conserv Recycl 54(12):1084–1094

Yin Y, Wang F, Sun P (2009) Landslide hazards triggered by the 2008 Wenchuan earthquake, Sichuan China. Landslides 6(2):139–152

Chapter 4
Hydrosphere—The Water Realm Which Supports Human Life

Taikan Oki

Abstract The Earth is sometimes called as an aqua planet, even though the total amount of water is only 0.02% of the planet, because the existence of water substantially controls the energy budget and the environment of its surface. The roles of storages and flows under the context of global hydrological cycles and Earth system science, and interactions among them are discussed, and estimated volumes and fluxes are quantitatively illustrated. Terrestrial, atmospheric, and combined water balance concepts are described and applications of them are presented. Representative estimates of global water balance consist of global mean precipitation, evapotranspiration, and runoff by various researchers are introduced and compared.

Keywords Water balance · Energy balance · Snow · Groundwater · Multi-model ensemble

4.1 Introduction

The Earth system is unique in that water exists in all three phases, that is, water vapor, liquid water, and solid ice, compared to the situations in other planets. The total volume of water on the Earth does not change on shorter than geological timescales (Oki et al. 1999), and is approximately estimated as 1.4×10^{18} m^3 which corresponds to a mass of 1.4×10^{21} kg (Fig. 4.1, revised from Oki and Kanae 2006). Although the water mass only constitutes 0.02% of the total mass of the planet (5.974×10^{24} kg), it is a critical matter for all organisms including humans in their survival (Oki et al. 2004).

Hydrosphere includes water vapor circulation in the atmosphere. Terrestrial hydrologic cycles are driven by atmospheric circulation and its consequences particularly precipitation. On the contrary, hydrologic conditions over land form boundary conditions for atmospheric processes and influences its variations. Biospheric processes interact with hydrologic cycles particularly through interception and transpiration by vegetation (Oki et al. 2013).

T. Oki (✉)
Institute for Future Initiatives, The University of Tokyo, Tokyo, Japan
e-mail: oki@ifi.u-tokyo.ac.jp

Y. Himiyama et al. (eds.), *Human Geoscience*, Advances in Geological Science,
https://doi.org/10.1007/978-981-32-9224-6_4

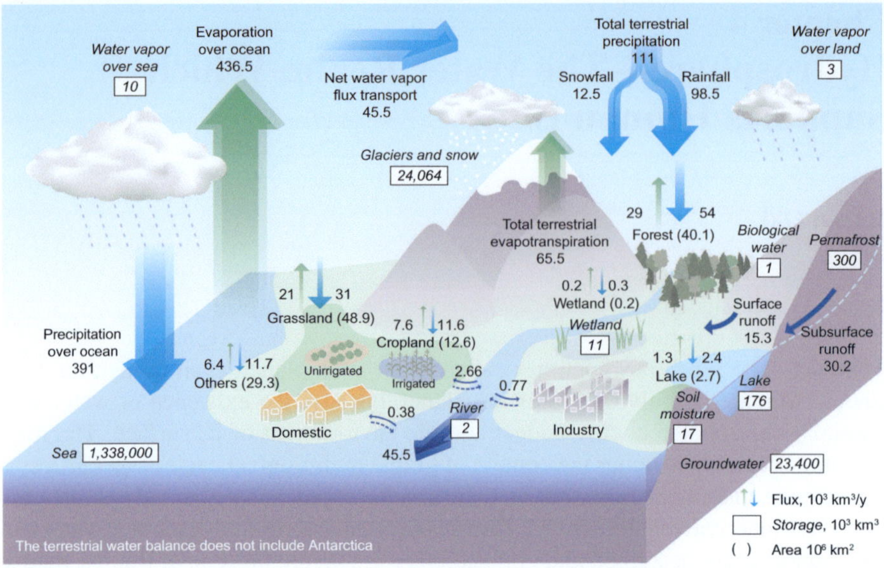

Fig. 4.1 From Oki and Kanae (*Science*, 2006); Global hydrological fluxes (1000 km^3 yr^{-1}) and storages (1000 km^3) with natural and anthropogenic cycles are synthesized from various sources. Vertical arrows show annual precipitation and evapotranspiration over ocean and land with major landscapes (1000 km^3 yr^{-1}). Parentheses indicate the area (million km^2)

Hydrologic cycles are also determined by geological/topographical conditions in the lithosphere, however, there are strong interactions between hydrosphere and lithosphere in terms of the formation of land through geomorphologic processes. Cryosphere and oceanosphere are parts of hydrosphere but due to their massive volume and longer time scales, they are often discussed apart from hydrosphere.

4.2 Major Water Reservoirs on the Earth

Water exists on the Earth surface in various forms, such as liquid, ice, and vapor, and is stored in various storages (sometimes called reservoirs), such as oceans, groundwater, lakes, rivers, and atmosphere. Water flows among these storages, and the water flow through unit area per unit time is called water flux. One of major issues in global hydrology is to quantify how much water is stored in each storage, how long water stays on average, and how much water flows among storages. Comprehensive estimates of water storages and fluxes are schematically illustrated in Fig. 4.1 (Revised from Oki and Kanae 2006), in which stored water are mainly from Korzun (1978) and fluxes are from Dirmeyer et al. (2006). The presented numbers in Fig. 4.1 have differences within approximately ±10% compared with the latter estimates by Rodell et al. (2015). Various estimates on global mean annual precipitation, evapotranspiration, and runoff are compared in Table 4.1. Mean precipitation overland is explained in detail in Park et al. (2017).

The proportion of water mass in the ocean is large (96.5%). The global hydrologic cycle never closes without including oceanic circulation. The ocean circulations carry large amounts of energy and water. The surface ocean currents are driven by surface wind stress, and the atmosphere itself is sensitive to the sea surface temperature. Temperature and salinity determine the density of ocean water, and both factors contribute to the overturning and deep ocean general circulation.

Water over land is stored in lakes, marshes, rivers, soils, aquifers (groundwater), and even in plant body (vegetation). Artificial reservoirs are estimated to have total capacity of approximately 7000–8000 km^3 in the world. They have an ability to suppress global sea level by a few centimeters (Chao et al. 2008; Pokhrel et al. 2012), sometimes explicitly considered in global hydrologic and water resources model (Hanasaki et al. 2008, 2010), however, excluded from the Fig. 4.1.

Solid water, namely ice and snow, covers land and water body in cold regions. Water vapor in atmosphere is invisible and the concentration is approximately 0.3% by mass and 0.5% by volume, however, clouds (with optical depth > 0.1) always cover 0.68 ($\pm$0.03) % of the Earth surface(Stubenrauch et al. 2013).

Major reserves other than the ocean are solid waters on the continent, including glaciers and permanent snow cover. Glaciers are ice accumulations originated from the atmosphere, and they move slowly on land over a long period. Glaciers form *U*-shaped valleys over land and leave moraine deposits when they retreat. If a glacier "flows" into an ocean, it often turns to be an iceberg. Glaciers evolve in a relatively

Table 4.1 Global water balance in mm yr^{-1} by different studies

	Type	Period	P	ET	R	ET/P
Oki and Kanae (2006)[a]	Model-based	1986–1995	826.5	**487.7**	**338.8**	0.59
GSWP2 (Dirmeyer et al. 2006)	Model-based	1986–1995	836.4	**488.4**	**348.0**	0.58
Trenberth et al. (2007)[b]	Observations	Varies by datasets	762.0	492.2	269.7	0.65
WaterMIP (Harding et al. 2011)	Model-based	1985–1999	872.0	**499.0**	**375.0**	0.57
Kim and Oki (2014)	Model-based	1979–2013	**757.0**	469.0	288.0	0.62
MEaSUREs (Rodell et al. 2015)	Observations and model-based	2000–2010 (mostly)	795.7	**481.8**	313.9	0.61

[a]Original values in $km^3\ yr^{-1}$ is divided by $1.34 \times 10^8\ km^2$ (global land area excluding Antarctica)
[b]Original values in $km^3\ yr^{-1}$ is divided by $1.48 \times 10^8\ km^2$ (global land area including Antarctica)
Boldface numbers indicate ensemble estimations

longer timescale in comparison to climatic change. They can also induce isostatic responses of continental-scale upheavals or subsidence in even longer timescales. Even though it was believed that the thermal expansion of oceanic water dominated the anticipated sea-level rise due to global warming, glaciers over land are also a major concern as the cause of sea-level rise associated with global warming in the coming decades.

Groundwater is the subsurface water in the saturated zone. It contributes to the runoff in the low-flow regime between storm events, i.e., during a dry spell. Deep groundwater may also reflect the long-term climatological condition. The groundwater quantity in Fig. 4.1 considers both gravitational and capillary water, but groundwater in the Antarctica (roughly estimated as 2×10^6 km^3) is excluded. Gravitational water is the water in the unsaturated zone (vadose zone) which moves downward by gravity. Capillary water is the water which moves upward due to capillary diffusion. Implementing a macro-scale groundwater dynamics, Koirala et al. (2014) estimated groundwater recharge flux as 31,789 km^3 yr^{-1} which is close to the flux of sub-surface runoff in Fig. 4.1 (30,200 km^3 yr^{-1}).

Soil moisture is the water being held above the groundwater table. It influences the energy balance at the land surface in the way of affecting evapotranspiration (which consists of soil evaporation, plant transpiration, and interception-loss) and changing surface albedo. Soil moisture also alters the fraction of precipitation partitioned into direct runoff and infiltration. When the temperature of the soil column keeps at or below 0 °C for more than two consecutive years, the condition is called permafrost. During the summer season, the upper part of the soil column thaws and the melting water infiltrates downwards, but the permafrost layer is still impermeable like a bedrock. Approximately one-third of precipitation turns to runoffs and one-third of the runoff is estimated as surface runoff. The shares of transpiration, canopy infiltration, and bare soil evaporation are close on global average.

4.3 Flows in Terrestrial Hydrological Cycles

Precipitation consists of rain, snow, and any other water falling from sky to the Earth surface, such as hail and graupel. Precipitation drives the hydrological cycles over land, and also over ocean through changing the ocean surface salinity and temperature, which affect its thermohaline circulation. In vegetated regions, precipitation is intercepted by canopy, and through fall will not occur until the interception storage will be full. Precipitation has large temporal variability and spatial heterogeneity, compared to other major hydrological quantities, and it is complicated and challenging to have reliable estimates of precipitation amount based merely on in situ rain-gauge network. Therefore hybrid products, such as GPCP (Adler et al. 2003) and GSMaP (Ushio et al. 2009), are operationally developed and provided by merging satellite-based estimates and in situ observations.

Snow and rain have different characteristics, and water and energy balances over land differ significantly with and without snow cover. Once the land surface is covered by snow, the surface temperature keeps 0 °C or below until the completion of snow melt. Snow surface also increases the land surface albedo up to 0.9 for new snow, and albedo decreases typically down to 0.6 or below by aging, covered with dust. Aerodynamic roughness is generally reduced with snow cover, and momentum exchange between land and atmosphere is also influenced by snow.

Vaporization of liquid water over land and its transfer into atmosphere is called evaporation. Direct vaporization of solid water (ice or snow) and its transfer is called sublimation. Evaporation and sublimation absorb heat (energy) and water vapor releases the heat when it condensates. Therefore transfer of water vapor is virtually equivalent to the transfer of heat, and evaporation and sublimation are called latent heat flux. Evaporation of water through stomata of leaves is specifically called transpiration, and sometimes distinguished from evaporation from free water surface since transpiration is controlled biogeochemical processes. The aperture of stomata, which dominates the transpiration rate, is related to the soil wetness in the root zone, not at the land surface, and also to the speed of photosynthesis at leaves. Vegetated land has different albedo compared to other land cover, such as bare soil, and has different energy and water balance, in addition to the influence of interception by leaves. Evapotranspiration is the terminology used to call evaporation and transpiration together.

Evapotranspiration is determined by both atmospheric and hydrological conditions, and soil wetness at land surface, in particular, significantly controls the partition of net radiation (energy input to land surface) into sensible and latent energy fluxes. Under wet condition of land surface with enough water to be evapotranspirated compared to available energy, evapotranspiration is mainly controlled by net radiation (energy input) to land surface and classified as radiation-driven (atmosphere-controlled). Under dry condition with limited water compared to available energy at land surface, actual evapotranspiration is smaller than potential evapotranspiration and sometimes formulated by surface resistance, and the condition is classified as hydrology-driven (soil-controlled).

Runoff is the water which flows out from the area, typically as river discharge. River discharge, passing through a stream flow station, accumulates total runoff generated in the upper watershed.

Without considering the river discharge, global hydrologic cycles will not meet the balance. In addition to carrying water back to the ocean from land, river discharge carries various materials, such as sediment, nutrients, organic matters, and chemicals, from land to ocean. River discharge also changes the surface salinity and temperature of ocean, at least close to the coast, and influences the circulation of ocean and the formation of sea ice.

Runoff occurs when the intensity of rainfall or snowmelt exceeds the infiltration capacity of the soil (Hortoian runoff), or rainfall directly runs-off when it falls on the saturated surface ("source area") generated mainly by saturated lateral flow over base rock (Dunne runoff).

It is believed that water movement in unsaturated and saturated soil layers can be formulated by Richard's Equation with detailed information on topography and soil properties, such as hydraulic conductivity and porosity.

4.4 Global Water Balance

The water balance over land is described as Eq. (4.1) where P, E, R, and S are precipitation, evapotranspiration, runoff, and terrestrial water storage, respectively, within an arbitrary boundary as illustrated in Fig. 4.2a:

$$\frac{\mathrm{d}S}{\mathrm{d}t} = P - E - R \tag{4.1}$$

Atmospheric water vapor flux convergence contains water balance information in addition to the traditional hydrological elements such as precipitation, evapotranspiration and discharge. The basic concepts as well as the application of atmospheric data to estimate terrestrial water balance were first presented by Starr and Peixöto (1958). The atmospheric water balance for a column of atmosphere from the bottom at land surface to the top of the atmosphere is described by the equation,

$$\frac{\mathrm{d}W}{\mathrm{d}t} = Q + (E - P) \tag{4.2}$$

where W represents the precipitable water (i.e., column integrated water vapor), and Q is the water vapor flux convergence in the atmosphere. Since the atmospheric water content in both solid and liquid phases is generally small, only the water vapor is considered in Eq. (4.2). Figure 4.2b shows that the water storage in an atmospheric column is increased by the lateral convergence of water vapor and evapotranspiration from the bottom of the column (i.e. land surface), and decreases due to the precipitation falling from the bottom of the atmosphere column to the land (Oki et al. 1995).

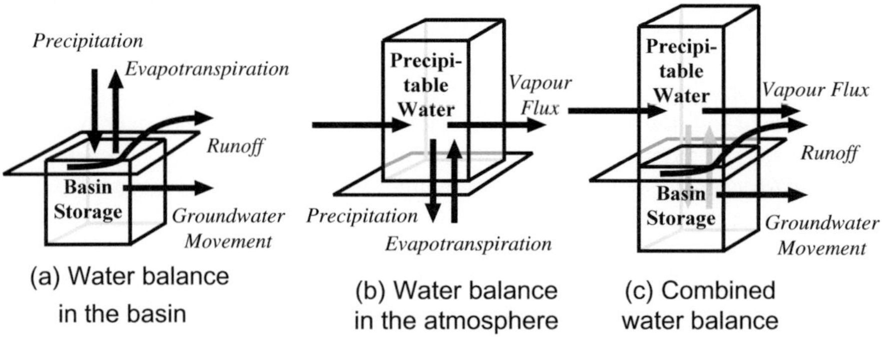

Fig. 4.2 Schematic diagram for **a** terrestrial water balance, **b** atmospheric water balance, and **c** combined atmosphere–land surface water balance corresponding to Eqs. (4.1), (4.2), and (4.3), respectively (Oki et al. 1995)

Since there are common terms in Eqs. (4.1) and (4.2), they can be combined into:

$$-\frac{dW}{dt} + Q = (P - E) = \frac{dS}{dt} + R \tag{4.3}$$

Figure 4.2c illustrates the balance in this equation. The difference of precipitation and evapotranspiration is equal to the sum of the decrease of atmospheric water vapor storage and lateral (horizontal) convergence, and is also equal to the sum of the increase of water storage over the land and runoff. Theoretically, Eq. (4.3) can be applied for any control volume of the land area combined with the atmosphere above, however, the practical applicability depends on the accuracy and availability of atmospheric and hydrologic information.

Global off-line hydrological modeling framework (Oki et al. 1999) has been used to estimate large scale water cycles since this is still the only available methodology which covers global area for a sufficient timespan without any gap. To reduce or estimate simulations uncertainties, approaches such as data model integration (e.g., data assimilation) and multi-model ensemble (MME) have been proposed. Making Earth System Data Records for Use in Research Environments (MEaSUREs) project compiled various source of dataset including remote sensing, atmospheric reanalysis and model simulations and optimized sets of flux terms using a data assimilation technique (Rodell et al. 2015). MME approach has been performed as a community effort frequently. International model intercomparison projects (MIPs) such as Global Soil Wetness Project (GSWP; Dirmeyer et al. 2006; Dirmeyer 2011) and Water Model Intercomparison Project (WaterMIP; Haddeland et al. 2011) are good examples adopting MME approach to quantify the fluxes of water cycles globally. Table 4.1 (Kim and Oki 2016) compares recent studies which estimate global water balance in different approaches.

References

Adler RF, Huffman GJ, Chang A, Ferraro R, Xie P-P, Janowiak J, Rudolf B, Schneider U, Curtis S, Bolvin D, Gruber A, Susskind J, Arkin P, Nelkin E (2003) The version-2 global precipitation climatology project (gpcp) monthly precipitation analysis (1979–Present). J Hydrometeor 4:1147–1167

Chao BF, Wu YH, Li YS (2008) Impact of artificial reservoir water impoundment on global sea level. Science 320:212–214

Dirmeyer PA, Gao XA, Zhao M, Guo ZC, Oki T, Hanasaki N (2006) GSWP-2 multimodal analysis and implications for our perception of the land surface. Bull Am Meteor Soc 87:1381–1397

Dirmeyer PA (2011) A history and review of the global soil wetness project (GSWP). J Hydrometeor 12:729–749

Haddeland I, Clark DB, Franssen W, Ludwig F, Voß F, Arnell NW, Bertrand N, Best M, Folwell S, Gerten D, Gomes S, Gosling SN, Hagemann S, Hanasaki N, Harding R, Heinke J, Kabat P, Koirala S, Oki T, Polcher J, Stacke T, Viterbo P, Weedon GP, Yeh P (2011) Multimodel estimate of the global terrestrial water balance: setup and first results. J Hydrometeor 12:869–884

Hanasaki N, Kanae S, Oki T, Masuda K, Motoya K, Shirakawa N, Shen Y, Tanaka K (2008) An integrated model for the assessment of global water resources—part 1: model description and input meteorological forcing. Hydrol Earth Syst Sci 12:1007–1025

Hanasaki N, Inuzuka T, Kanae S, Oki T (2010) An estimation of global virtual water flow and sources of water withdrawal for major crops and livestock products using a global hydrological model. J Hydrology 384:232–244

Harding R, Best M, Blyth E, Hagemann S, Kabat P, Tallaksen LM, Warnaars T, Wiberg D, Weedon GP, Van Lanen H, Ludwig F, Haddeland I (2011) WATCH: current knowledge of the terrestrial global water cycle. J Hydrometeor. 12:1149–1156

Kim H, Oki T (2014) [Global climate] River discharge [in "State of the Climate in 2013"]. Bull Amer Meteor Soc. 95(7):S23–S24

Kim H, Oki T (2016) Macroscale hydrological modeling and global water balance, in terrestrial water cycle and climate change: natural and human-induced impacts, geophysical monograph. In: Tang Q, Oki T (eds) vol 221, Amer. Geophys. Union, Wiley

Koirala S, Yeh PJF, Hirabayashi Y, Kanae S, Oki T (2014) Globalscale land surface hydrologic modeling with the representation of water table dynamics, J Geophys Res Atmos 119(1):75–89

Korzun VI (1978) World water balance and water resources of the earth—studies and reports in hydrology, 25th edn. UNESCO, Paris

Oki T, Blyth EM, Berbery EH, Alcaraz-Segura D (2013) Land use and land cover changes and their impacts on hydro climate, ecosystems and society. In: Asrar GR, Hurrell JW (eds) Climate science for serving society: research, modeling and prediction priorities. Springer Science + Business Media, Dordrecht

Oki T, Entekhabi D, Harrold TI (2004) The global water cycle. In: Sparks RSJ, Hawkesworth CJ (eds) State of the planet: frontiers and challenges in geophysics. geophysical monograph series. vol 150, 414 pages. AGU Publications, 225–257

Oki T, Kanae S (2006) Global hydrological cycles and world water resources. Science 313(5790):1068–1072

Oki T, Musiake K, Matsuyama H, Masuda K (1995) Global atmospheric water balance and runoff from large river basins. Hydrol Process 9:655–678

Oki T, Nishimura T, Dirmeyer P (1999) Assessment of annual runoff from land surface models using total runoff integrating pathways (TRIP). J Meteor Soc Japan 77:235–255

Park KJ, Yoshimura K, Kim H, Oki Taikan (2017) Chronological development of terrestrial mean precipitation. Bull Am Meteor Soc 98(11):2411–2427

Pokhrel YN, Hanasaki N, Yeh PJ-F, Yamada TJ, Kanae S, Oki T (2012) Model estimates of sea-level change due to anthropogenic impacts on terrestrial water storage. Nature Geosci 5:389–392

Rodell M, Beaudoing HK, L'Ecuyer TS, Olson WS, Famiglietti JS, Houser PR, Adler R, Bosilovich MG, Clayson CA, Chambers D, Clark E, Fetzer EJ, Gao X, Gu G, Hilburn K, Huffman GJ, Lettenmaier DP, Liu WT, Robertson FR, Schlosser CA, Sheffield J, Wood EF (2015) The observed state of the water cycle in the early twenty-first century. J. Climate 28:8289–8318

Starr VP, Peixöto J (1958) On the global balance of water vapor and the hydrology of deserts. Tellus 10:189–194

Stubenrauch CJ, Rossow WB, Kinne S, Ackerman S, Cesana G, Chepfer H, Di Girolamo L, Getzewich B, Guignard A, Heldinger A, Maddux BC, Menzel WP, Minnis P, Pearl C, Platnick S, Poulsen C, Riedi J, Sun- Mak S, Walther A, Winker D, Zen S, Zhao G (2013) Assessment of global cloud data sets from satellites: project and database initiated by the GEWEX radiation panel, Bulletin of the American Meteorological Society. Ame Meteor Soc 94(7):1031–1049

Trenberth KE, Smith L, Qian T, Dai A, Fasullo J (2007) Estimates of the global water budget and its annual cycle using observational and model data. J Hydrometeor. 8:758–769

Ushio T, Sasashige K, Kubota T, Shige S, Okamoto K, Aonashi K, Inoue T, Takahashi N, Iguchi T, Kachi M, Oki R, Morimoto T, Kawasaki Z (2009) A Kalman filter approach to the global satellite mapping of precipitation (GSMaP) from combined passive microwave and infrared radiometric data. J Meteor Soc Japan 87A:137–151. https://doi.org/10.2151/jmsj.87A.137

Chapter 5
Human Sphere—The Earth Surface Realm Created by Human

Yukio Himiyama

Abstract The earth realms, such as atmosphere, hydrosphere, lithosphere and biosphere, have offered the main framework for the major international sustainability programmes for over half a century. However, human sphere is now emerging as an important realm to be added to them. It provides a broad range of academic disciplines with a common ground where and for which to work together for sustainability. Human sphere comprises of human, its various products and environments of both natural and artificial characters. It now surpasses its biological limit, affecting inorganic earth realms such as atmosphere, hydrosphere and lithosphere. A new model chart of the earth realms containing human sphere is proposed. Human sphere has not only been expanding vigorously, but also evolving with large scale changes in land use, life style, agriculture, industrial and other activities, altering its inner subsystems and affecting the whole earth system. They form the background of various global environmental problems and large scale geo-disasters, as well as development of human society. Appropriate monitoring of the changes of human sphere together with its interaction with the other spheres, analyses of the related data and information, and enhancement of understanding in human sphere are essential for global sustainability.

Keywords Human sphere · Earth realm · Earth system · Sustainability · Global environment

5.1 Introduction

In the traditional view of physical geography, inorganic earth realms consist of atmosphere, hydrosphere and lithosphere, partly overlapping with each other, while biosphere is organic realm evolved at their interface (Strahler and Strahler 1976). The concept of the earth realm has been used and developed widely since the 1960s as a useful tool or framework to grasp the complex earth system and to coordinate

Y. Himiyama (✉)
Hokkaido University of Education, Sapporo, Japan
e-mail: himiyamay@kkd.biglobe.ne.jp

Y. Himiyama et al. (eds.), *Human Geoscience*, Advances in Geological Science,
https://doi.org/10.1007/978-981-32-9224-6_5

various disciplines concerned when global environmental problems were becoming increasingly serious.

In 1971, UNESCO launched the *Man and the Biosphere Programme* (MAB) as an international scientific programme that aimed at establishing scientific basis for improvement of relationships between people and their environments. It was intended to combine natural and social sciences, economics and education to improve human livelihoods and equitable sharing of benefits, and safeguard natural and managed ecosystems, thus promoting innovative approaches to economic development that are socially and culturally appropriate, and environmentally sustainable (UNESCO 2017).

In 1987, International Council for Science (ICSU) started the *International Geosphere-Biosphere Programme* (IGBP) (Seitzinger et al. 2015). It is noted that "geosphere", which was not addressed in MAB, was placed at the centre of the programme. It established core projects covering the key global environmental issues related with hydrosphere, atmosphere, land and biosphere, namely BAHC (hydrological cycle), GCTE (terrestrial ecosystems), IGAC (atmospheric chemistry), PAGES (past global changes), LOICZ (land-ocean interactions), LUCC (land use/cover change), GAIM (modeling), and DIS (data & information), in order to implement the programme.

In June 1992, Agenda 21, the United Nations Programme of Action for Sustainable Development, was adopted at the United Nations Conference on Environment and Development (UNCED) at Rio de Janeiro (United Nations 1992). It consisted of 40 chapters in four sections, namely (i) social and economic dimensions; (ii) conservation and management of resources for development; (iii) strengthening the role of major groups; (iv) means of implementation. UNESCO reviewed related activities and achievements in "Encyclopedia of Life Support Systems" (UNESCO/EOLSS 2002), with the opening chapter written by N. Glazovsky under the title "Environmental structure and function—Earth system" with reference to the earth's spheres, namely the atmosphere, hydrosphere, pedosphere, lithosphere, and living matter, which are closely linked with each other by flows of matter and energy (Grazovsky 2002). The "Global Environment Outlook" series of the United Nations Environment Programme (UNEP) also has a similar composition, with chapters on atmosphere, land, water, biodiversity, and earth system, e.g., GEO_5 (UNEP 2012).

The earth realms, or spheres, have provided the main framework not only for physical geography, but also for the major international sustainability programmes for over half a century. It took some time, however, before the concept of human sphere, or humanosphere, was put forward explicitly. It was Takafumi Matsui, a Japanese geo-physicist, who coined the term "ningen-ken" in Japanese, meaning human sphere in English, in 1998. He was deeply involved in the study of the earth realms and the earth system with focus on the energy and material flows and stocks. He considered human sphere as a subsystem of the earth system just like the other physical spheres, which seemingly could be identified more specifically in physical scientific terms than "man" or "human". By so doing he warned the danger of rapid expansion of human sphere (Matsui 1998).

For more in-depth discussion on the inorganic realms, namely atmosphere, lithosphere, and hydrosphere, see Chaps. 2–4 in the present book, respectively.

5.2 Human Sphere

Human sphere refers to the earth realm which comprises of human, its various products and environments of both artificial and natural characteristics in human use or under human impact. It is the core realm for the life support system of human on the earth. Human sphere in this sense covers much of the earth's surface. It characteristically consists of both organic and inorganic matters. The term is relatively new even in the global change research community, probably because it was considered to be marginal in comparison to the other realms, and partly because the "man-and-environment" concept, which was sometimes equated with "man-and-biosphere", was considered to be good enough. However, it is becoming increasingly clear that it may provide a timely common ground for a wide range of sciences in general, and geosciences in particular, to work together towards the common goal of sustainable society. In Japan, a new committee on "Human Geoscience", which addresses the issues related with human sphere, was established in 2005, with participation from such fields as geography, hydrology, soil science, land science, geology, geomorphology, ocean science, climatology, seismology, volcanology, and hazard study (see Chap. 1).

Figure 5.1 is a schematic representation of the earth realms including inorganic realms such as atmosphere, hydrosphere, and lithosphere, together with the organic realm called biosphere, and human sphere. The three inorganic realms are shown as intersecting ellipses, with each one overlaps the other two in a limited area, indicating that some of the substance of each realm is included in the other two. The green circle

Fig. 5.1 The earth realms shown as intersecting eclipses and circles

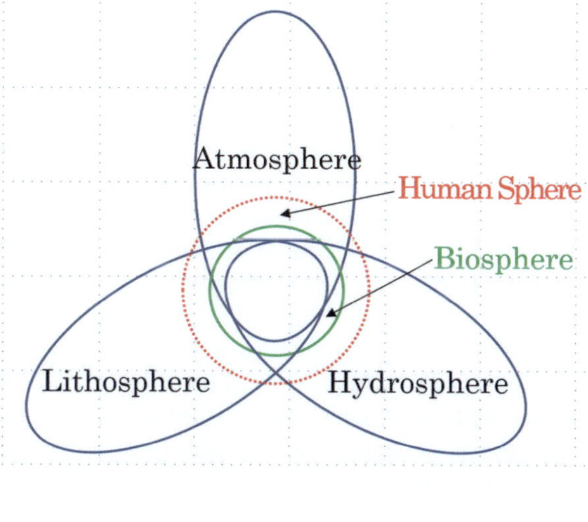

overlapping into the three surrounding realms represents biosphere, which draws its substance from its inorganic environment. The smaller size of the biosphere circle signifies that the total mass of the matter held within the biosphere is smaller than the other three inorganic realms. The human sphere is expressed by a dotted red circle, which surpasses the biosphere circle. It represents not only human and their properties and products of all kinds, but also the environment used, modified or influenced by human, or supporting or accommodating human. Dotted line is used for the human sphere circle in order to emphasize variation or flexibility of the content and extent of human sphere. It can be larger or smaller than the biosphere circle depending on the detail of definition and the historic period it represents. It was originally a marginal existence contained within biosphere, but it grew gradually, and then at increasingly high speed, surpassing its biological limit and altering the inorganic realms, especially since the time of industrial revolution.

It is now widely recognized that human sphere is no longer a tiny trivial existence on the earth, but a non-negligible actor which may disturb the earth system considerably. Many scientists now believe that we are no longer in Holocene Epoch but instead in *Anthropocene Epoch* which is characterized by great human disturbance of the earth system which in turn may endanger human existence. The relative size of the human sphere circle changes by time and definition, but its expansion by time would be irrevocable.

5.3 Unprecedented Expansion of Human Existence

Human depends on its environment heavily, e.g., for breezing air, burning fuel, trapping water from rivers, lakes and underground, catching fish, growing crops, plants, trees and livestock, building and constructing structures, extracting minerals, leisure and sports, and dumping waste. Therefore, human sphere represents the realm far more than human body itself. However, the mass and the number of human on the earth do have significant effect on all those human use of, and impact on, the environment and the earth's life support systems.

The expansion of human existence can be seen quantitatively in the increase of human population (see Chap. 18) and in the increase of more intensive use of the land (see Chap. 6, soil (see Chap. 7) and the water surfaces such as the ocean, rivers and lakes (see Chap. 8). The History Database of the Global Environment (HYDE) developed under the authority of the Netherlands Environmental Assessment Agency and the leadership of Klein Goldewijk, which is still in progress, is an ambitious challenge to offer an overview of the global change of human existence (NEAA 2010; Klein Goldewijk et al. 2011). HYDE has been closely associated with LUCC (Land Use and Land Cover Change Programme) and its successor GLP (Global Land Project). It presents time series of population and land use for the last 12,000 years, together with some other indicators including GDP, livestock, agricultural areas and yields, private consumption, greenhouse gas emissions and industrial production data for the last century.

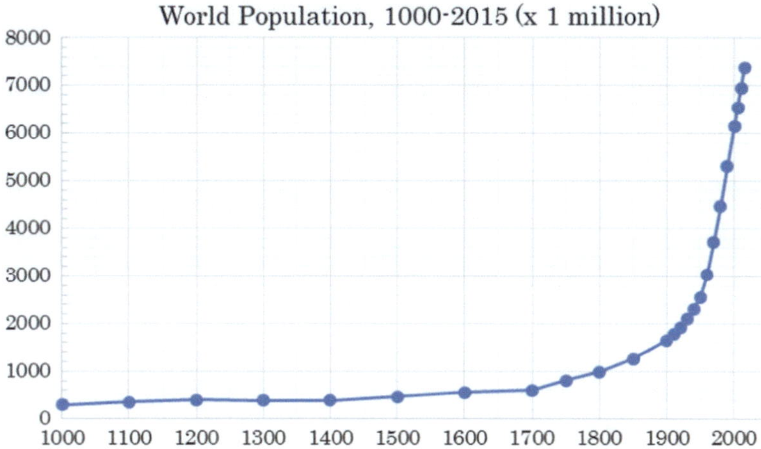

Fig. 5.2 Trend of world population, 1000–2015 [based on HYDE 3.1 and UN (2017)]

Figure 5.2 shows part of the long-term population trend of the world from the year 1000 onward based on the HYDE 3.1 database and the United Nations World Population Prospects (UN 2017). World population was estimated to be about 391 million in 1000 according to HYDE 3.1, while it became 7,383 million in 2015 according to the UN, i.e., 20 times larger than in 1000. The twentieth century witnessed a particularly high growth of population. The population became 3.7 times larger during the twentieth century, with the highest annual increase rate of 2.1% observed in 1962. Although human as a biological existence occupies only small part of human sphere, it is undoubtedly the core of human sphere, and it has been bringing about systemic changes to the earth system. The expansion of human sphere is far more than the increase of human population, as it is amplified greatly by the increasing use and consumption, modification, and exploitation of its environments of both proximate and distant nature by human, various human activities including residing, eating, drinking, moving, studying, gathering and entertaining, as well as economic and industrial activities.

Bulk of these activities takes place on the land, which is the surface of the earth that is not covered by water, in the form of various land uses and land cover. In the history of human civilization, expansion of agricultural land has been one of the greatest events resulting in the expansion of human sphere, both directly and indirectly, through population increase. Figure 5.3 shows the trend of the world pasture area during 1000–2005. The world pasture area was estimated to be about 142.9 million ha in 1000 (1% of land area), while it was 3,428.9 million ha (23% of land area) in 2000, i.e., 23-fold increase within a millennium, according to the HYDE 3.1 database. Unlike population, however, the increase slowed down in the latter half of the twentieth century, and even started to decline in the twenty-first century.

Figure 5.4 shows the trend of the world cropland area during 1000–2005. It was estimated to be about 153.325 million ha in 1000, while it became 1,532.11 million

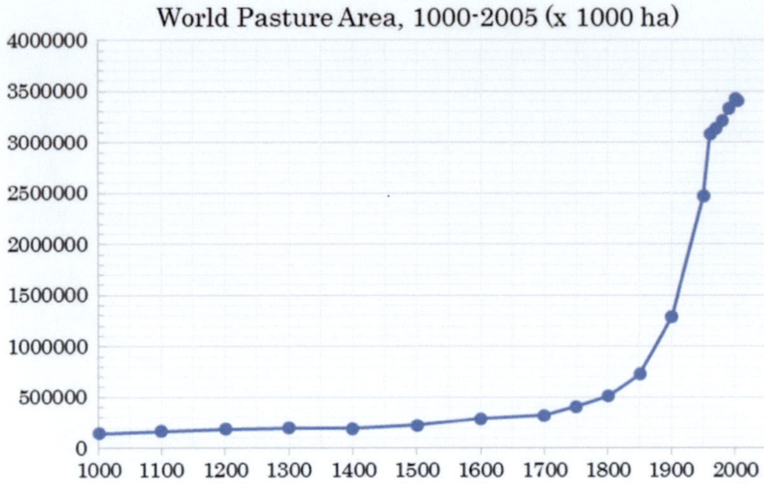

Fig. 5.3 Trend of world pasture area, 1000–2005 (×1000 ha) (based on HYDE 3.1)

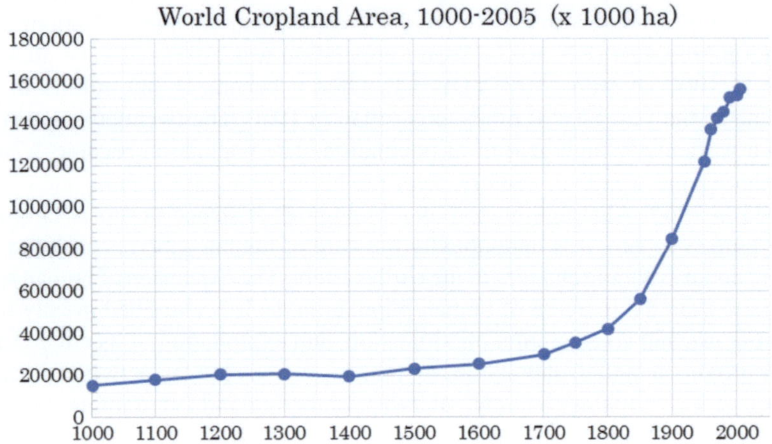

Fig. 5.4 Trend of world cropland area, 1000–2005 (based on HYDE 3.1)

ha in 2000 according to the HYDE 3.1 database, i.e., 10 times larger. Unlike population, however, the increase slowed down in the latter half of the twentieth century, and even started to decline in the early twenty-first century. It was doubled during the nineteenth century, but during the twentieth century, the increase rate went down to 1.8-fold. These facts indicate that although the increase of the world human population is likely to continue for some decades, cropland and pasture land together may have hit its peak at around year 2000 and entered a declining mode already. Thus, the dramatic increase of cropland and pasture land have had immense effects on global biogeochemical cycles and the whole earth system.

Human activities and their effects are now spreading higher in the air (atmosphere), deeper in the oceans (hydrosphere) and underground (lithosphere), and wider on the land and water surfaces, both in visible and invisible ways, and they are intensifying. In another words, human sphere has been expanding into atmosphere, hydrosphere, and lithosphere. Since the industrial revolution, atmosphere has been affected by increasing amount of gaseous and particulate wastes emitted by human action, and some of them have been causing global climatic change, human health and other problems. The emission of CO_2, which is known as greenhouse gas, was estimated to be 0.03 billion ton in 1800, 2 b. ton in 1900, 6 b. ton in 1950, 14.9 b. ton in 1970, 22.3 b. ton in 1990, 36.17 b. ton in 2014, and 36.18 b. ton in 2015, i.e., it increased 1,200-fold during 1800–2015 (Ritchie and Roser 2018). Increase of air transport has also been dramatic. In 2016 airlines carried as many as 3.8 billion passengers on scheduled services with an increase of 7% over 2015 (IATA 2017).

Figure 5.5 shows *Chikyu*, a gigantic scientific drilling vessel 210 m long, 130 m high from bottom, weighing 56,752 ton, which is capable of drilling 3000 m below the deep ocean floor (JAMSTEC 2018). The ship is a front-runner of the expansion

Fig. 5.5 Riser drilling vessel "Chikyu"

of human sphere into lithosphere, accommodating a number of research projects on such themes as (i) *Investigating the "whys" of great earthquakes*, by drilling at the Nankai Trough; (ii) *The origin of life*, by exploring in deep subseafloor coalbeds; (iii) *Quest for the mantle*, by drilling into the mantle of the earth through the oceanic crust; (iv) *Cores—history books of the earth's past*, by using the many geological sample core from the formations drilled by *Chikyu* as a record of past global environmental changes.

The most noticeable expansion of human sphere into hydrosphere might regrettably be the man-made pollution of the world waters by various chemicals, plastics, wastes, garbage, etc. A disturbing fact discovered recently is the pollution by microplastics, i.e., small plastic particles (<1 mm) originating from the degradation of larger plastic debris. These microplastics have been accumulating in the marine environment for decades and have been detected throughout the water column and in sublittoral and beach sediments worldwide. The pollution spread throughout the world's seas and oceans, even into the remote and largely unknown deep seas (Van Cauwenberghe et al. 2013).

5.4 Increasing Stress in Human Sphere

The expansion of human sphere is not a simple spatial enlargement, but a process with its structural and qualitative changes such as large scale changes in land use, life styles, agriculture, industrial and other economic activities, which involves systemic changes, and it affects the whole earth system. These changes form the background of many global environmental problems and large scale geo-disasters as well as the development of human society.

Human has been increasing its population, expanding human sphere, and intensifying its activities, but it cannot continue them endlessly within the limited space and physical conditions on the earth. In fact expansion of agricultural land is becoming extremely difficult, as seen above. Furthermore, shortage of irrigation water and land degradation are likely to become tenser, and farmland abandonment is feared to increase. Lester Brown warns the danger of falling underground water tables, eroding soils, expanding deserts, rising temperatures, and melting ice to food security with many examples from all over the world (e.g. Brown 2011).

Figure 5.6 shows desertification of the former grassland with some standing trees in northern Tarim Basin, China. Mainly due to overgrazing by cattle, Taklimakan Desert overtook the once green ground nearby. There are water pipes on the sandy ground for dripping salty underground water to re-establish vegetation, but it looks an extremely difficult task. This is a symbolic scene of interaction between lithosphere, biosphere and human sphere.

Land is where lithosphere or biosphere meets with atmosphere, and where terrestrial biosphere and human sphere are under strong pressure from cultivation, grazing,

Fig. 5.6 Desertification in Tarim Basin, China

urbanization, mining, and other human activities. Land is not for agriculture alone, and the demand for non-agricultural use of land, such as forest, residential use, urban use and nature conservation, are also likely to increase. *Global Environment Outlook 5* (UNEP 2012) warns that competing demands for food, feed, fuel, fibre and raw materials are intensifying pressure on land, and that globalization and urbanization are aggravating competing demands on land.

Urbanization stands for the growth of cities, expansion of urban land use, concentration of people and commodities, concentration of urban-type industries and functions, development of urban systems, and urban-type cultures (see Chap. 19). In developing countries or regions, urbanization is under a strong influence of economic development, motorization, and modernization, while in developed countries and regions, changes backed by affluence are more evident. Rapid urbanization threatens the surrounding rural areas and push urbanization to more vulnerable areas. Urbanization also separate people from the land and nature, and may promote more human-centric thoughts.

The present book addresses global environmental problems such as those related to desertification, urbanization, population increase, climate change and pollution. For detail, see Chaps. 18–21. The book also focuses on major geo-disasters, such as those related to active faults, earthquakes, tsunamis, volcanic eruptions, flooding, typhoons, land slide, and accident of nuclear power plant. For detail, see Chaps. 9–17.

5.5 Concluding Remarks

Human sphere is expanding and changing rapidly by increasing human population and growing human activities, and is altering its inner subsystems and affecting the whole earth system. It is in no doubt a threatening situation. However, recognition of human sphere as an earth realm in fact offers a common ground for cooperation to a wide range of geosciences specializing in different kinds of geo-disasters and global environmental threats and to those involved in sustainability research and education, a place where and for which to work together for sustainability in general and for reduction of geo and environmental disasters in particular. Relevant monitoring of the changes of human sphere and its interaction with the other realms, analyses of the data and information, and its understanding and governance are essential for global sustainability.

Human tends to put priority on shorter-term economic benefit, convenience or comfort, and longer-term benefit such as longer-term economic benefit, sustainability and safety tend to be put aside, resulting in worsening global environmental problems, increasing geo-disasters, and declining socio-economic stability and resilience. It is hoped that the concept of *human sphere* is used to remind people of the pressing situation of the *Anthropocene Epoch* where we are now in, and to help them understand it in the context of the earth system. Chapters 22–24 discusses the strategy towards sustainable society with these in mind.

References

Brown L (2011) World on the edge—how to prevent environmental and economic collapse. W. W Norton, London

Grazovsky N (2002) Environmental structure and function: Earth system. Knowledge for sustainable development—an insight into the Encyclopedia of life support systems, vol 1. UNESCO/EOLSS Publishers, Paris and Oxford, pp 1–27

IATA (2017) 2017 edition of IATA world air transport statistics. https://www.iata.org/pressroom/pr/Pages/2017-10-09-01.aspx

JAMSTEC (2018) About Chikyu: https://www.jamstec.go.jp/chikyu/e/about/

Klein Goldewijk K, Beusen A, de Vos M, van Drecht G (2011) The HYDE 3.1 spatially explicit database of human induced land use change over the past 12,000 years. Glob Ecol Biogeogr 20(1):73–86. https://onlinelibrary.wiley.com/doi/abs/10.1111/j.1466-8238.2010.00587.x

Matsui T (1998) What is human sphere? In: Earth and planetary science—ser. 14 social earth science. Iwanami Shoten, Tokyo, pp 1–12. (J)

Netherlands Environmental Assessment Agency (2010) History Database of the Global Environment (HYDE): http://themasites.pbl.nl/tridion/en/themasites/hyde/index.html

Ritchie H, Roser M (2018) CO_2 and other greenhouse gas emissions. Published online at OurWorldInData.org. Retrieved from: https://ourworldindata.org/co2-and-other-greenhouse-gas-emissions

Seitzinger SP, Gaffney O, Brasseur G, Broadgate W, Ciais P, Claussen M, Erisman JW, Kiefer T, Lancelot C, Monks PS, Smyth K, Syvitski J, Uematsu M (2015) International geosphere-biosphere programme and earth system science—three decades of co-evolution. Anthropocene 12:3–16

Strahler AN, Strahler AH (1976) Elements of physical geography. Wiley, New York
UNEP (2012) Global environment outlook 5: environment for the future we want. Progress Press, Valletta, Malta
UNESCO (2017) Man and the biosphere programme. http://www.unesco.org/new/en/natural-sciences/environment/ecological-sciences/
UNESCO, EOLSS Publishers (2002) Knowledge for sustainable development—an insight into the Encyclopedia of life support systems. UNESCO/EOLSS Publishers, Paris and Oxford
United Nations (1992) Agenda 21: The United Nations Programme of Action from Rio. United Nations, New York
United Nations (2017) World population prospects: the 2017 revision, DVD Edition: https://esa.un.org/unpd/wpp/Download/Standard/Population/
Van Cauwenberghe L, Vanreusel A, Mees J, Janssen CR (2013) Microplastic pollution in deep-sea sediments. Environ Poll 182:495–499

Part III
Natural Resources and Society

Chapter 6
World Land Resources at Stake

Yukio Himiyama

Abstract Land is an essential resource for human, which offers places for living, food production and bulk of other activities of human, and its ways of use and management are of paramount importance for welfare, wealth, security and sustainability for human. However, despite substantial research and management on land, particularly on its use, rapid population increase, economic development, globalisation, etc. are creating various challenges in land and its use. Land degradation and other land-related problems are seen in many parts of the world, particularly in developing countries and regions. This chapter reviews studies and management practices useful for understanding and elaborating them, and explores the proper ways for using and managing land resources.

Keywords Land use · Land resource · Land management · Urbanisation · Disaster

6.1 Introduction

Land is the solid part of the surface of the earth, or in other words the part of the surface of the earth other than the ocean. Total land area on the earth is 147,244,000 km^2, which occupies 28.9% of the earth's surface according to the latest calculation by the Geospatial Information Authority of Japan (Suga et al. 2013). Much of human sphere and large part of biosphere are on land, including its immediate subsurface. It is one of the most essential resources for human, which is naturally provided, except some man-made land such as reclaimed land, and is used by human as place for living, production, mining mineral and energy resources, supply of water, food, timber and other biological substances, consumption, waste disposal and most other human activities.

Land also provides human with its environment which contains wild life, forest, lakes, rivers, and deserts. People adapt themselves to the given natural conditions of land such as land form, land cover and soil, and socio-economic conditions of land such as its economic value or locational advantage, overcoming less favourable

Y. Himiyama (✉)
Hokkaido University of Education, Sapporo, Japan
e-mail: himiyamay@kkd.biglobe.ne.jp

Y. Himiyama et al. (eds.), *Human Geoscience*, Advances in Geological Science,
https://doi.org/10.1007/978-981-32-9224-6_6

conditions. The desire for more land and better land might have been the nature of human since its early days of evolution, and it might have pushed human to expand or move its territory, or to modify the ways to use or manage it, or to protect it. The ways how land resource is used or managed are of paramount importance for welfare, wealth, security and sustainability for human.

As land resource is such a big issue, there are many ways to see it. Mather and Chapman (1995) saw forests, biodiversity and environmental conservation as main issues of land resource in the context of environmental resources. In the present book, it is discussed in the context of natural resources and society, with soil resource and water resource discussed in separate chapters. Urbanisation and geo-disasters, which are both very important issues of land resource but haven't been treated well enough for their importance in land science, are given special attention here.

The twentieth century saw an unprecedented increase in human population of the world, as shown in Fig. 6.1 (NEAA 2010; Klein Goldewijk et al. 2011; United Nations 2017). The estimated world population in 1900 was 1654 million, and that in 2000 was 6145 million, i.e., 3.7-fold increase in a century. The increase of population, together with the rapid industrialisation, urbanisation, and pervasion of materialistic cultures, resulted in the increase of burden to land by human. Accumulation of waste and pollutants and environmental degradation has become common on land as well as in the sea and the atmosphere. Human has been expanding and intensifying its activities on land, but it cannot continue it forever ignoring the limits to growth on the earth. Further increase of agricultural land in the world is unlikely in many parts of the world, and soil degradation, shortage of irrigation water, farmland abandonment due to collapse of rural communities, competing demands of different land uses, environmental degradation by development etc. will accelerate unless governed properly (Brown 2009).

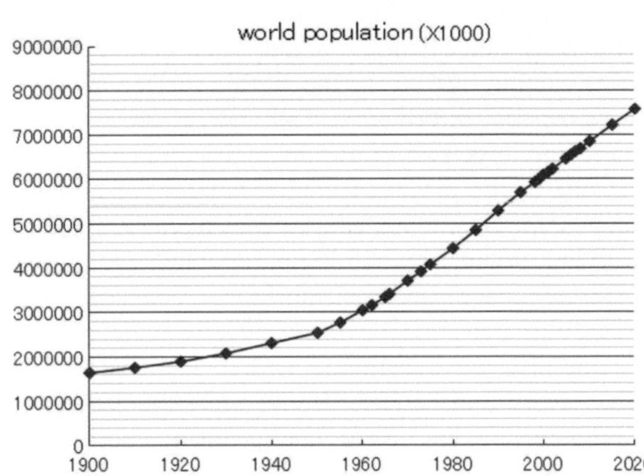

Fig. 6.1 World population trend since 1900 [based on HYDE 3.1 and UN (2017)]

6.2 Land Use

Before World War II, the main concern of the society about land was rural land use for food production. In Europe, where conflicts for land resources became increasingly serious through the twentieth century, rural land use became an important national issue for survival of the country. In Britain in the early twentieth century, there existed topographic maps made by the Ordnance Survey, but they had only limited information of land use. So, Dudley Stamp conducted a land-use survey of the whole of Britain in the 1930s, mainly in 1931 to 1934, in order to map the food production capacity of the land of Britain (Stamp 1948). The survey was significant not only because it helped the country to survive the naval blockade by Germany during World War II, but also because it initiated national-level land-use survey in the world, and offered the basic information and spatial data of the land use in Britain for analyses, evaluation, and planning.

One of the outcomes of the Stamp's survey was the establishment of a planning system in Britain with the prime aim of protecting good agricultural land. However, the planning authorities thus established were bitterly criticised by Alice Coleman, the Director of the Second Land Utilisation Survey of Britain carried out in the 1960s, on the ground that they failed to protect good agricultural land and instead even accelerated its decline (Coleman 1977). By comparing her survey results with the Stamp's, Coleman discovered that 5% of total improved farmland in England and Wales was lost between 1933 and 1963 mainly due to urban sprawl, and warned food crisis which may arise as a result of rapid increase of the world population. It was a justifiable proclamation indicating the basic principle of land-use planning relevant in the era of global environmental crisis, which may be shared by 'Limits to Growth' (Meadows et al. 1972).

Figure 6.2 is part of the 1:25,000 land use map sheet named 'Merthyr Tydfil' (SO00 and SO10) produced by the Second Land Utilisation Survey by adding detailed information of land use and vegetation on to the Ordnance Survey base map (Coleman and Maggs 1968). It shows the area around a small village in South Wales called Aberfan. 1 km national grid lines of UK and some types of land use and vegetation can be identified, namely, field vegetables (purple); grassland (light green); woodland (dark green); settlement (grey); heath and rough land (yellow); derelict land (black stipple). The map shows a typical landscape of South Wales which once flourished by coal mining but was then struggling to live in the narrow valley. Aberfan is known as the site of the disaster in 1966 in which a colliery spoil tip created on the mountain slope above the village, which overlaid a natural spring, slid downhill due to the heavy rain, killing 116 children and 28 adults in and near the local junior school. The location of the tip, which is classified as derelict land, and the school are shown by T and S, respectively, and a narrow stream at the southern edge of the tip leading to the school is visible. This case shows the power of large scale land-use map with fine information of topography provided by the Ordnance Survey base map for disaster prevention, i.e. more sustainable land use.

Fig. 6.2 Part of 1:25,000 land use map 'Merthyr Tydfil' produced by the Second Land Utilisation Survey of Britain

The works of Stamp and Coleman became a model of study of land use not only in the UK but also in other countries. A remarkable example of it is the publication of 'Land Use in China' edited by Wu and Guo (1994) that was written in connection with the '1:1,000,000 Land Use Map of China' (Wu 1988). The map sheets themselves had rich information on natural and social characteristics, land use, problems of land use and the countermeasures in each covered area, and they, together with the book, offer invaluable information of land use in China. Thus, land use studies armed with land use maps have shown their effectiveness in many countries.

It is noted that official statistical data of land use in China, as well as in many other countries, is quite unreliable, although efforts have been made in recent years to improve the past data in China and elsewhere.

Land use had been a major focus of geography till the early 1980 s, but as urbanisation, globalisation and other societal changes became prominent, geographical study of land use linked with land use maps was gradually broadened, if not scattered. It got a new focus and momentum in 1996, when the International Geographical Union (IGU) established a Study Group (later Commission) on Land Use/Cover Change called IGU-LUCC in order to contribute to the newly born Land Use/Cover Change Programme (LUCC) created in 1995 as a core project jointly by the International Geosphere/Biosphere Programme (IGBP) and the International Human Dimensions Programme (IHDP).

6.3 Land Use/Cover Change in Global Research Context

LUCC played an important role in adding land science into IGBP and IHDP, i.e. into the global environmental research framework. 'Land-Use and Land-Cover Change—local processes and global impacts' (Lambin and Geist 2006) summarises the achievements of the programme and shows direction for its successor, namely the Global Land Project (GLP). The book consists of the following parts: (1) Introduction: local processes with global impacts; (2) Global land-cover change: recent progress, remaining challenges; (3) Causes and trajectories of land-use/cover change; (4) Multiple impacts of land-use/cover change; (5) Modelling land-use and land-cover change: scenarios from the local to global scale; (6) Searching for the future of land; (7) Linking land-change science and policy: current lessons and future integration; (8) Conclusion.

'Our Earth's Changing Land—an encyclopedia of land-use and land-cover change' (Geist 2006) is another notable output of LUCC. It is a resource book full of information on land use/cover change, which was supported by IGU-LUCC and other related international bodies. The book deals with land change with focus on the changes of land use and land cover, where land use refers to the use or meaning assigned to land, while land cover refers to biophysical state of the earth's surface and its immediate subsurface. There are over 300 items under such headings as 'conventions and environmental agreements', 'cropland change', 'dryland change', 'forest change', 'land-cover conversion and modification', 'land-use dynamics', and 'models and scenarios'.

LUCC was merged with Global Change and Terrestrial Ecosystems (GCTE) Programme in 2005, and was restructured as Global Land Project (GLP, 2005–2015), and continued its important role as a global research project under the IGBP-IHDP framework. GLP assumed a leading role in the development of database and indices that are useful for the global research projects. HYDE (History Database of the Global Environment), which was developed by Netherlands Environmental Agency (Klein Goldewijk et al. 2011), is an example of such database, which offers those of long term change of population, land use, and other related items. Consolidation of such database could be done only by the joint work of researchers and database authority. GLP was restructured again in 2015 under the new platform of Future Earth.

Despite of the remarkable achievements of LUCC and GLP, considerable part of the mission of LUCC and GLP had not been completed as expected. Under such circumstances, IGU-LUCC played a unique role in filling some of the gaps caused by these transformations of core projects (Himiyama 2018). One of the notable schemes of IGU-LUCC has been the publication of the 'Atlas—Land Use/Cover Changes in Selected Regions in the World' (e.g. Himiyama and Bicik 2017). This scheme has promoted production and use of historical information bases and maps of land use/cover changes in over twenty countries/regions to date [e.g. 'Land Cover Change Mapping by Applying the CORINE Land Cover Database: Regional Example' (Otahel et al. 2005)]. Each article normally covers a substantially large area, includes large A3-size maps of national/regional scales, has information on data source and preparation; land use/cover classification; study methods; important facts, processes, drivers and mechanisms of land use/cover changes, etc. Other publications of IGU-LUCC include, among many others, the following two books, both consisting of five parts each as follows:

- 'Understanding Land-use and Land-cover Change in Global and Regional Context' (Milanova et al. 2005): Global trends of land-use and land-cover change and their interaction with global environmental changes; Regional diversity and trends of land-use and land-cover changes; Integration of environmental, socio-economic and historical knowledge of land-use and land-cover change; Documentation, classification and mapping issues; Land-use management and policy in various regions.
- 'Land-use Changes in Comparative Perspective' (Himiyama et al. 2002): Data and methods; Urbanisation; Sustainability of rural land use; Land reclamation and coastal change; Land-use change in peripheral areas.

6.4 Urbanisation and Land Use

Growth of cities, both in population and area terms, is a common phenomenon in the world in the twentieth and twenty-first century. It characteristically accompanies concentration of population, economic, social, cultural, and other activities, changing lifestyles, and expansion of urban areas, and is inclusively called urbanisation. It is an indicator of regional development, and it is rapid in the countries or regions at developing stage. It is important for land use not only because it is itself a major change of land use which has great socio-economic effect as well as safety implication for the people living or working there, but also because it is a major driving force of the changes of the whole land system including land-use system.

However, accurate identification of urban expansion, which is necessary for urban and regional planning, is often a challenge. Globally, urbanisation of different countries is usually compared by percentage of population residing in what are called cities, while the definition of a 'city' does not always reflect the level of urbanizsation or concentration of population properly, and varies greatly from country to country or even from city to city within a country. 'Urbanization in Japan since 1960 viewed from the DID statistics' (Himiyama and Morishita 2003) is an example of showing the use of the DID (Densely Inhabited District) statistics in Japan in grasping urbanisation since 1960, and concluded that the population density in the urban area represented by the DID declined sharply, as against commonly held expectation, during the 1960s when Japan was experiencing rapid economic growth and urbanisation, and the decline gradually levelled, off afterwards. DID was proved to be an operational indicator of urbanisation, but it is not in common use internationally.

Rapid urbanisation threatens sustainability in many developing countries, with serious pollution of air, water, and land, and there has been numerous research on the issue. However, they mostly focus on major cities, overlooking the others, and the time period only since 1990, which is not enough for understanding urbanisation and its consequences. Moreover, difficulty in obtaining usable reliable data of urban expansion, namely, lack of relevant maps or statistics of the past as well as the present, makes it inevitable to rely on satellite images heavily, which makes it difficult to grasp urbanisation accurately or in longer time range going back to the 1980s and before in many developing countries.

Himiyama et al. (2018) studied urbanisation in Jiangsu Province and Zhejiang Province in China since ca. 1930 by using the 1:50,000 topographic maps produced by Japan, 'Atlas of Cities of China' (Zhou and Chu 1994) and Landsat ETM+ satellite images, and showed that the urban expansion in the 1990s was about four-fold in 10 years, i.e., at the speed twice as fast as that during the rapid economic growth in Japan in the 1960s, that expansion was faster in medium to smaller cities than in big cities, and that the agricultural land around cities gave way to urban expansion rapidly. For more detailed discussions on urbanisation, see Chap. 19.

6.5 Geo-Disasters and Land Resources

In recent years, people tend to favour short-term benefit or economic advantage, easiness, or comfort at the cost of sustainability or safety of the use of land or ocean, which makes global environmental problems more serious, increases disasters, and accelerates socio-economic fragility and instability. The high concentration of human activities in the coastal areas, particularly in the coastal plains, further accelerates the danger.

Disasters caused by geo-hazards, such as earthquakes, tsunamis, volcanic activities, and extreme weather, affect land and its use seriously, and they are sometimes generated or amplified by the existing land use. 'Global Environment Outlook' (GEO), a comprehensive report on global environment issued every five years or so by the United Nations Environment Programme (UNEP), repeatedly warned the increase of economic costs of great disasters, and suggested that global warming, destruction of the natural environment or inappropriate land uses for short-term economic benefit, migration of population to urban and coastal areas, etc., may be the reasons for the increase (UNEP 2002, 2007). However, unlike IGU-LUCC, both LUCC and GLP put relatively low priority on disasters (Himiyama 2018). Ironically, the recent increase of great geo-disasters is gradually convincing sustainability research communities the importance of incorporating disasters into their programmes.

Japan has many disasters affecting, or being caused by, the existing land use, so the importance of incorporating disaster into the study of land use/cover has been known for many years. It was recognised again in Japan on March 11, 2011, when the Great East Japan Earthquake of M 9.0 and the gigantic tsunami that followed attacked the Pacific coast in eastern Japan, devastating huge coastal area. According to the Geospatial Information Authority of Japan (2011), the area inundated by the tsunami

Fig. 6.3 Newly developed residential area (in front) on the hill side for the people who were moved from the tsunami-stricken coastal lowland (right back), Miyako City, Japan, April 2017

was 561 km^2. By land-use type, paddy field occupied 38.4%, other agricultural land 5.4%, forests 7.0%, built-up areas 32.9% of the inundated area.

Figure 6.3 shows a newly developed residential area (in front) on the hill side about 60 m above sea level in Miyako City in North-East Japan for the people who were moved from the tsunami-stricken coastal lowland (right back) about 5 m above sea level. This low land was densely populated until the tsunami, but all the residents there have moved to a safer place like the one in this photograph. The emptied lowland is now used as a baseball stadium, solar panels, and some commercial facilities. The twenty or so houses in the photograph are among the 15 thousand houses built on the higher ground by or for those who lost their houses by the tsunami (Reconstruction Agency 2018). The impact of the March 11 tsunami does not stop at the inundated coastal low land, but it stretches further deep inland.

Figure 6.4 shows the abandoned paddy field near Tokyo Electric Power Company (TEPCO) Fukushima Daiichi Nuclear Power Plant, which has little chance of recovery as paddy field because of the heavy contamination by radioactive substances originating from the accident of the plant caused by the earthquake and tsunami of March 11, 2011. Seven years later, there are still tens of thousands of former residents evacuated from the contaminated area who cannot, or not willing to, return home because of still remaining radiation and collapsed local communities. Efforts have been made to reduce contamination of less contaminated paddy fields or dry fields, but thousands of hectares of agricultural land is still in serious difficulty. The accident of nuclear power plant and the resultant radioactive contamination are extremely complex and delicate issues, and even official data and information are not always reliable. For more detailed discussion about nuclear disaster, see Chap. 17, also ‘Unravelling the Fukushima Disaster’ (Yamakawa and Yamamoto 2017a) and ‘Rebuilding Fukushima’ (Yamakawa and Yamamoto 2017b).

Fig. 6.4 Abandoned paddy field near TEPCO Fukushima Daiichi Nuclear Power Plant, Futaba Town, Japan, April 2017

'The Impact of the Great East Japan Tsunami of 2011 on Land Use' (Himiyama et al. 2014) is an article written jointly by those involved in IGU-LUCC and in GLP. It is meaningful to understand the Great East Japan Disaster not only in the context of the established disaster studies, but also in the framework of GLP or the other global change research programmes such as those in Future Earth, bearing in mind that disasters are increasing in frequency and severity under the tense interactions between humans and nature worldwide.

The present book addresses global environmental problems such as those related to desertification, urbanisation, population increase, climate change and pollution. For detail, see Chaps. 18–21. The book also focuses on major geo-disasters, such as those related with active faults, earthquakes, tsunamis, volcanic eruptions, flooding, typhoons, land slide, and accident of nuclear power plant. For detail, see Chaps. 9–17.

6.6 UNEP's Global Environment Outlook

The United Nations Environment Programme (UNEP) issues a comprehensive report on global environment entitled 'Global Environment Outlook' (GEO) every five years or so, and it usually has a separate chapter on land as well as on atmosphere, water, and biodiversity. In fact, land is referred to in several other chapters, and the chapter on land is known as a comprehensive periodic outlook on land resources in the world. The main messages on land as a resource in GEO5 (UNEP 2012) are as follows:

- Pressure on land resources has increased during recent years despite international goals to improve their management.

- Economic growth has come at the expense of natural resources and ecosystems.
- Competing demands for food, feed, fuel, fibre, and raw materials are intensifying pressures on land.
- Globalisation and urbanisation are aggravating competing demands on land.
- Improved governance and capacity building are crucial to achieving sustainable land management.
- Potential exists to create more sustainable land systems.

'GEO-6, the latest GEO, was issued on June 2016, not in print but electronically (UNEP 2016). UNEP proclaims that the assessment for the six UNEP regions paints a comprehensive picture of the environmental factors contributing to human health and well-being at the regional level and that it also highlights the complexity of the interlinked environmental, social and economic challenges now confronting decision-makers. A great advantage of this new system is that it can give more region-specific account to the region's priority issues. The key message on the state of land resources in each region written in GEO-6 (UNEP 2016) is briefly summarised below:

- Africa: Land is the most valued asset and a critical resource for all aspects of life and development in Africa, but it faces new challenges from changing use, including urbanisation, mining, deforestation, agricultural expansion, and infrastructure development.
- Asia Pacific: Preserving land resources in their best condition is of paramount importance for Asia and Pacific. Despite interventions to conserve them, however, land resources are continuously put under pressure by rapidly emerging social, demographic and economic changes in the region.
- Europe: Land resources show the interaction between development trends, globalisation, landscape frangmentation, urbanisation, coastal over-development and a set of drivers and pressures leading to land degradation processes and changes in land use.
- Latin America and the Caribbean: Demand for food crops, livestock, wood, oil, and mining, coupled with adverse socio-economic conditions and the need for foreign investment, exerts pressure on decision-makers to prioritise short-term goals that may result in land degradation.
- North America: Land resources are generally in good shape. Developed landscapes largely meet society's needs for housing, economic development, and community well-being. Natural landscapes satisfy many food and fibre needs of the region, while contributing to global supplies.
- West Asia: The spread of land degradation and desertification and its economic and environmental consequences are the most critical challenges. Shrinkage of agricultural lands due to population growth, urbanisation, land degradation, and desertification will jeopardise regional food security.

Thus, the regional messages of GEO-6 are much more region-specific and concrete than those of GEO5. GEO-6 also has the summary of the six regional assessments, but it has some drawbacks. Firstly, it is a mere collection of the summaries of the six regional reports and lacks global assessments. Second, some parts of the summaries oversimplify, ignore or misinterpret some important items in the main regional reports, probably because of the bias or lack of understanding of those concerned.

6.7 Towards Sustainable Use and Management of Land Resource

Land on the earth is huge, but it is likely to be over-populated already and human population is still increasing rapidly, with increasing per capita demand for land as well as many other resources. There have been substantial research and management practices on land, particularly on its use, as reviewed above. However, as the GEO-6 regional reports indicate, the state of land, its use, and management differ greatly from region to region, from one stage of development to another, and from one viewpoint to another. In order to keep land and its use more sustainable, it is necessary to pursue research on land and its use and management at different scales, different regions, and in multidisciplinary perspectives.

The United Nations adopted the '2030 Agenda for Sustainable Development' (UN 2015) in 2015, and approved 17 Sustainable Development Goals (SDGs) as an urgent call for action by all countries for peace and prosperity for people and the planet. It offers a great opportunity and responsibility to the human geoscience community in general, and to the land science community in particular, to contribute to this global initiative for sustainability. The following two goals are particularly closely related to land resource:

- Goal 11. Make cities and human settlements inclusive, safe, resilient and sustainable
- Goal 15. Protect, restore and promote sustainable use of terrestrial ecosystems, sustainably manage forests, combat desertification, and halt and reverse land degradation and halt biodiversity loss.

However, land is related with most other goals as well, so efforts should be made to promote research and management practices contributing to Goal 11 and Goal 15 and the other Goals, bearing in mind the importance of their coordination and consolidation of basic land science research and database making. There is always a danger of 'failing to see the wood for the trees'.

Land is close to everyone but maybe too close to be recognised by human. Land is a very precious resource, but may have too diverse a value to be understood by human. It may look calm and stable, but it is not so always, or everywhere. In fact, land faces various serious challenges in the world, and proper use and management of land are of paramount importance for welfare, wealth, security and sustainability for human.

References

Brown L (2009) Plan 4.0—mobilizing to save civilization. W. W. Norton, London
Coleman A (1977) Land use planning—success or failure? Archit J 3(165):91–134
Coleman A, Maggs KRA (1968) Land use survey handbook. Second land use survey, London
Geist H (ed) (2006) Our earth's changing land—an encyclopedia of land-use and land-cover change. Greenwood Press, Westport, USA
Geospatial Information Authority of Japan (2011) Land use area by type and municipality in the tsunami inundation zone. http://www.gsi.go.jp/common/000060371.pdf (in Japanese)
Himiyama Y (2018) Towards sustainable land use in Asia. In: Himiyama Y (ed) Exploring sustainable land use in Monsoon Asia. Springer, Singapore, pp 3–11
Himiyama Y, Bicik I (eds) (2017) Land use/cover changes in selected regions in the world volume XII, IGU-LUCC
Himiyama Y, Morishita Y (2003) Urbanization in Japan since 1960 viewed from the DID statistics. Rep Taisetsuzan Inst Sci 37:37–52
Himiyama Y, Hwang M, Ichinose T (eds) (2002) Land-use changes in comparative perspective. Science Publishers, Enfield, USA
Himiyama Y, Watanabe T, Hara K (2014) The impact of the great east Japan tsunami of 2011 on land use. Glob Environ Res 18(2):191–206
Himiyama Y, Ikeshita M, Shinde T (2018) Urbanization in Jiangsu Province and Zhejiang Province in China since ca. 1930. In: Himiyama Y (ed) Exploring sustainable land use in monsoon Asia. Springer, Singapore, pp 15–32
Klein Goldewijk K, Beusen A, de Vos M, van Drecht G (2011) The HYDE 3.1 spatially explicit database of human induced land use change over the past 12,000 years. Glob Ecol Biogeogr 20(1):73–86
Lambin EF, Geist HJ (eds) (2006) Land-use and land-cover change—local processes and global impacts. Springer, Berlin
Mather A, Chapman K (1995) Environmental resources. Longman Sci. & Tec, Harlow
Meadows DH, Meadows DL, Randers J, Behrens WW III (1972) The limits to growth. Universe Books, New York
Milanova E, Himiyama Y, Bicik I (eds) (2005) Understanding land-use and land-cover change in global and regional context. Science Publishers, Enfield
Netherlands Environmental Assessment Agency (2010) History database of the global environment (HYDE): http://themasites.pbl.nl/tridion/en/themasites/hyde/index.html
Otahel J, Feranec J, Cebecauer T, Pravda J, Husar K (2005) Land cover change mapping by applying the CORINE land cover database (regional example). In: Himiyama Y, Mather A, Bicik I, Milanova E (eds) Land use/cover changes in selected regions in the world volume III & IV. IGU-LUCC, Asahikawa
Reconstruction Agency (2018) State of construction of new housing on higher land. Reconstruction Agency home page, Japan. http://www.reconstruction.go.jp/ (in Japanese)
Stamp D (1948) The land of Britain—its use and misuse. Longmans & Geographical Publications, London
Suga M, Ando A, Ubukawa A, Ueda M (2013) Calculation of sea and land area and their ratio of each latitude band by Global Map version 2. GSI J 124:95–98 http://www.gsi.go.jp/common/000085617.pdf (in Japanese)
UNEP (2002) Global environment outlook 3: environment for development. Progress Press, Valletta, Malta
UNEP (2007) Global environment outlook 4: environment for development. Progress Press, Valletta, Malta
UNEP (2012) Global environment outlook 5: environment for the future we want. Progress Press, Valletta, Malta

UNEP (2016) GEO-6: Global environment outlook: all regional assessments. UNEP, Nairobi, Kenya https://europa.eu/capacity4dev/unep/blog/geo-6-global-environment-outlook-all-regional-assessments
United Nations (2015) Transforming our world: the 2030 agenda for sustainable development. https://sustainabledevelopment.un.org/post2015/transformingourworld
United Nations (2017) World population prospects: The 2017 Revision, DVD Edition: https://esa.un.org/unpd/wpp/Download/Standard/Population/
Wu C (ed) (1988) 1:1,000,000 land use map of China. Science Press, Beijing (in Chinese)
Wu C, Guo H (eds) (1994) Land use in China. Science Press, Beijing (in Chinese)
Yamakawa M, Yamamoto D (2017a) Unravelling the Fukushima disaster. Routledge, New York
Yamakawa M, Yamamoto D (2017b) Rebuilding Fukushima. Routledge, New York
Zhou G, Chu C (1994) Atlas of cities of China

Chapter 7
World Soil Resources at Stake

Tsuyoshi Miyazaki and Chihiro Kato

Abstract After human beings initiated soil cultivation and food production, civilizations in the world thrived thanks to fertile soils but they went to ruin due to the harsh treatment of soils. To keep a sustainable society and to nourish the earth population more than 10 billion in the future, soils must be maintained in good conditions suitable for growing crops and providing safe lives. However, during about 20 years at the end of twentieth century, 2 billions hectare of land degraded because of human activities and soil resources in the world are now standing on the edge of precipice. The four actions, namely, prevention of soil degradations, global stores of soil organic matter, global limits of nitrogen and phosphorus uses, and improving soil observation systems, are the greatest priorities. Global adoption of sustainable soil management is a mutual issue for all human beings.

Keywords Soil degradation · ANTHROSOLS · Climate change · Fukushima

7.1 Soil and Human Beings

Soil and human relations started about 10,000 years ago when human beings initiated soil cultivation and food production in their lands. According to the description by Carter and Dale (1974), who published a famous book entitled "Topsoil & Civilization", civilizations in the world thrived thanks to fertile soils but they went to ruin due to the harsh treatment of soils.

Modernization of agriculture has spawned new earth crisis. Some of large-scale irrigation and drainage facilities in arid area have promoted salinization of soils. Overgrazing and over cultivation under the increasing population pressures have accelerated salinization and alkalization of soils. Figure 7.1 shows salt crust formation in north-east area of China, where the overgrazing of livestock have accelerated salinization and alkalization of soils (Wang et al. 2009).

T. Miyazaki (✉)
The University of Tokyo, Tokyo, Japan
e-mail: t_miya8519_1102@yahoo.co.jp

C. Kato
Hirosaki University, Hirosaki, Japan

Y. Himiyama et al. (eds.), *Human Geoscience*, Advances in Geological Science,
https://doi.org/10.1007/978-981-32-9224-6_7

Fig. 7.1 Surface salt crust formation in north-east China

Excessive irrigations in arid area have caused water loggings, where the groundwater levels exceeded the land surfaces, resulting in barren lands. Some of the deforestations in the mountain forest slopes and large scale land developments have been promoting soil erosions. Wind erosions and dust storms from single-crop farming lands after cutting such agricultural crop as wheat have been threatening farmers. Figure 7.2 shows a famous dust storm approaching Stanford, Texas, 1935, generated from croplands, where the farmers had been suffered from drought and wind erosion.

Spreading of excess fertilizers and pesticides has caused chemical degradations of soils and ground water contaminations. Introductions of heavy machineries to agricultural lands have been causing such soil physical degradations as hard pan formations.

Figure 7.3 shows the degradation of soils in the world, where soils are classified into four categories, namely, very degraded soil, degraded soil, stable soil, and soil without vegetation. According to the report presented by GLASOD (Global Assessment of Soil Degradation) study in 1997, 2 billions hectare of land degraded because of human activities during about 20 year at the end of twentieth century.

To keep a sustainable society and to nourish the earth population more than 10 billion, soils must be maintained in good conditions suitable for growing crops and to live safely.

Fig. 7.2 A dust storm in 1935. Reprinted from NOAA Photo Library

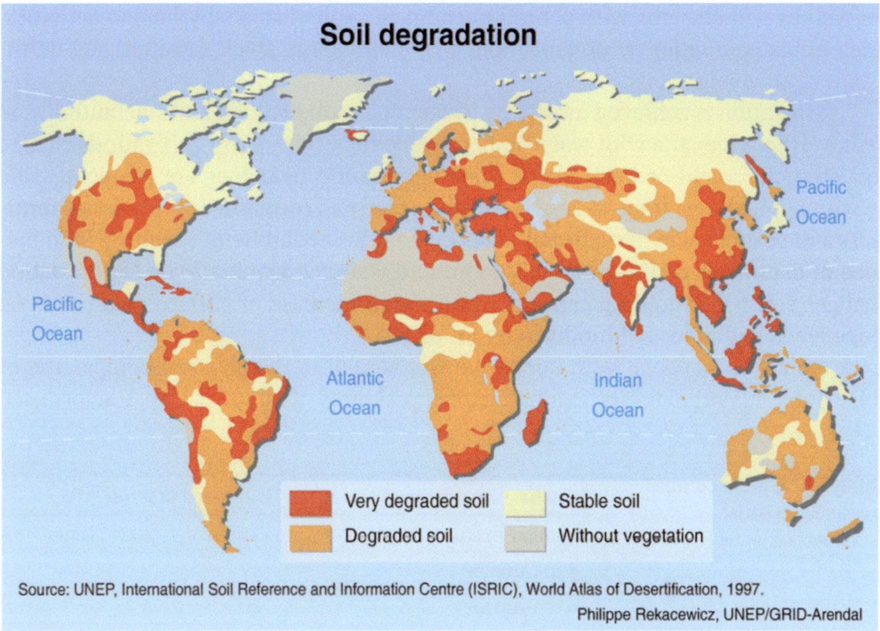

Fig. 7.3 Soil degradation of the world. Reprinted from Rekacewicz (1997)

7.2 Soil Changes with Natural Processes

Soil changes naturally through such processes as soil formation with chemical weathering, physical weathering, and biological weathering. Additional soil changes occur by transportation of soils through soil erosion, soil deposition, and transfiguration under the effects of biological activities.

Soils were initially formed under the effects of climate, resulting in chemical and physical weathering of country rocks. Afterward, soils were lost from lands through water erosion and wind erosion. Natural sedimentations formed alluvial plains, diluvial plains, and fan slopes. Natural disasters, such as slope failures, landslides, tsunami deposits, and eruption of volcanos, reformed fresh soils on the earth. Biological activities, such as microbial metabolism, nest formation by ant, and root net formation by plants, changed soil conditions. In summary, soils could be formed and could change naturally without human beings, while human beings couldn't and cannot exist without soils.

7.3 Soil Changes with Human Impacts

As well as soil changes with natural phenomena, soil changes by human activities. Such human activities as urbanization, crop cultivation, stock-farming, and deforestation, all change soils.

Urbanization is achieved mostly by tremendous soil banking and soil cutting, and by covering wide area with water-impermeable materials. Crop cultivations change soil physical conditions especially when agricultural machines are frequently and heavily applied, resulting in very hard subsurface pan formation. Overuses of fertilizers and pesticides in agricultural fields degrade soil conditions, resulting in imbalance of chemicals and occasionally in ground water contaminations. Deforestation changes soil water holding capacity, resulting in increase of surface water flow on slope lands and in loss of biodiversity in forests.

A typical soil change with human impact is seen in the physical compaction of soils. Table 7.1 provides the estimated tread pressures of human's step, tractors used

Table 7.1 Estimated tread pressures on soils

Causes of tread pressure	Estimated pressure kPa
Human's step	15
4 ton-tractor	17
6 ton-tractor	20
23 ton-tractor	49
Automobile	98
Hoof of cattle	172

Note: 1 Pa is equal to 1 N/m^2

in agricultural fields, cars, and hoof of cattle. It is noted that the larger the tractor, the larger the estimated tread pressure. Larger tractor may give higher efficiency of soil treatment, while it affects more on soil compaction. The severely compacted soils are not suitable for crop growth and decline the soil hydraulic permeability, resulting in poor efficiency of excess water drainage.

7.4 International Year of Soils (IYS) 2015 and International Decade of Soils (IDS) 2015–2024

The Intergovernmental Technical Panel on Soils (ITPS), the main scientific advisory body to the Global Soil Partnership (GSP) hosted by the Food and Agriculture Organization of the United Nations (FAO), issued "Status of the World's Soil Resources" report responding to the International Year of Soils (2015) declared by the General Assembly of the United Nations.

After the declaration, in consideration of the importance of world's soil resources, International Union of Soil Science (IUSS) proclaims the International Decade of Soils 2015–2024. In addition, they are going to move towards the Centenary of the IUSS formation in 2024. These activities from the soil scientists' side are all generated from their belief that to maintain healthy life and environment, the significance of soils remains continuously at the forefront of political and scientific planning and decision making.

7.5 Possible Soil Changes Under the Climate Change

7.5.1 Earth's Global Warming Raises Soil Temperature

Hartman et al. (2013) showed that the globally-averaged land and ocean surface temperature has risen 0.85 °C over the period of 1880 to 2012. Bai et al. (2014) showed that soil temperature at depth of 50 cm has increased at an average rate of 0.79 °C per decade from 1982 to 2000 in Mojav Desert region in the USA, and pointed out that the temporal changes of soil temperature were correlated to temporal changes of air temperature.

Changes in soil temperature are rather more complex in snow zone. The observed data from 30 climate stations across Canada showed that a significant warming trend of mean soil temperature below depth of 5–150 cm was detected in spring and summer, while such a warming trend was not detected in winter and autumn (Qian et al. 2011). They pointed out that, due to the reduced snow cover in winter, heat insulation effect of snow cover on underlying soil probably decreased in winter and resulted in increase of soil temperature in spring after snow melt. The same tendency has also been found in simulated future soil temperature in forests sites in Sweden (Jungqvist et al. 2014).

7.5.2 Global Warming Promotes Biochemical Reaction in Soil (Promote or Alter)

Various kinds of chemicals exist in soil and thus biochemical reactions perpetually occur in vadose zone. Single biochemical reactions generally depend on temperature. Arrhenius equation describes this temperature dependence, where the reaction rate increases with temperature rising. It is reasonably estimated that the increase of soil temperature accompanied by global warming accelerates the bio-chemical reactions in soil.

Soil respiration, the sum of the respiration of root and soil microbes, is another important biochemical reaction in soil. For example, Bond-Lamberty and Thomson (2010) reported, based on the literature review published over the past 40 years, that soil respiration has a positive correlation with air temperature.

The temperature coefficient Q10 is defined as a factor which describes the increase rate of soil respiration with 10 °C rise of temperature. Q10 is widely used for quantitative expression of chemical reaction in biological processes related to temperature (Hashimoto 2005). Q10 values vary with climate zones, soil type, land uses, (forest, agricultural lands, grasslands, etc.) and so on. For example, the value of Q10 is 1.31 for bare Andosol in temperate zone of Japan (Choi and Inoue 2004), 2.3–2.4 for crop lands of maize, and 2.8 for that of wheat (Nonaka et al. 2004). Q10 values range from 1.30 to 3.45 and the mean value is 2.18 in Japanese forests (Hashimoto 2005). Note that Q10 value also increases with temperature.

The increase in atmospheric CO_2 affects soil respiration, too. Under higher atmospheric CO_2 condition, fine roots of plants may grow more and their respirations may increase, resulting in increase in substrates of soil microbes. Niinisto et al. (2004) reported that higher air temperature, higher atmospheric CO_2, and their combination increased the annual total soil respiration in boreal pine forests.

It is important to describe here the different results presented by several researchers. Wang et al. (2012) and Wan et al. (2007) showed that annual soil respiration rates definitely increased under elevated CO_2 concentration condition but they did not necessarily increase under higher temperature. Tingey et al. (2006) reported that neither elevated CO_2 concentration nor warming stimulated soil respiration.

The reason why the effects of global warming on soil respiration did not agree among researchers lies on the fact that other natural factors are dominating than biotic and abiotic factors. Note that such natural factors as nitrogen enrichment, soil moisture depletion, labile soil C pool decrease, and changes in the soil microbial community structure are likely occurring under warming condition (Tingey et al. 2006). When these factors are dominating, there is some possibility that both biotic effects and abiotic effects are negligibly small (Wan et al. 2007), resulting in obtaining different data among researchers. This is why further researches are required in the future.

7.5.3 Global Warming Alters Soil Communities

1 g of soil contains up to 1 billion bacterial cells consisting of tens of thousands of taxa, and thus soils are highly diverse (Wagg et al. 2014). Changes in air temperature and rainfall pattern accompanied with global warming will alter the relative abundance and function of soil communities since the temperature sensitivity and the growth rate of soil community members vary with species (Classen et al. 2015). For example, DeAngelis et al. (2015) mentioned that 5 °C of temperature rising for 20 years in a temperate forest increased the bacterial-to-fungal ratio of the community.

Taking both the changes in soil moisture and the variations of rainfall pattern into account, response of microbial community to climate change might be far complex (Classen et al. 2015). For example, microbial communities are estimated to be stimulated rapidly with moisture changes, while fungal communities are estimated to be stimulated slowly due to their slower-growing characteristics. In addition, soil fungal communities may shift from one dominant member to another, but bacterial communities remain constant (Kaiserman et al. 2015).

Changes in microbial community structure are likely to lead to changes in ecosystem function (Classen et al. 2015). For example, microbial community structure in soil horizons where organic matter is relatively rich changed more. Warming and elevated atmospheric CO_2 may have altered the quantity and quality of soil organic matter, resulting in the change of soil biodiversity.

Soil biodiversity and soil community composition are deeply related with such ecosystem multi-functionality as decomposition of plant litter and nitrogen (N) turnover. For example, soil biodiversity was less in a sieved soil than in raw and unsieved soil. The sieving of raw soil, by which soil particles of similar size prepared simpler living spaces for ecosystem, generated a simpler soil community composition, and resulted in lower ecosystem multi-functionality (Wagg et al. 2014).

Based on the field experiments across five terrestrial and aquatic locations ranging from the subarctic to the tropics, it was shown that the rates of litter decomposition, carbon cycle, and nitrogen cycle decreased as the diversity of microbial community decreased (Handa et al. 2014).

Though numerous reports mentioned the consequences of sifts in soil biodiversity for ecosystem functions, there is still much to be learned to reveal the impacts of soil biodiversity change caused by global warming (FAO and Intergovernmental Panel on Soils 2015).

7.5.4 Climate Change Alter Soil Moisture Conditions and Physical Properties

Soil stores water, mainly supplied by precipitation, in its unsaturated zone and approximately 60% of the whole terrestrial precipitation returns back to the atmosphere by terrestrial evapotranspiration (Oki and Kanae 2006). Solar energy consumed by latent heat flux during terrestrial evapotranspiration is more than half of the total solar energy absorbed by land surface (Trenberth et al. 2009). However, soil moisture conditions are strongly dependent on regional climate conditions and the regionally dependent soil moisture conditions constrain plant transpiration and photosynthesis, which affects regional water, energy and biogeochemical cycles (Seneviratne et al. 2010).

Thus soil moisture is involved in a number of complex feedbacks from local to global scales and it plays a major role in climate change project. Soil moisture itself is also important for crop production and soil biodiversity, and predicting future soil moisture condition is useful for discussing adaptation of climate change.

Seneviratne et al. (2006, 2010) predicted the future soil moisture with three IPCC-AR4 GCMs (Global Climate Models), which van Ulden and van Oldenborgh (2006) showed the most accurate representation of present-day North-hemispheric circulation pattern, and reported that, in North Hemisphere summer (June/July/August), mean soil moisture in 2080–2099 will decrease in Europe and the Southern U.S. compared with the period of 1970–1989. They mentioned that it mainly causes the decreases in mean JJA precipitation, especially in central Europe, but it is also due to the accumulated precipitation deficits and in some regions evapotranspiration excess in preceding months. IPCC-AR5 (2014) summarized that decrease in soil moisture in certain region is high confidence and moister surface soils in anywhere is little-to-no confidence at the end of twenty-first century under the Representative Concentration Pathway (RCP)2.6, 4.5, 6.0 and 8.5 scenarios.

7.5.5 Prediction of Future Soil Moisture by Downscaling Global Climate Model, GCM

When discussing impacts of climate change including soil moisture at regional and/or finer scale, applying downscaling of GCM projections into optimal temporal and spacial scales are recommended (Prudhomme et al. 2002; Wilby et al. 2004; Kato and Nishimura 2015; etc.). For example, Masaki et al. (2011) conducted the prediction of hydraulic condition of North-east Thailand to reveal the impacts of climate change on agriculture with downscaled projections which were obtained from 10 GCMs. They reported that the annual soil moisture may decrease over the region, but the degree of water deficit may vary with geographical condition. Wada et al. (2013) predicted the future Irrigation Water Demand (IWD), which was calculated with surface water balance, soil water balance and the difference between potential and

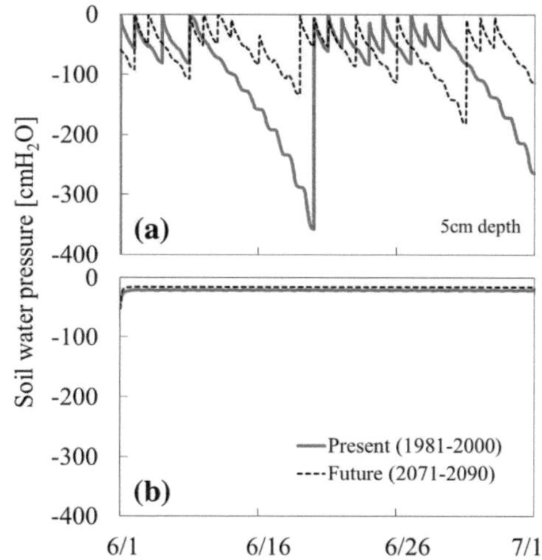

Fig. 7.4 Comparison of simulated present and future soil moisture condition in Toyama city, Japan, with (**a**) and without (**b**) temporally downscaled GCM projection under the condition that projected monthly precipitation depths in June were the largest. Reprinted and modified from Kato and Nishimura (2015) with permission from The Japanese Society of Irrigation, Drainage and Rural Engineering

actual crop evapotranspiration of 23 regions across the world with downscaled GCM projections. They showed that IWD will considerably increase (>20% by 2100) from May to September over North and Central America and most of Asia. Applications of GCM projections to phenomena of not only region or basin scales but also field scales have also been attempted.

Kato and Nishimura (2015) investigated several types of downscaling the GCM projections for the purpose of adaptation of climate change. Especially, they focused on downscaling rainfall data into hourly scale in order to predict soil water condition of agricultural lands where soil hydrological properties are heterogeneous. Finally, they exemplified comparisons of present (dashed lines) and future (solid lines) soil water pressures at the depth of 5 cm in Toyama city, Japan, as shown in Fig. 7.4. It is noted that when we use temporally downscaled GCM projection we can obtain a realistic soil water pressure prediction as shown in Fig. 7.4a, while when we do not adopt downscaling procedure we obtain a nonrealistic and too much-simplified soil water pressure prediction as shown in Fig. 7.4b. Downscaling is crucial for predicting future soil moisture conditions under the climate change.

7.6 Soil Restoration and Hazard Regulation Techniques

7.6.1 Soil Hazard Due to Earthquake Followed by Radioisotope Contamination in 2011, Japan

Owing to the steady and persevering researches on the effects of Chernobyl accident, 1986, the behavior of cesium 137 after falling out on ground surface have been clarified in many aspects. Vidal et al. reported, 1995, that 75–90% of radioactive cesium 137 felt out on land surface were partitioned into the molecular layers of clay minerals resulting in strong adsorption by these minerals, while 10–25% of them remained as the mobile and exchangeable cation. Komamura et al. claimed, 2006, based on the continuous measurements of 137Cs included in soils all over Japan from 1959 to 2000, that 82–85% of radioactive cesium 137 felt out on land surface was partitioned into the molecular layers of clay minerals, while 15–18% of them remained as the mobile and exchangeable cation. Table 7.2 shows the fate of 137Cs reported by Vidal et al. (1995) and by Komamura et al. (2006).

Shiozawa et al. 2011, reported the radioactive Cs concentration profiles in Fukushima soil at May 24th, 2011, two months after the Fukushima Daiichi nuclear disaster, in undisturbed paddy field which are shown by solid lines and in plowed paddy field which is given by dashed lines in Fig. 7.5. It was clearly shown that 96%

Table 7.2 Fate of Cesium 137 after falling out on the ground surface

Origin of 137Cs	Chernobyl accident		Behaviors of 137Cs after falling out on the ground surface
Form of 137Cs	Vidal et al. (1995) (%)	Komamura et al. (2006)	
Exchangeable	10–25	15 ± 4% (paddy fields)	Being transported with water movement
		18 ± 4% (upland fields)	
Adsorptive	75–90	85% (paddy fields)	Stay at the ground surface, being brew away by wind, or being swept away by water together with fine soil particles
		82% (upland fields)	

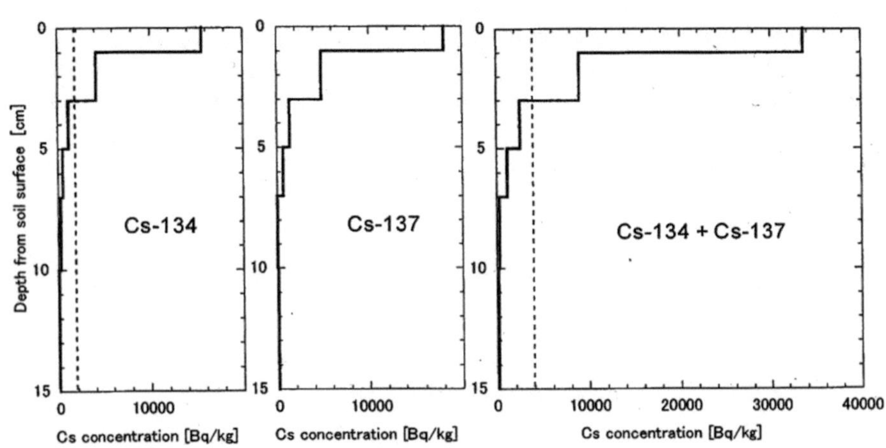

Fig. 7.5 Radioactive Cs concentration profiles in soil on May 24th, 2011 in undisturbed paddy field (solid line) and plowed paddy field (dashed line). Reprinted from Shiozawa et al. (2011) with permission from Japan Radioisotope Association

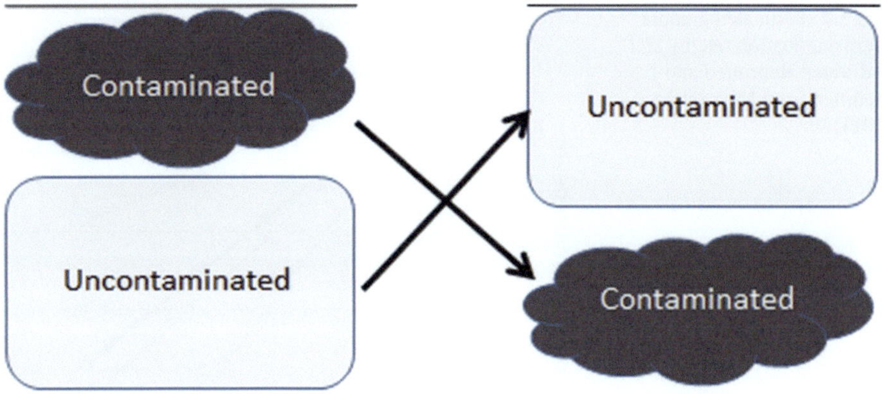

Fig. 7.6 Turning over procedure of contaminated and uncontaminated soils

of Cs134 and Cs137 remained in the upper 5 cm soil layer under uncultivated paddy field.

Since materials like soils and water can attenuate gamma beams, it is recommendable to cut the surface contaminated soils and replace them with the uncontaminated subsurface soils. Figure 7.6 illustrates the soil-turning over procedure.

By using the mass attenuation coefficient of gamma beam as shown in Table 7.3, we can estimate the degree of gamma beam attenuation as a function of the thicknesses of materials (Miyazaki et al. 1985).

The attenuation of one-dimensional gamma beam by a shading material is described by

$$\frac{I}{I_0} = e^{-\rho\mu x}$$

where I is the gamma count through a shading material, I_0 is the originally generated gamma count, ρ is the density of shading material, μ is the mass attenuation coefficient of shading material, and x is the thickness of the shading material. When the contaminated and uncontaminated soils are two-dimensionally placed one upon another, the gamma beam emission is regarded to be vertically one dimensional. Therefore, by knowing the soil bulk density, the mass attenuation coefficient of the

Table 7.3 Mass attenuation coefficients of gamma beam through several materials

Materials	Dry bulk density (g/cm^3)	Mass attenuation coefficient (cm^2/g)
Water	1	0.0835
Volcanic ash soil	0.569	0.0764
Alluvial clayey soil	1.277	0.0756

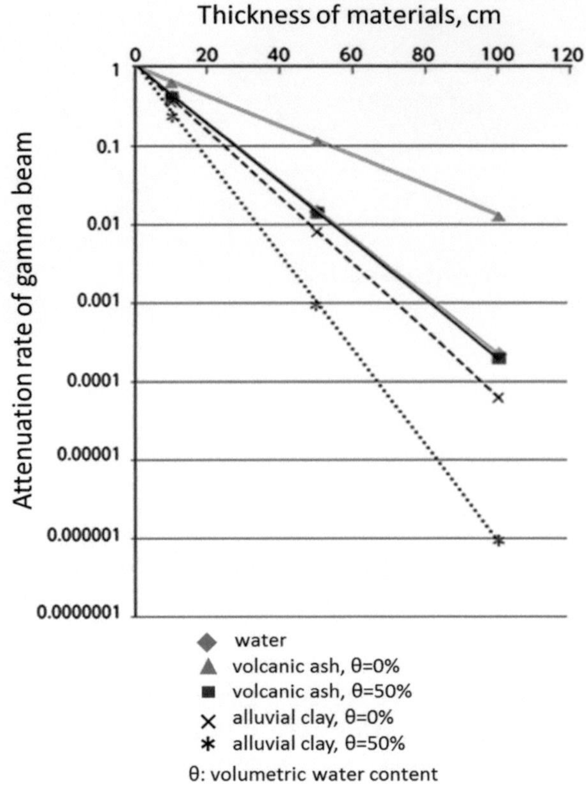

Fig. 7.7 Estimated gamma beam attenuation rate by soil and water. Reprinted and modified from Miyazaki (2011)

soil, and the uncontaminated soil thickness, we can estimate the degree of gamma beam attenuation generated from the buried contaminated soil.

Figure 7.7 represents the estimated degree of gamma beam attenuation. For example, if we cover the contaminated soil by 55 cm-thick volcanic ash soil containing volumetric water content of 50%, we can expect the gamma beam emission to be 1/100 of the originally generated count at the ground surface.

The adsorption of Cs137 by clay minerals is irreversibly strong and stronger more with time duration. This means that buried Cs137 is hardly transported to the groundwater underneath or into the plant roots.

The decontamination by physical removal of the contaminated soil brings about another subject to be solved where and how to find the final disposal area.

7.6.2 Recommendation by ITPS

Integrating all the investigation of soils in the world and realizing that world soil resources are now at stake, FAO and Intergovernmental Technical Panel on Soils

(ITPS) (2015) strongly recommended the following four actions, namely, prevention of soil degradations, global stores of soil organic matter, global limits of nitrogen and phosphorus uses, and improvement of soil observation systems.

Soil resource in the world is thus standing on the edge of a precipice. Global adoption of sustainable soil management is a mutual issue for all human beings.

References

Bai Y, Scott TA, Min Q (2014) Climate change implications of soil temperature in the Mojave Desert, USA. Front Earth Sci 8(2):302–308

Bond-Lamberty B, Thomson A (2010) Temperature-associated increases in the global soil respiration record. Nature 464:579–583

Carter VG, Dale T (1974) Topsoil & civilization, Revised Edition, University of Oklahoma Press, Norman, 292p

Choi EN, Inoue Y (2004) Relationship between dynamic change of CO_2 flux over Andisol upland field and remotely sensed surface temperature. J Agric Meteorol 60(1):33–41 (in Japanese with English abstract)

Classen AT, Sundqvist MK, Henning JA, Newman GS, Moore JAM, Cregger MA, Moorhead LC, Patterson CM (2015) Direct and indirect effects of climate change on soil microbial and soil microbial-plant interactions: what lies ahead? Ecosphere 6(8):1–21

DeAngelis KM, Pold G, Topcuoglu BD, van Diepen LTA, Varney RM, Blanchard JL, Melillo J, Frey SD (2015) Long-term forest soil warming alters microbial communities in temperate forest soils. Front Microbiol 6:1–13

FAO and Intergovernmental Technical Panel on Soils (ITPS) (2015) Status of the world's soil resources (SWSR)—main report. Italy, Rome

Handa IT, Aerts R, Berendse F, Berg MP, Bruder A, Butenschoen O, Chauvet E, Gessner MO, Jabiol J, Makkonen M, McKie BG, Malmqvist B, Peeters ETHM, Scheu S, Schmid B, van Ruijven J, Vos VCA, Hattenschwiler S (2014) Consequences of biodiversity loss for litter decomposition across biomes. Nature 509:218–221

Hartmann DL, Klein Tank AMG, Rusticucci M, Alexander LV, Brönnimann S, Charabi Y, Dentener FJ, Dlugokencky EJ, Easterling DR, Kaplan DR, Soden BJ, Thorne PW, Wild M, Zhai PM (2013) Observations: atmosphere and surface. In: Climate Change (2013) The physical science basis. Contribution of Working group I to the fifth assessment report of the intergovernmental panel on climate change [Stocker TF, Qin D, Plattner G-K, Tignor M, Allen SK, Boschung J, Nauels A, Xia Y, Bex V, Midgley PM (eds)]. Cambridge University Press, Cambridge, United Kingdom and New York, NY, USA

Hashimoto S (2005) Q10 values of soil respiration in Japanese forests. J For Res 10:409–413

Jungqvist G, Oni SK, Teutschbein C, Futter MN (2014) Effect of climate change on soil temperature in Swedish boreal forests. PLoS ONE 9(4):1–12

Kaisermann A, Maron PA, Beaumelle L, Lata LC (2015) Fungal communities are more sensitive indicators to non-extreme soil moisture variations than bacterial communities. Appl Soil Ecol 86:158–164

Kato C, Nishimura T (2015) Temporal downscaling methods of GCM projections for predicting soil moisture in agricultural land. Trans JSIDRE 295:11–19 (in Japanese with English abstract)

Komamura M, Tsumura A, Yamaguchi N, Fujiwara H, Kihou N, Kodaira K (2006) Long-term monitoring and analysis of 90Sr and 137Cs concentrations in rice, wheat and soils in Japan from 1959 to 2000. Bull Natl Inst Agro Environ Sci No 24:1–21 (in Japanese with English summary)

Masaki Y, Ishigooka Y, Kuwagata T, Goto S, Sawano S, Hasegawa T (2011) Expected changes in future agro-climatological conditions in Northeast Thailand and their differences between general circulation models. Theor Appl Climatol 106(3):383–401

Miyazaki T, Hasegawa S, Kasubuchi T (1985) A comparison of single-, double-, and dual-gamma ray measurement of soil and water. Trans JSIDRE 119:29–37 (in Japanese with English abstract)
Miyazaki T (2011) To revive water, soil and green. Bull Agric Acad Jpn 16:83–91 (in Japanese)
Niinisto SM, Silvola J, Kellomaki S (2004) Soil CO_2 efflux in a boreal pine forest under atmospheric CO_2 enrichment and air warming. Glob Chang Biol 10:1363–1376
NOAA Photo Library (1935) theb1365, NOAA's National Weather Service (NWS) Collection, http://www.photolib.noaa.gov/htmls/theb1365.htm
Nonaka K, Aoki M, Horie K (2004) CO_2 flux from soil surface under crop rotation practice. J Agric Meteorol 60(3):197–207 (in Japanese with English abstract)
Oki T, Kanae S (2006) Global hydrological cycles and world water resources. Science 313(5790):1068–1072
Prudhomme C, Reynard N, Crooks S (2002) Downscaling of global climate models for flood frequency analysis: where are we now? Hydrol Process 16:1137–1150
Qian B, Gregorich EG, Gameda S, Hpkins DW, Wang XL (2011) Observed soil temperature trends associated with climate change in Canada. J Geophys Res 116:D02106
Rekacewicz, P. (1997) UNEP/GRID-Arendal, International Soil Reference and Information Centre (ISRIC), World Atlas of Desertification. https://www.grida.no/resources/5507
Seneviratne SI, Corti T, Davin EL, Hirschi M, Jaeger EB, Lehner I, Orlowsky B, Teuling AJ (2010) Investigating soil moisture-climate interactions in a changing climate: a review. Earth Sci Rev 99:125–161
Seneviratne SI, Lüthi D, Litschi M, Schär C (2006) Land-atmosphere coupling and climate change in Europe. Nature 443:205–209
Shiozawa S, Tanoi K, Nemoto K, Yoshida S, Nishida K, Hashimoto K, Sakurai K, Nakanishi T, Nihei N, Ono Y (2011) Vertical concentration profiles of radioactive caesium and convective velocity in soil in a paddy field in Fukushima. Radioisotopes 60(8):323–328
Tingey DT, Lee EH, Waschmann R, Johnson MG, Rygiewicz PT (2006) Does soil CO_2 efflux acclimatize to elevated temperature and CO_2 during long-term treatment of Douglas-fir seedlings? New Phytol 170:107–118
Trenberth KE, Fasullo JT, Kiehl J (2009) Earth's global energy budget. Bull Ame Meteorol Soc 90:311–323
van Ulden AO, van Oldenborgh GJ (2006) Large-scale atmospheric circulation biases and changes in global climate simulations and their importance for climate change in Central Europe. Atmos Chem Phys 6:863–881
Vidal M, Roig M, Rigol A, Llarado M, Rauret G, Wauters J, Elsen A, Cremers A (1995) Two approaches to the study of radiocaesium partitioning and mobility in agricultural soils from the Chernobyl area. Analyst 120:1785–1791
Wada Y, Wisser D, Eisner S, Flörke M, Gerten D, Haddeland I, Hanasaki N, Masaki Y, Portmann FT, Stacke T, Tessler Z, Schewe J (2013) Multimodel projections and uncertainties of irrigation water demand under climate change. Geophys Res Let 40(17):4626–4632
Wagg C, Bender SF, Widmer F, van der Heijden MGA (2014) Soil biodiversity and soil community composition determine ecosystem multifunctionality. Proc Natl Acad Sci USA 107(4):1443–1446
Wan S, Norby RJ, Ledford J, Weltzin JF (2007) Responses of soil respiration to elevated CO_2, air warming, and changing soil water availability in a model old-field grassland. Glob Chang Biol 13:2411–2424
Wang L, Seki K, Miyazaki T, Ishihama Y (2009) The causes of soil alkalinization in the Songnen Plain of Northeast China. Paddy Water Environ 7:259–270. https://doi.org/10.1007/s10333-009-0166-x
Wang X, Nakatsubo T, Nakane K (2012) Impacts of elevated CO_2 and temperature on soil respiration in warm temperate evergreen Quercus glauca stands: an open-top chamber experiment. Ecol Res 27:592–602
Wilby RL, Charles SP, Zorita E, Timbal B, Whetton P, Mearns LO (2004) Guidelines for use of climate scenarios developed from statistical downscaling methods. IPCC, (online), http://www.ipcc-data.org/guidelines/dgm_no2_v1_09_2004.pdf

Chapter 8
World Water Resources at Stake

Taikan Oki

Abstract Overviews on the basic human needs on water and world water withdrawals are introduced, and the reason why we should concern on the availability is explained by its uneven distribution seasonally and geographically. The relationship between human society and water is also illustrated with an example of the dawn of civilization why all of them are located in semi-arid regions downstream of moderately large rivers, not gigantic rivers, with their head waters in wet region. The global estimates of withdrawals for irrigation and artificial reservoirs to support the stable supply of water, and their impacts on sea-level rise or suppress are discussed. The special characteristics of water compared to energy and food are articulated and the reason why water-intensive commodities, such as food, are traded internationally as "virtual water trade", instead of water itself is elucidated.

Keywords Green water · Civilization · Artificial reservoirs · Sea level · FEW nexus

8.1 Introduction

Life on the Earth evolved in water, and utilizes water as a media to support necessary chemical reactions in their bodies. Therefore, all organisms, including humans, require water for their survival, and ancient civilizations were also developed on large rivers.

Even though fluid intake requirements are 1.5–3.0 L per person per day (Laurent et al. 2010), between 50 and 100 L of water per person per day are required to ensure basic human health (i.e., water for sanitation; UN OHCHR 2015). In addition, approximately 2–3 m^3 per person per day of water including rain water is consumed in the cropland during cropping period to produce food for humans (e.g., Hanasaki et al. 2010). The required amount of water is comparatively larger than food itself (Oki et al. 2017): Humans are supplied with 82 g of fat and 81 g of protein per capita per day (FAO 2014), and 2.0 L of petroleum per person per day (or a total energy

T. Oki (✉)
Institute for Future Initiatives, The University of Tokyo, Tokyo, Japan
e-mail: oki@ifi.u-tokyo.ac.jp

Y. Himiyama et al. (eds.), *Human Geoscience*, Advances in Geological Science,
https://doi.org/10.1007/978-981-32-9224-6_8

consumption of 3.4 kg of oil equivalent, including oil, coal, natural gas, nuclear, hydro, biofuels and waste, and other fuel types) was consumed daily in 2010 (US EIA 2015).

As a consequence of population growth and economic developments during the twentieth century, world water withdrawal has increased 6.3 times from 600 to 3,800 km^3/year, and it was more than the population growth of 3.7 times from 1.65 billion to 6.1 billion (Shiklomanov 1997). Even though water resources development has been the central issue of civilization, it was in the late twentieth century when water was recognized as one of the resources economically; water resources research of American Geophysical Union (AGU) was launched in 1961, UNESCO (United Nations Educational, Scientific and Cultural Organization) defined hydrology in 1964, and UNESCO and WMO started the International Hydrological Decade (IHD) in 1965 in order to facilitate education and capacity building, and enhance water resources management and governance.

Yet, on the global scale, current water withdrawals, 4000 km^3/year (Rockstörm et al. 2009), are well below the theoretically maximum of available freshwater resources, approximately 40,000 km^3/year.

Conventional engineers of water resources consider the water withdrawn from surface and groundwater as water resources, and evapotranspiration as a loss of water from the precipitated water. In that sense, precipitation minus evapotranspiration over land is a measure of the maximum available freshwater resources. The major part of this is surface water, particularly, river discharge. In contrast to the conventional view, it has been noted that evapotranspiration from non-irrigated cropland also is a water resource that is beneficial to society (Postel et al. 1996). To distinguish between this kind of resource and conventional resources, evapotranspiration flow has been named green water (Falkenmark 1997), and conventional withdrawal from rivers and groundwater has been named bluewater. Approximately 3800 km^3/year of blue water is currently withdrawn by human beings, and that accounts for less than 10% of the maximum available freshwater resources in the world (Oki and Kanae 2006). Evapotranspiration is estimated to be 7080 km^3 during cropping period from non-irrigated and irrigated croplands, and 12,960 km^3/year from permanent grazing land (Hanasaki et al. 2010). The total evapotranspiration from cropland and permanent grazing land, 20,040 km^3, corresponds to approximately 30% of the total terrestrial evapotranspiration.

Why should we be concerned about water scarcity when presently only 10% of maximum available blue water and 30% of green water resources are used? The reason is the high variability of water resource availability in time and space (Oki and Kanae 2006). For example, the monthly mean discharge at the Obidos station in the Amazon River differs by a factor of 2 between the highest and the lowest months, even for climatologically averaged values. River discharge is more variable in smaller river basins in general, and daily river discharge is, of course, more variable than monthly river discharge. Because of this temporal variability, it is impractical to use 100% of the available renewable freshwater resources for human society. Flow during floods and wet seasons cannot be used during the low flow seasons unless storage systems are in place. That is why there are millions of artificial reservoirs, lakes, and ponds in the world, most of the major rivers are regulated, and enormous lengths of canals and channels, sometimes tans-boundary, are developed to physically transport water. Water resources developments are the human activities to interfere hydrologic cycles and averaging the spatial (geographical) heterogeneities and temporal (seasonal) variations and to meet water demand where and when needed. Consequentially, real hydrologic cycles on the current Earth is not natural anymore and we see a lot of traces of human interventions on hydrologic cycles even on the global scale (Oki et al. 2013). Modeling efforts have been made to incorporate anthropogenic activities within global hydrologic cycles (Hanasaki et al. 2008; Pokhrel et al. 2012a, b).

8.2 Dawn of Civilization and Water

As Herodotus, an ancient Greek historian, called Egypt as "the gift of the Nile", the development of the civilization of Ancient Egypt was tightly linked with the Nile River. Ancient Mesopotamian civilization, where the wheat cultivation first started in the Fertile Crescent, evolved in the fertile land of the Tigris/Euphrates River. Further, the Indus civilization and Yellow River civilization are called by the names of the major rivers, and big rivers seem indispensable for the rise of ancient civilization. There is no wonder that the existence of large river was the necessary condition for building civilization because water is indispensable for the survival of animals including humans, and also for the production of food.

However, on the contrary, known ancient civilizations did not necessarily evolve in the largest rivers in the world, such as the Amazon, the Congo, the Mississippi, and the La Plata Rivers. The common hydro-climatological characteristic of the ancient civilizations is that they are located in semi-arid regions downstream of moderately large rivers with their head waters in water-rich regions, such as tropical rainforest. Large rivers with water sources in rainy areas are all over the world, but large rivers flowing into the sea through arid areas are very few, limited to these ancient civilization areas.

Dry and arid area may not be perceived as a suitable habitat in ancient days, however, more stable and rich crop yield can be expected in arid regions provided that water can be supplied stably. Also, in a low wetland, it should be hard to transport, and always be needed to face with the dangers of waterborne infectious diseases at all times. As a consequence, people who were able to develop irrigation technology in arid areas could develop ancient civilizations, and irrigation and water resource development have always been with thc history of human civilization.

8.3 Irrigation

Even though small ponds and reserves have been commonly used for irrigation throughout the world, a lot of large reservoirs were constructed particularly in the latter half of the twentieth century. The increase of the artificial reservoirs supported the rapid increase of irrigation withdrawals, and resulted the associated increase of crop yield per area, which accounted approximately 2.9 times from 1961 to 2010. Even though the area of cropland increased by only 10% during the period and the harvested area increased less than that, the total crop production increased by a factor over 3.0, more than the rate of population growth (2.4), and the calorie supply per capita increased 1.3 times during the period (Fig. 8.1). This phenomenal growth was to a large extent due to a doubling of the irrigation area and the corresponding increase of water withdrawal for irrigation in addition to the increased usage of fertilizer (Oki and Kanae 2006).

The total capacity of reservoirs is estimated to be more than 8000 km^3 globally, and it corresponds to that global sea level will be reduced by approximately 25 mm if all the reservoirs are filled (Chao et al. 2008). Lettenmaier and Milly (2009) estimate that sea-level rise, which over the last 50 years has averaged about 3 mm/year, would have been 15–20% larger in the middle of the last century were it not for the reduction in freshwater flux to the oceans associated with filling of manmade reservoirs (they also note that the rate of filling has since decreased substantially, perhaps to a global net less than zero due to infilling of reservoirs with sediment and slowing of reservoir construction).

On the other hand, diffusion of electric pump, and probably the subsidies for the electricity in the agricultural sector, accelerated the ground water withdrawals and caused ground water depletion in some regions of the world (e.g., Famiglietti 2014).

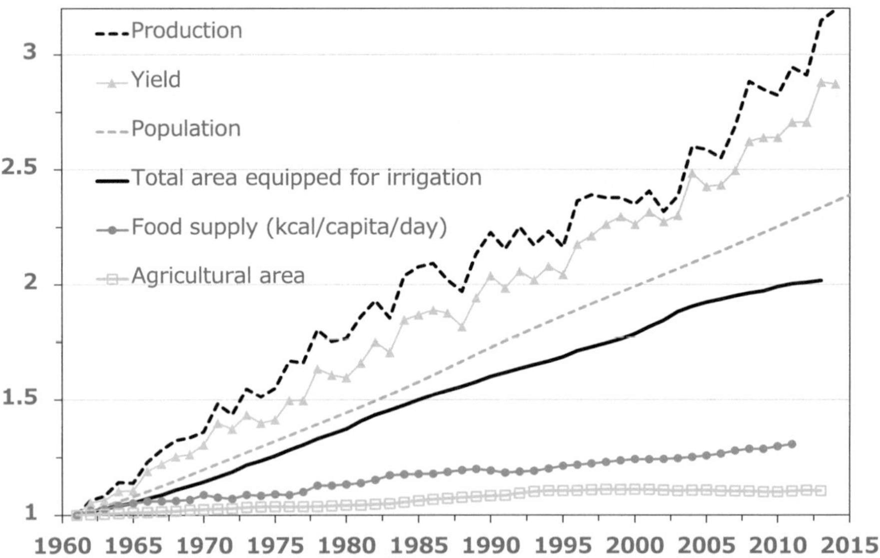

Fig. 8.1 Global trends in total crop production, average crop yield, world population, total area equipped for irrigation, average food supply, and total agricultural area with normalized by the values in 1961 as 1.0. Based on data from the Food and Agriculture Organization of the United Nations (2014)

Pokhrel et al. (2012a, b) estimated on the basis of an integrated modeling framework that artificial reservoir water impoundment caused a sea-level change (SLC) of −0.39 mm/year, while unsustainable groundwater use (groundwater depletion), climate-driven terrestrial water storage (TWS) change, and the net loss of water from endorheic basins contributed +1.05, +0.09, and +0.03 mm/year of the SLC, respectively. Therefore, the net TWS contribution to SLC during 1961–2003 is + 0.77 mm/year. Their result for the anthropogenic TWS contribution to global SLC partially fills the gap in the global sea-level budget reported in Chap. 5 of the Fourth Assessment Report (AR4) by the Working Group I of IPCC (Bindoff et al. 2007). Even though their estimates of ground water depletion are claimed to be larger than other estimates (Konikov 2013; Pokhrel et al. 2013), certain amount of food is produced using non-renewable water resources, depleting groundwater in some areas over the world, and traded internationally (Hanasaki et al. 2010; Dalin et al. 2017).

Individual farmers simply need water, and they merely use the water from reservoirs or groundwater aquifer, and individual consumers just get the necessary food for their survivals. However, such actions as a whole affect the global water cycles, depleting groundwater and influencing sea-level fluctuations.

8.4 FEW Nexus and Remarks

Food, energy, and water nexus (FEW Nexus) should also be noted when social-ecological system and nature-human interactions are considered. There are tight connections, inter-dependencies, complementarities, trade-offs, and competitions among food, energy, and water uses, securities, and their changes under climate change (Fig. 8.2; Oki 2016). However, there is tremendous difference in weight among the necessary quantity for human's survival, and also the unit price. A tremendous volume of water is consumed per person and water has an extremely low unit price compared to food and energy (Oki et al. 2017). Therefore water is rarely traded over long distances by itself, there is no international market price of water, but

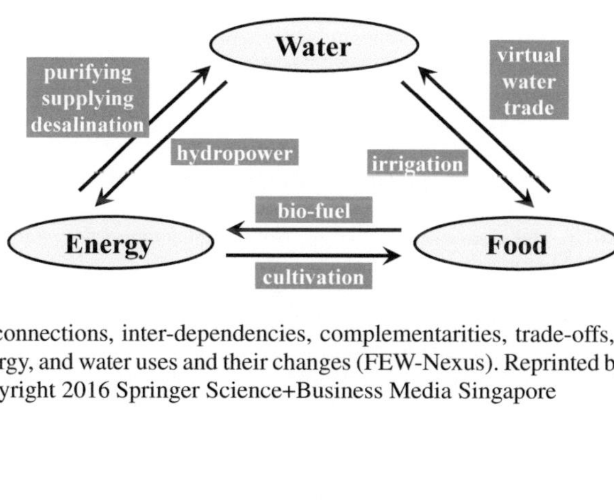

Fig. 8.2 Tight connections, inter-dependencies, complementarities, trade-offs, and competitions among food, energy, and water uses and their changes (FEW-Nexus). Reprinted by permission from Oki (2016). Copyright 2016 Springer Science+Business Media Singapore

the total weight of the water consumed to produce traded commodities exceeds the weight of any other commodity traded in the world. It is known that economically rich but water-scarce countries tend to reduce local water consumption by importing water-intensive commodities, and the concept is called "virtual water trade" (Allan 1998; Oki et al. 2017).

The major common factor controlling the sustainability and security of the FEW Nexus is land, therefore land use/land cover management should also be integrated into the global risk management framework of FEW Nexus and climate change. A more detailed introduction can be found in the cross-chapterbox "The Water–Energy–Food/Feed/Fibre Nexus as Linked to Climate Change" in AR5/WGII (Arent et al. 2014). Water cannot be secured by its own and should be placed in a wider context, such as climate change adaptation, disaster risk management, and sustainable development. Finally, recent evolution of "socio-hydrology" deals with the coupled human-water system, tries to understand the feedback mechanisms in long-term and short-term and to have a capability to assess and predict possible future changes in the system considering both geophysical and sociological aspects (Wagener et al. 2010; Sivapalan et al. 2012; Pande and Sivapalan 2016).

References

Allan JA (1998) Virtual water: a strategic resource global solutions to regional deficits. Groundwater 36:545–546

Arent DJ, Döll P, Strzepek KM, Jiménez Cisneros BE, Reisinger A, Tóth FL, Oki T (2014) Cross-chapter box on the water–energy–food/feed/fiber nexus as linked to climate change. In: Field CB, Barros VR, Dokken DJ, Mach KJ, Mastrandrea MD, Bilir TE, Chatterjee TE, Ebi KL, Estrada YO, Genova RC, Girma B, Kissel ES, Levy AN, MacCracken S, Mastrandrea PR, White LL (eds) Climate change 2014: impacts, adaptation, and vulnerability. Part A: global and sectoral aspects. contribution of working group II to the fifth assessment report of the intergovernmental panel on climate change. Cambridge University Press, Cambridge, United Kingdom and New York, NY, USA, pp. 163–166

Bindoff NL, Willebrand J, Artale V, Cazenave A, Gregory J, Gulev S, Hanawa K, Le Quéré C, Levitus S, Nojiri Y, Shum CK, Talley LD, Unnikrishnan A (2007) Observations: oceanic climate change and sea level. In: Solomon S, Qin D, Manning M, Chen Z, Marquis M, Averyt KB, Tignor M, Miller HL (eds) Climate change 2007: the physical science basis. Contribution of working group I to the fourth assessment report of the intergovernmental panel on climate change. Cambridge University Press, Cambridge, United Kingdom and New York, NY, USA

Chao BF, Wu YH, Li YS (2008) Impact of artificial reservoir water impoundment on global sea level. Science 320:212–214

Dalin C, Wada Y, Kastner T, Puma MJ (2017) Groundwater depletion embedded in international food trade. Nature 543:700–704

Famiglietti J (2014) The global groundwater crisis. Nat Clim Change 4:945–948

Falkenmark M (1997) Meeting water requirements of an expanding world population. Phil Trans R Soc Lond B 352:929–936

Food and Agriculture Organization of the United Nations (2014) FAOSTAT. http://faostat.fao.org/

Hanasaki N, Kanae S, Oki T, Masuda K, Motoya K, Shirakawa N, Shen Y, Tanaka K (2008) An integrated model for the assessment of global water resources—Part 1: model description and input meteorological forcing. Hydrol Earth Syst Sci 12:1007–1025

Hanasaki N, Inuzuka T, Kanae S, Oki T (2010) An estimation of global virtual water flow and sources of water withdrawal for major crops and livestock products using a global hydrological model. J Hydrol 384:232–244

Konikow LF (2013) Over estimated water storage. Nat Geosci 6:3

Laurent LB, Christine J, Liliana J, Christel M, Wei T, Isabelle B (2010) Understanding fluid consumption patterns to improve healthy hydration. Nutr Today 45:S22–S26

Lettenmaier DP, Milly PCD (2009) Land waters and sea level. Nat Geosci 2:452–454

Oki T (2016) Integrated water resources management and adaptation to climate change. In: Biswas AK, Tortajada C (eds) Water security, climate change and sustainable development, water resources development and management. Springer, Science+Business Media Singapore

Oki T, Kanae S (2006) Global hydrological cycles and world water resources. Science 313(5790):1068–1072

Oki T, Blyth EM, Berbery EH, Alcaraz-Segura D (2013) Land use and land cover changes and their impacts on hydroclimate, ecosystems and society. In: Asrar GR, Hurrell JW (eds) Climate science for serving society: research, modeling and prediction priorities. Springer Science+Business Media, Dordrecht

Oki T, Yano S, Hanasaki N (2017) Economic aspects of virtual water trade. Environ Res Lett 12:044002

Pande S, Sivapalan M (2016) Progress in socio-hydrology: a meta-analysis of challenges and opportunities. WIREs Water. https://doi.org/10.1002/wat2.1193

Pokhrel Y, Hanasaki N, Koirala S, Cho J, Yeh PJ-F, Kim H, Kanae S, Oki T (2012a) Incorporating anthropogenic water regulation modules into a land surface model. J Hydrometeorol 13:255–269

Pokhrel YN, Hanasaki N, Yeh PJ-F, Yamada TJ, Kanae S, Oki T (2012b) Model estimates of sea-level change due to anthropogenic impacts on terrestrial water storage. Nat Geosci 5:389–392

Pokhrel YN, Hanasaki N, Yeh PJ-F, Yamada TJ, Kanae S, Oki T (2013) Reply to 'Overestimated water storage'. Nat Geosci 6:3–4

Postel SL, Daily GC, Ehrlich PR (1996) Human appropriation of renewable fresh water. Science 271:785–788

Rockström J, Steffen W, Noone K, Persson Å, Chapin FS III, Lambin E, Lenton TM, Scheffer M, Folke C, Schellnhuber H, Nykvist B, De Wit CA, Hughes T, van der Leeuw S, Rodhe H, Sörlin S, Snyder PK, Costanza R, Svedin U, Falkenmark M, Karlberg L, Corell RW, Fabry VJ, Hansen J, Walker B, Liverman D, Richardson K, Crutzen P, Foley J (2009) Planetary boundaries: exploring the safe operating space for humanity. Ecol Soc 14(2):32

Shiklomanov IA (ed) (1997) Assessment of water resources and water availability in the world. Background report for the comprehensive assessment of the freshwater resources of the world, WMO/SEI, Geneva, Switzerland

Sivapalan M, Savenije HHG, Blöschl G (2012) Socio-hydrology: a new science of people and water. Hydrol Process 26:1270–1276

U.S. Energy Information Administration (2015) International energy statistics

UN Office of the High Commissioner for Human Rights (OHCHR) (2015) Fact sheet No. 35. The Right to Water. www.ohchr.org/Documents/Publications/FactSheet35en.pdf

Wagener T, Sivapalan M, Troch PA, McGlynn BL, Harman CJ, Gupta HV, Kumar P, Rao PSC, Basu NB, Wilson JS (2010) The future of hydrology: an evolving science for a changing world. Water Resour Res 46:W05301

Part IV
Natural Hazards and Society

Chapter 9
Active Fault and Earthquake Disasters

Yasuhiro Suzuki

Abstract An earthquake is a phenomenon caused by faulting; therefore, we focus on faults that could possibly be reactivated in the future, as active faults, to estimate potential earthquake disasters and their prevention. In order to assess the future activity of faults, it is worth noting the paleo-seismological history of faults wherein the tectonic regime has remained similar. Fault activity is subject to the geology and geomorphology of a region, but the concept and purpose of active fault research are also closely related to earthquake disaster prevention. Therefore, active faults are an important research theme in human geoscience, which is the study of how geoscience is related to human activity. Since the Kobe earthquake of 1995, the Japanese government began emphasizing active fault research as a factor in earthquake disaster prevention. Advances in research and technology have improved the quality and availability of active fault information and have made it possible to better understand co-seismic fault activity. The 2016 Kumamoto earthquake served as an opportunity to learn more about the relationship between active faults and earthquakes, and prompted some verification of the validity of the 21-year effort.

Keywords Active fault · Hazard · Earthquake · Disaster prevention · Resilience

9.1 Introduction

Earthquakes are a complex phenomenon triggered by the interaction between Earth and human activities. It is, therefore, one of the most widely researched topics in human geoscience. The goals of these studies are to gain a better understanding of the phenomena as well as to work towards mitigating the effects of earthquake disasters.

There are regions where earthquakes are likely to but seldom occur. Identifying the root cause of an earthquake is where most research begins. Plate tectonics explains that the Earth's surface is divided into multiple plates and interpolate earthquakes often occur due to the interaction between these plates at their boundaries. The stress generated at the plate boundary is transmitted through its vicinities, which sometimes

Y. Suzuki (✉)
Disaster Mitigation Research Center, Nagoya University, Nagoya, Japan
e-mail: ysz@nagoya-u.jp

Y. Himiyama et al. (eds.), *Human Geoscience*, Advances in Geological Science,
https://doi.org/10.1007/978-981-32-9224-6_9

cause intraplate earthquakes within the plate. The fault that reactivates and causes a future intraplate earthquake is called an active fault.

In this chapter, we outline the research history of active faults in Japan and discuss active faults from a sociological perspective. Active fault research has made great strides in the latter half of the twentieth century, deepening our understanding of active faults. However, when the Kobe Earthquake was triggered by an active fault reactivation in 1995, resulting in more than 6000 deaths, it became clear that current disaster prevention measures against this phenomenon were insufficient. Thereafter, the Japanese government started active fault assessment and the results were publicly announced. At the same time, it was emphasized that strong ground motion can occur everywhere. In 2016, when the Kumamoto Earth-quake was triggered by the reactivation of an active fault, we were confronted with the question of whether or not the countermeasures of the past 20 years were effective.

The 2011, the East Japan, great earthquake was said to be an "unexpected" earthquake, which made us realize that conventional earthquake prediction was inadequate. A reassessment on whether or not the safety of nuclear power plants against active faults is sufficient was carried out. The safety of some nuclear power plants was questionable. What we learned from this situation, as a basic question of disaster prevention theory, was how earthquake countermeasures should be taken against large-scale disasters of low-frequency occurrence, like earthquakes caused by active faults.

9.2 Overview of an Active Fault

9.2.1 What Is an Active Fault?

In 1891, the M8.0 Nobi earthquake occurred in the southern part of Gifu Prefecture in the Central Japan region. At that time, several earthquake faults appeared on the ground surface, whose lengths reached 80 km in total. The faults were recorded in photographs and this earthquake became famous worldwide (Fig. 9.1). Theoretical discussions of the cause of earthquakes began with this event, and the modern seismology started to develop. In addition, this earthquake made the Japanese Government to set up the Imperial Earthquake Investigation Committee and to start studies on earthquake damage reduction. The Nobi earthquake is generally considered the beginning of seismology and earthquake engineering.

The upper crust, which constitutes earth surface layer, may be deformed by stress that causes fault rupture. We cannot deny the possibility of future reactivation if the faults have repeated activity in the late Quaternary because the stress field is estimated to be constant in that period. For this reason, we also classify such faults as "active faults" for disaster prevention. This concept was first established in Japan after the Kita Tango Earthquake of 1927, which was generated by faulting (Tada 1927).

Fig. 9.1 Fault scarplet emerged with the Nobi Earthquake of 1891 in Gifu Prefecture, Central Japan. Photo was taken by Milne, J. and Murton, W. K.

Faulting, a destructive phenomenon, which is accompanied by strong ground vibration, is the earthquake itself. Therefore, paying attention to the existence of active faults leads to predicting the place where larger earthquakes occur, which is essential for disaster prevention. In addition, it is likely that heavy damage will occur due to the ground movement just on the fault, so the information on detailed position of the fault is important for disaster mitigation.

In order to recognize an active fault, it is necessary to find the evidence of repeating activities in the late Quaternary by using geomorphological, geological, or geophysical methods. By using them, if we find a deformation zone on the geomorphological surface or geological strata formed in the late Quaternary, an active fault is recognized along with it. The important thing is that each method has advantages and disadvantages and we should be careful about their methodological limitations. When one method indicates a high possibility of active fault, this suspect is not easily deniable by using another method. In active fault assessment for important structures, it is necessary to pay sufficient attention to the methodological limitations.

9.2.2 History of Active Fault Research in the Twentieth Century

After 1891, the inland earthquakes occurred one after another, such as the 1894 Shonai Earthquake, the 1896 Rikuu Earthquake, the 1925 Kita-Tajima Earthquake, the 1927 Kita-tango Earthquake, the 1930 Kita-Izu Earthquake, the 1943 Tottori Earthquake, the 1945 Mikawa Earthquake, and the 1948 Fukui Earthquake. The relationship between the earthquakes and the faults was investigated in each event and it was assumed that reactivation of active faults can cause an earthquake.

Therefore, it became necessary to know the active faults distribution. However, in the first half of the twentieth century, only geomorphological methods were available, in which the interpretation of linear valleys or triangular terminal faces resulted in repeated fault movements. Actually, as the terrain is formed by various reasons, such topographical features are only circumstantial evidence. Consequently, those days information on the active faults existence was not defined but presumed.

However, active fault studies since the 1960s have greatly developed for those that emphasize direct evidence. Aerial photographs throughout Japan became widely available since 1960, and it became possible to find traces of earthquake faults, which are direct evidence of paleo-seismic activities. It enabled us to clarify precisely the active faults distribution in Japan. Figure 9.2 is a schematic diagram of the deformed landform by faulting, drawn by Atsumasa Okada. In this figure, it is shown that the deformed landform caused by one earthquake event is observed on the newer geomorphic surface, and on the older surfaces, the deformation accumulation can be confirmed. At last, a methodology to recognize active faults was established by paying attention to the direct evidence of repeating activity (Watanabe and Suzuki 1999).

The work for recognizing active fault throughout Japan has been greatly advanced since the 1970s. Sohei Kaizuka led a cooperation project with many Japanese geomorphologists and geologists in which aerial photographs were interpreted based on standardized criteria for identifying active faults, and created the distribution map.

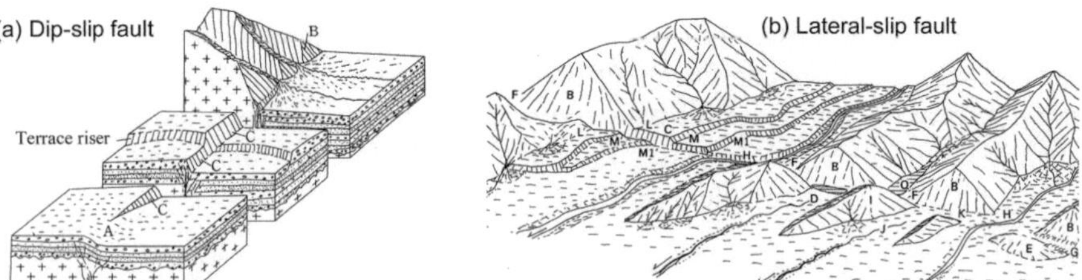

A: Flexure scarp. B:Truncated Spur, C:Fault scarplet, D:Sag pond, E:Tectonic bulge, F:Fault saddle, G:Fault graben, H:Offset drainage channel, I:Shutter ridge, J:Truncated river channel, K:Wind gap, L-L':Offset piedmont line, M-M':Offset terrace riser, Q:Dammed lake

Fig. 9.2 Schematic diagram of deformed landform by faulting (The Research Group for Active Faults 1980)

The active fault was classified into three categories, I to III, depending on the certainty of its existence, and three categories A to C were assigned depending on its slip-rates. The results were published as "Active Fault in Japan", and a revised edition was published in 1991 (Fig. 9.3, The Research Group for Active Faults 1980; The Research Group for Active Faults of Japan 1991).

In response to these research results, Matsuda (1975) proposed a relational expression between the active fault length and the earthquake scale, and the relation between fault length and the offset amount during the earthquake. These are empirical formulas from limited cases in the past, and as he himself points out, the formulas cannot predict accurately the future activities. However, as it is convenient to estimate roughly future earthquakes, it has often been used. In addition, Matsuda (1981) proposed the concept of "precaution fault", which is based on the "characteristic earthquake occurrence model" and states that active faults periodically repeat activity of the same characteristic scale. If a longer active fault does not have historical

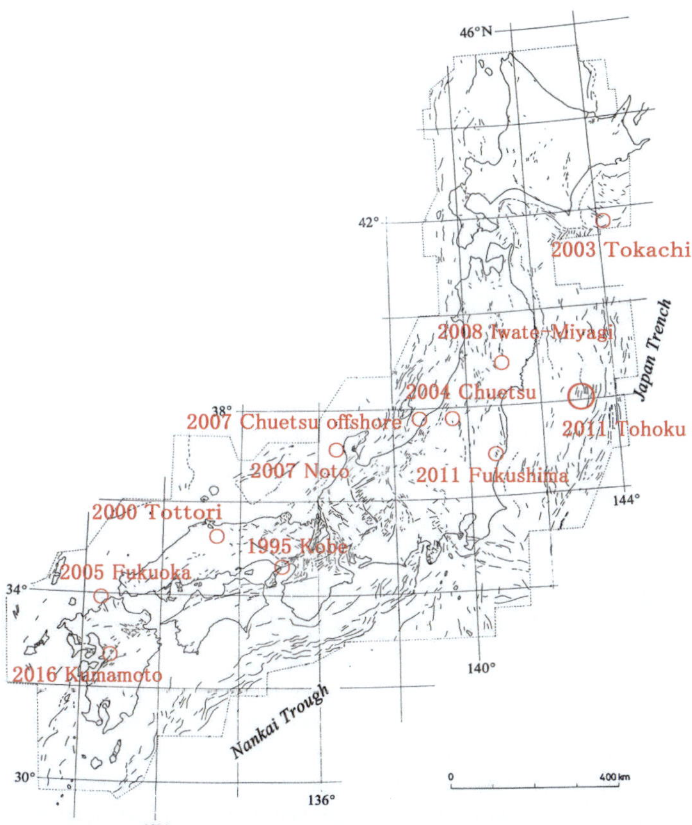

Fig. 9.3 Main active faults in and around Japan and major earthquakes occurred after 1995. Modified from The Research Group for Active Faults of Japan (1980)

activity records, it can be regarded as having a high potential of resulting in a large future earthquake. Twelve active faults in Japan are nominated to precaution faults, which has attracted attention after the 1995 Kobe Earthquake.

The Research Group for the Active Faults of Japan (1980) mentioned in a discussion paper that the work marked the final step in the “age of discovery” of active faults in Japan. However, according to a review on the active fault research progresses by Suzuki (2013b), paleo-seismological research started since 1980s and the study on fault evolution mechanism or on the geomorphological development have been conducted. It has become possible to use aerial photographs with scale of 1:10,000 in active fault research, and age determination and volcanic ash identification research have developed. In particular, excavation surveys and seismic reflection surveys have been intensively conducted since 1995. Detailed Digital Elevation Model (DEM) became available for geomorphological analysis after the 2000s. Under these circumstances, new active faults continued to be discovered even in the twenty-first century. Newly discovery of active faults is sometimes socially problematic, as it could have a considerable influence on nuclear power plant construction and urban development.

9.3 The 1995 Kobe Earthquake and the Active Fault

9.3.1 Outline of the Kobe Earthquake

During the early morning of January 17, 1995, an intra-plate earthquake of M 7.3 occurred in the southern part of Hyogo Prefecture. This earthquake was caused by rejuvenation of the Rokko-Awaji Fault, extending from the Awaji Island to the Kobe area. As the earthquake fault was located just under the city, strong ground motion occurred; this destroyed many buildings, railways, and highways, also liquefying the ground surface along the coast. The earthquake resulted in more than 6000 deaths and 43,000 injuries.

A 10-km long surface earthquake fault appeared on the northwest coast of Awaji Island (Fig. 9.4). The fault cut off paddy fields and roads and led to the collapse of the building just above it. Maximum displacement was 2 m in the right lateral direction. In contrast, no surface earthquake fault appeared in the vicinity of Kobe City; however, an “extremely severe damage zone” was formed with 1 km width in the city area (Fig. 9.5).

Seismic wave analysis provided insight into the movement of the source fault. This indicated that in the northern part of Awaji Island, displacement was large in shallow parts of the crust, whereas in the Kobe area, main displacement occurred not in shallower but in deeper parts. This was consistent with the unclear features of the surface fault.

Fig. 9.4 Nojima Fault, earthquake surface fault of the 1995 Kobe Earthquake. Photo taken by Nakata, T.

Fig. 9.5 Collapsed highway at Kobe area in 1995. Photo presented by Kobe City

9.3.2 Active Fault Study After the Kobe Earthquake

Surveys on the active fault that caused the Kobe earthquake started just after the event. First, the earthquake fault that appeared clearly on Awaji Island was investigated in detail. An excavation survey was conducted in February 1995 (Fig. 9.6). At the time, knowledge of active faults in Japanese society was insufficient; it was, therefore, requested that the cause of the earthquake and the repeating interval of earthquakes be clarified. The result of these surveys was reported, not only to academic society but also to the public; many citizens were thus able to understand the relationship between active faults and earthquakes.

In the Kobe area, surveys were conducted to clarify the reasons for the formation of the "extremely severe damage zone". The active Rokko fault, which was known before the earthquake, was located just along the mountain foot, with the "extremely severe damage zone" forming about 1 km away. The reason for its formation needed to be investigated. Specialists of strong ground motion emphasized the basin wedge effect, i.e., the unique basin structure of the Kobe area was regarded to be a factor amplifying vibrations. Meanwhile, seismic reflection surveys were conducted in the area, and revealed the existence of a concealed active fault under the "extremely severe damage zone". Moreover, the deformed landform was recognized by tectonic geomorphological survey there, and fault continuity was estimated (Fig. 9.7). In this figure, the star shows the point where the concealed fault was confirmed by the reflection survey, and the distribution of the stars is consistent with the fault line estimated geomorphologically.

Fig. 9.6 Excavation research for Nojima Fault in 1995 (Suzuki 2001)

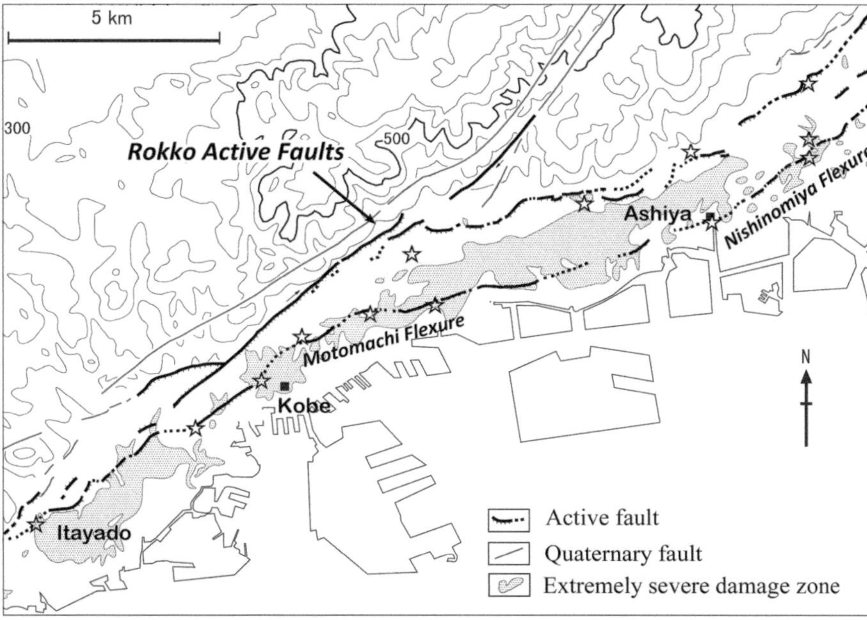

Star shows the point where active faults were found by reflection seismic survey

Fig. 9.7 Active fault distribution in Kobe area. Figure was revised from Watanabe et al. (1997)

The Active Fault Research Committee of Kobe City conducted boring at a number of sites to investigate the geological structure across the concealed fault (Fig. 9.8). Results showed that strata containing AT (25,000 years old) and K-Ah (7,000 years old) volcanic ashes had vertical flexure of 6 and 2 m, respectively. It could, therefore, be concluded that the fault has been repeatedly active in the late Quaternary. The question is whether or not the fault reactivated at the time of the Kobe earthquake. Unfortunately, geodesic observation technology of that time was not able to answer this question.

After the earthquake, seismic resistance standards for buildings and infrastructure and for seismic safety of nuclear power plants were carefully reconsidered by the government. Strong earthquake motions caused by faulting far exceeded expectations prior to the earthquake, and it was revealed that strong tremors with 1-s frequency tended to cause heavy damage to wooden houses. It was also noted that artificial modification of landforms, such as through cutting or filling, made the ground vulnerable. Considering the significant damages which result from such events, earthquake countermeasures were comprehensively reviewed.

Surveys of active faults since 1995 have followed the methodology used for the Kobe earthquake. Tectonic geomorphological surveys, excavation surveys, and seismic reflection surveys were conducted nationwide. The Headquarters for Earthquake Research Promotion of Japan was established within the government; this selected

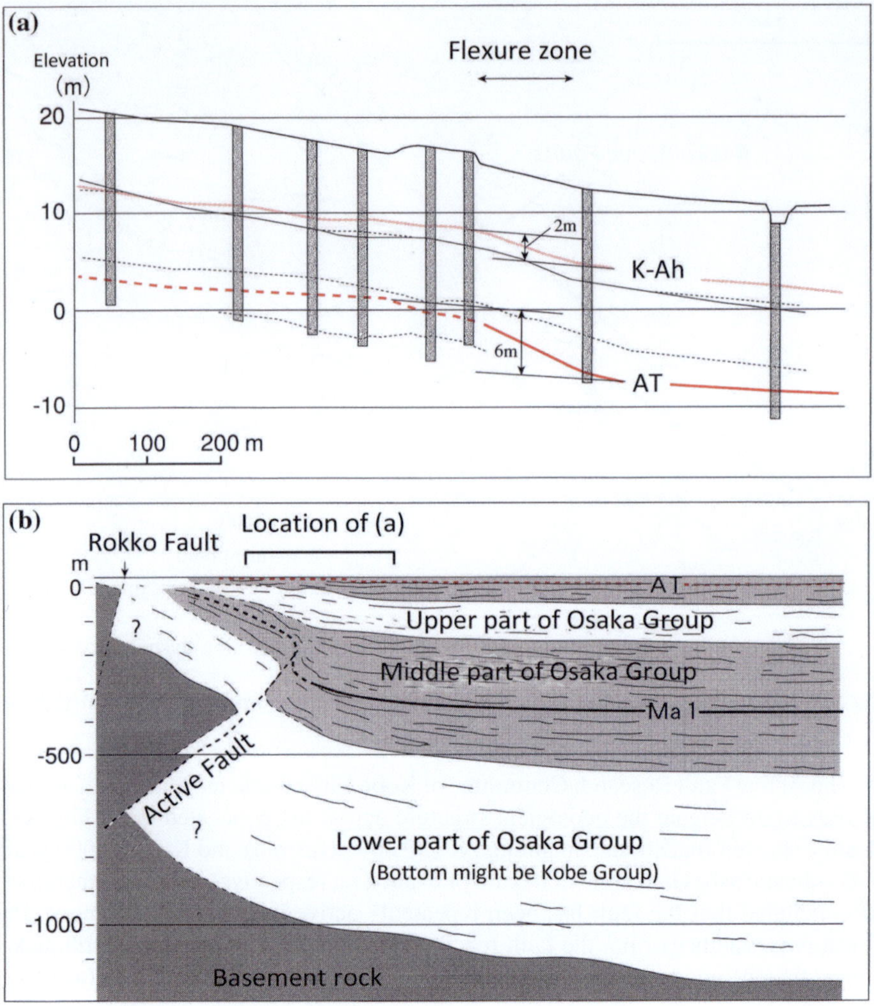

Fig. 9.8 Cross-section revealed by boring and seismic reflection survey (Kobe City and Construction Engineering Research Institute 1998)

about 100 major active faults and set up a basic research plan for disaster preparedness. Details of these Headquarters will be provided in the next chapter.

The notion of earthquake disaster prevention changed significantly with this earthquake. Prior to the earthquake, it was not believed necessary to inform the public of natural hazards, but after the earthquake, it was decided that such information must be released. The government started to provide people with information on the location of active faults and on the possibility of earthquake occurrence. The Geospatial Information Authority of Japan, in cooperation with active fault researchers, began to develop "1:25000 Active Fault Maps in Urban Areas" (Fig. 9.9) and has continued

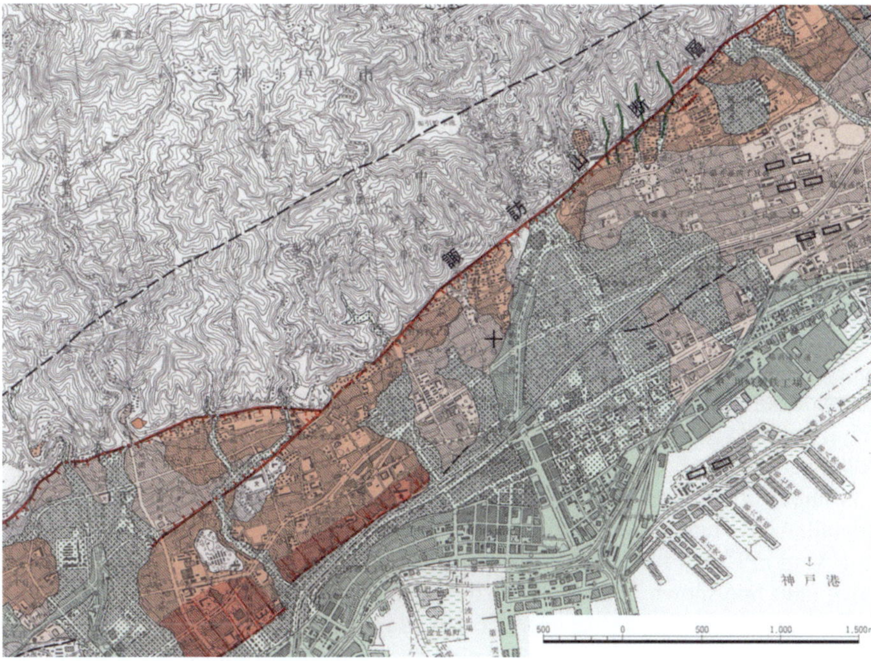

Fig. 9.9 "1:25,000 Active Fault Map in Urban Areas" for Kobe area (Watanabe et al. 1997)

this work for more than 20 years since. Excavation surveys to clarify paleo-seismicity and to evaluate the possibility of earthquake occurrence in the near future have also been continuously carried out.

The impact of the Kobe earthquake can be described by the expression "collapse of a safety myth." There was concern about the seismic safety of buildings and infrastructure. At the same time, local citizens mistakenly thought that "there is no possibility of a large earthquake in the Hanshin area," increasing the damage. For this reason, the Headquarters began to inform the public of the existence of active faults and to educate people regarding the fact that large earthquakes can occur anywhere in the country.

9.4 The 2016 Kumamoto Earthquake and Active Fault Research

9.4.1 Overview of the Kumamoto Earthquake

Two large earthquakes occurred in Kumamoto Prefecture, Japan, on 14 and 16 April 2016 with a magnitude of 6.8 and 7.3, respectively. Although the Headquarters for

Earthquake Research Promotion announced that the two earthquakes arose from different faults, the surface earthquake faults overlapped each other (Sugito et al. 2016), resulting in classification of a series of faults, together called the Futagawa-Hinagu fault, which is concluded to have activated twice. Because this long active fault generated an earthquake of smaller-than-expected size, it should be considered at risk for larger earthquakes in the future; this was one of the most important lessons learned from the Kumamoto earthquake (Suzuki et al. 2016a).

In Mashiki town and Nishihara village, which are both located near the fault, strong ground motion of Japanese seismic intensity 7 occurred, resulting in an "extremely severe damage zone." The death toll from the earthquake reached 200, with 2,700 people injured and 180,000 evacuated. In Minami-Aso village, mountain slopes collapsed in a number of places. Surprisingly, at least five automobiles overturned onto their sides in the same direction under the influence of strong ground motion in the vicinity of the surface fault, which indicated that Minami-Aso likely suffered severe shaking corresponding to seismic intensity 7 (Fig. 9.10).

A survey after the earthquake revealed the distribution of surface earthquake faults in detail. Active faults had been mapped before the earthquake to the "1:25,000 Active Fault Map in Urban Areas." Most surface earthquake faults appeared just along the active fault line shown in the map. In the foreshock on 14 April, a maximum offset of 8 cm appeared over a distance of 6 km (Sugito et al. 2016), whereas offsets of up to 30 cm occurred at the same location during the main shock on 16 April. The displacement during the main shock extended further northeast, measuring 31 km long and reaching the inside of the Aso caldera (Kumahara et al. 2016, Fig. 9.11). A maximum displacement of 2 m, right lateral, was observed in Mashiki and Nishihara. Additionally, clear normal faulting appeared parallel to the main fault, which indicates a slip-partitioning phenomenon in which the lateral and longitudinal offset appeared separately (Toda et al. 2016).

Surface ruptures were found both along and away from the Futagawa-Hinagu fault. A clear rupture appeared in the northern part of the Aso caldera, and its cause was discussed. Ruptures also appeared on the west slope of Caldera and within the city of Kumamoto (e.g., Goto et al. 2017). The Geospatial Information Authority analyzed SAR (Synthetic Aperture Radar) Interferometry using the Advanced Land Observing Satellite (ALOS-2) and clearly revealed details of the surface changes mentioned above (Fig. 9.12, Fujiwara et al. 2016). Additionally, this analysis revealed the earthquake source fault models, which indicate in turn that the southern area, where the foreshock of 14 April occurred, moved slightly with the foreshock and reactivated to a higher degree with the main shock. This result is consistent with the surface rupture observations mentioned above. Moreover, the National Research Institute for Earth Science and Disaster Prevention (NIED) analyzed the fault rupture process during the main shock (Fig. 9.13, Kubo et al. 2016). With these new technological developments, the relationship between underground faulting and surface rupture during the earthquake can be understood more clearly.

Fig. 9.10 Surface fault and damages with the 2016 Kumamoto Earthquake (Suzuki et al. 2016a)

9.4.2 Issues Raised by the Kumamoto Earthquake

The Kumamoto earthquake was the first representative M7 earthquake caused by reactivation of the major active faults in Japan that was selected by the Headquarters for Earthquake Research Promotion. The Headquarters assessed the active fault in 2002 and 2013 and announced that the probability of earthquake occurrence was at a "slightly higher level." The 2016 earthquake disaster prompted inquiry into whether this evaluation was correct and effective for disaster preparedness.

Probability of occurrence is calculated using occurrence interval data and the elapsed time after the last event. The probability for the Futagawa-Hinagu fault should be carefully reconsidered using data from the many excavation studies still being conducted after the earthquake, as a previous study indicated that the segmentation of the fault might have been incorrect (Suzuki et al. 2016a). According to this study,

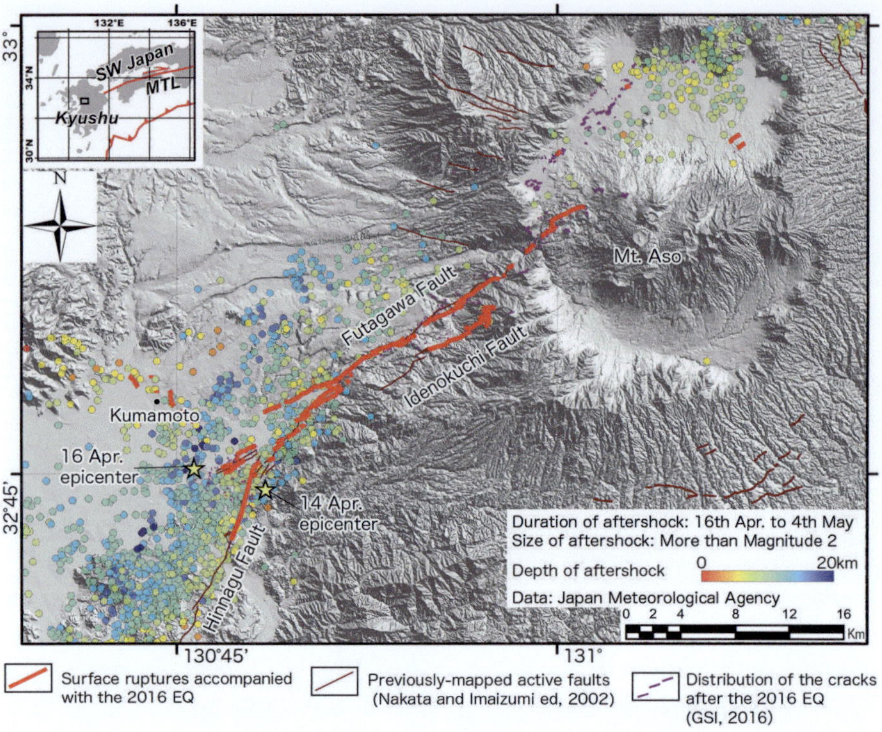

Fig. 9.11 Distribution of earthquake fault of the 2016 Kumamoto Earthquake (Kumahara et al. 2016)

the Headquarters divided the fault, which generated the 2016 main shock, into two different faults named the Futagawa fault and the Hinagu fault in 2013; however, they moved concurrently in the 2016 earthquake on 16 April. The Headquarters announced another evaluation in 2002 indicating that these two faults formed a series of faults (the Futagawa-Hinagu fault) which were likely to move together and generate one large earthquake; this evaluation is consistent with observations of the 2016 Kumamoto earthquake.

Because segmentation of a fault is essential for risk evaluation, the methodology for segmentation determination needs to be reconsidered. In addition, this earthquake posed issues regarding methods of ground motion evaluation. According to the fault rupture process revealed by NIED, significant displacement in a relatively shallow area 2–3 km underground caused strong motion at the surface. However, the conventional prediction method for ground motion does not forecast strong earthquake motion in areas shallower than 2–3 km. Moreover, an “extremely severe damage zone” was formed within several hundred meters of the earthquake fault. Conventional theory does not explain such strong motion in this limited linear area.

After the Kobe earthquake, the government announced the existence of active faults and related hazards, but has never recommended special safety measures in

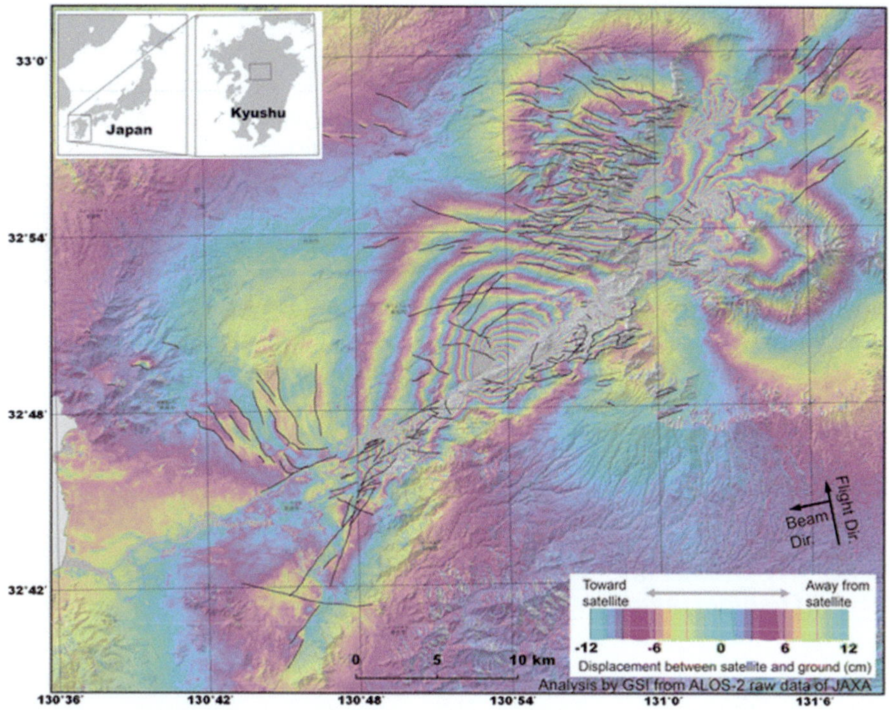

Fig. 9.12 SAR interferograms showing ground surface displacement associated with the 2016 Kumamoto earthquake, analyzed by using data of April 15 and April 29, 2016 (Fujiwara et al. 2016)

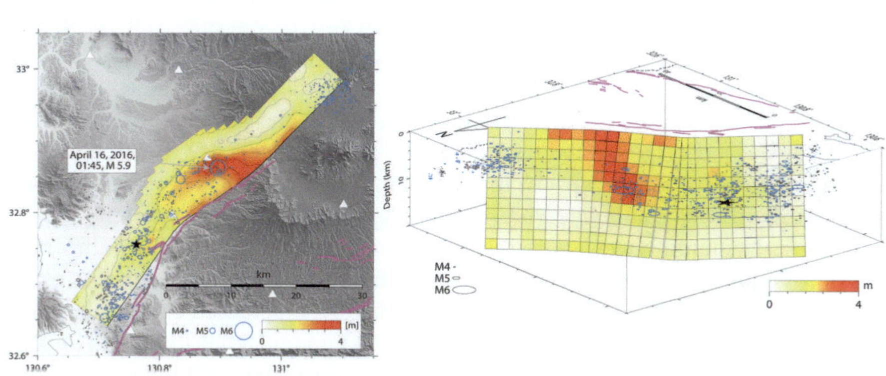

Fig. 9.13 Source rupture process of the M7.3 Kumamoto Earthquake of 16 April 2016, derived from strong-motion waveforms using multiple-time-window linear waveform inversion (Kubo et al. 2016)

the vicinity of active faults. Many local residents of Kumamoto said that they knew of the existence of active faults, but did not take any particular safety measures. There were several new buildings in the town of Mashiki that incorporated the latest earthquake resistance standards but were heavily damaged. It indicates the necessity for special safety measures in the vicinity of active faults.

The Japanese government was aware of the necessity of earthquake countermeasures nationwide after the Kobe earthquake, asserting repeatedly that "strong shakes can happen anywhere." This assertion could be misleading, as most strong tremors have a seismic intensity in the upper-5 to lower-6 range, with earthquakes of seismic intensity in the upper-6 to 7 range occurring only in limited areas. It is therefore necessary to better evaluate earthquake risk and strengthen earthquake resistance in areas where extremely strong ground motion is predicted.

9.5 Social Problems on Active Faults

9.5.1 Measures Against Strong Ground Motion and Fault Displacement

In New Zealand and the state of California, USA, there are regulations concerning land use near the active fault (Nakata, 1990). After the Kobe earthquake, there was debate as to whether such regional designations should be made in Japan. In that case, there were opposing arguments, including: (1) the position of a fault is difficult to determine exactly; (2) the appropriate setback distance from a fault is difficult to determine; and (3) many areas unconnected to an active fault may still experience earthquake risks. After the 1999 Chi-chi earthquake in central Taiwan caused the collapse of numerous buildings on fault lines, the Taiwanese government passed a regulation governing land use on active fault lines. With the Kumamoto earthquake, severe damage occurred on the earthquake fault line, especially in Minami-Aso, Mashiki, and Nishihara. In view of this disaster, it is necessary to once again comprehensively discuss countermeasures against damage from active faults.

In 1997, Nishinomiya city established an ordinance that obliged developers to examine the precise location of active faults if they plan to build houses near active faults. In 2008, Fukuoka prefecture began to encourage increased earthquake resistance for building in the vicinity of the active Kego fault, which is located in the city area. Moreover, Tokushima Prefecture passed an ordinance in 2013 that specified a special area for active fault measurement. If a public building is planned in the area, a geological survey is obligatory, and the building should be constructed a given minimum distance from the fault. While the Japanese government does not enforce legal restrictions, local government ordinances and independent judgments comprise a new direction for the future of earthquake disaster prevention in Japan.

In this discussion, it is important to consider countermeasures against ground shaking and ground deformation separately. It may also be necessary to separate discussions of countermeasures for public buildings and private houses. Active fault research needs to verify how accurately the location of the fault can be confirmed before an earthquake, and the Kumamoto earthquake provides a rich subject for this verification.

9.5.2 *Countermeasures Against Large, Low-Frequency Disasters*

The low frequency of fault activity presents great difficulty in reaching a consensus regarding building countermeasures against active faults. The 2011 Great East Japan earthquake raised similar issues, as tsunamis of this scale are expected to occur only once in 500–1000 years. Discussion of how to prepare for a large but low-frequency disaster is required. It could be said that the serious accident at the Fukushima Daiichi Nuclear Power Plant of TEPCO (Tokyo Electric Power Company Holdings, Incorporated) was due to inadequate countermeasures against large, low-frequency disasters, and this should prompt discussion of how to determine an appropriate disaster prevention level.

The Nuclear Regulation Authority, which was established after the 2011 earthquake, prohibits placing important structures in nuclear power plants directly over active faults. However, some oppose uniform prohibition of such building, arguing that these structures may not necessarily be broken by fault activity. Moreover, there is concern that strict enforcement of such regulations will render it impossible to construct nuclear power plants in Japan, a seismically active country, which would hinder economic development.

These arguments must be weighed with an understanding of the tectonic features of the Japanese archipelago before discussing the appropriate level of long-term disaster prevention against large, low-frequency disasters. The mission of human geoscience study is to contribute to objective discussions without being bound by short-term economic consequences or vested interests (Suzuki, 2013a).

9.6 Conclusions

In this chapter, topics concerning active faults were reviewed with an emphasis on earthquake disaster prevention. The information herein should inspire us to consider the appropriate disaster prevention level against large, low-frequency disasters. When discussing disaster prevention perspectives, an understanding of regional geographical characteristics is indispensable. Active fault research has been performed for many years as part of geomorphological development research.

When an active fault reactivates, both the ground surface and the geological sedimentary environment change, finally forming new landform such as mountains and basins. Fault activity continued throughout the Quaternary period with various effects on humankind. It is important to understand and discuss issues related to active faults from a long-term point of view, considering the results of geomorphic development research (e.g., Yonekura et al. 2001).

The ability to survive and thrive in the face of sudden disaster is sometimes called 'resilience' and is important to community sustainability. Both concepts of resilience and sustainability are included in the Sustainable Development Goals (SDGs) adopted by the United Nations in 2015, which seeks a vision for the future that requires both points of view. Resilience should be considered comprehensively, including social consensus on disaster prevention as well as individual disaster prevention measures (Suzuki et al. 2016b). Therefore, human geoscience can benefit greatly from becoming more trans-disciplinary, as the mission of human geoscience is to contribute to sustainability and the resilience of humankind.

References

Fujiwara S, Yarai H, Kobayashi T, Morishita Y, Nakano T, Miyahara B, Nakai H, Miura Y, Ueshiba H, Kakiage Y, Une H (2016) Small-displacement linear surface ruptures of the 2016 Kumamoto earthquake sequence detected by ALOS-2 SAR interferometry. Earth Planets Space 68:160. https://doi.org/10.1186/s40623-016-0534-x

Goto H, Tsutsumi H, Toda S, Kumahara Y (2017) Geomorphic features of surface ruptures associated with the 2016 Kumamoto earthquake in and around the downtown of Kumamoto City, and implications on triggered slip along active faults. Earth Planets Space 69:26. https://doi.org/10.1186/s40623-017-0603-9

Kobe City and Construction Engineering Research Institute (1998) Great Hanshin-Awaji Earthquake and geology and active fault of Kobe (in Japanese)

Kubo H, Suzuki W, Aoi S, Sekiguchi H (2016) Source rupture processes of the 2016 Kumamoto, Japan, earthquakes estimated from strong-motion waveforms. Earth Planets Space 68:161. https://doi.org/10.1186/s40623-016-0536-8

Kumahara Y, Goto H, Nakata T, Ishiguro S, Ishimura D, Ishiyama T, Okada S, Kagohara K, Kashihara S, Kaneda H, Sugito N, Suzuki Y, Takenami D, Tanaka K, Tanaka T, Tsutsumi H, Toda S, Hirouchi D, Matsuta N, Moriki H, Yoshida H, Watanabe M (2016) Distribution of surface rupture associated with the 2016 Kumamoto earthquake and its significance. Japan Geoscience Union Meeting 2016, MIS34-05 (in Japanese)

Matsuda T (1975) Magnitude and recurrence interval of earthquakes from a fault. Zishin 2(28):269–283 (in Japanese)

Matsuda T (1981) Active faults and damaging earthquakes in Japan-Macroseismic zoning and precaution faults zones. M. Ewing Series 4 (American Geophysical Union) 279–289

Nakata T (1990) Alquist-Pliolo Special Studies Zones Act of California and prevention of earthquake hazards. J Geogr 99:81–90 (in Japanese)

Sugito N, Goto H, Kumahara Y, Tsutsumi H, Nakata T, Kagohara K, Matsuta N, Yoshida H (2016) Surface fault ruptures associated with the 14 April foreshock (Mj 6.5) of the 2016 Kumamoto earthquake sequence, southwest Japan. Earth Planets Space 68:170. https://doi.org/10.1186/s40623-016-0547-5

Suzuki Y (2001) Prevention against earthquakes by active faults. Kokon-shoin (in Japanese)

Suzuki Y (2013a) Nuclear power plants and active faults. Iwanami-shoten (in Japanese)

Suzuki Y (2013b) Tectonic geomorphological active fault studies in Japan after 1980. Geogr Rev Jpn Ser B 86:6–21

Suzuki Y, Watanabe M, Nakata T (2016a) Lessons learned from the 2016 Kumamoto Earthquake and suggestion toward active fault prevention. Kagaku 86:839–847 (in Japanese)
Suzuki Y, Hayashi Y, Tsukahara K (2016b) Why is resilience lost? In: Hayashi Y et al (eds) Disaster resilient cities: concepts and practical examples. Elsevier, Amsterdam, pp 3–10
Tada F (1927) Two types of active faults. Geogr Rev Jpn 3:980–983 (in Japanese)
The Research Group for Active Faults (eds) (1980) Active faults in Japan: sheet maps and inventories. University of Tokyo Press, Tokyo (in Japanese)
The Research Group for Active Faults of Japan (eds) (1991) Active faults in Japan: sheet maps and inventories (Revised edn). University of Tokyo Press, Tokyo (in Japanese)
The Research Group for the Active Faults of Japan (1980) Active faults in and around Japan: the distribution and degree of activity. J Nat Disaster Sci 2:66–99
Toda S, Kaneda H, Okada S, Ishimura D, Mildon Z (2016) Slip-partitioned surface ruptures for the Mw 7.0 16 April 2016 Kumamoto, Japan, earthquake. Earth Planets Space 68:188
Watanabe M, Suzuki Y (1999) Active fault identification by aerial photograph interpretation. Kokon shoin (in Japanese)
Watanabe M, Suzuki Y, Okada A (1997) Active faults in the urbanized cities of Kobe, Ashiya and Nishinomiya, and the severely damaged zone associated with Hyogoken Nambu Earthquake. Trans Jpn Geomorphol Union 18:223–232 (in Japanese)
Yonekura N, Kaizuka S, Nogami M (2001) Landforms of Japan 1. University of Tokyo Press, Tokyo (in Japanese)

Chapter 10
Earthquake Disasters and Government Committees

Kenji Satake

Abstract After major devastating earthquakes, the Japanese government has set up committees to promote seismological research and to reduce future earthquake damage. Two committees have made efforts to assess possible precursors for large earthquakes with particular emphasis in Tokai region, but such short-term deterministic prediction is now considered difficult in the seismological community. The Headquarters of Earthquake Research Promotion, established after the 1995 Kobe earthquake, promoted to expand the geophysical observation network, made long-term and probabilistic forecast of earthquakes on active faults and subductions zones, predicted strong ground motion from such earthquakes, and made national seismic hazard maps including probabilistic estimation of ground motion. Such long-term forecast and ground motion prediction was made before the 2016 Kumamoto earthquake, but the results were not fully utilized by local governments or the general public to prepare for potential earthquake. The Central Disaster Management Council estimated the maximum possible earthquakes along the Nankai Trough and beneath the Tokyo metropolitan area, assessed the human, building and economic losses from such earthquakes, and made policies to reduce such damage.

Keywords Earthquake · Disaster · Committee · Japan

10.1 Introduction

Earthquakes have repeatedly caused damages in Japan. After major devastating earthquakes, the Japanese government has set up committees to promote seismological research and to reduce future earthquake damage. Four current committees directly related to earthquake research are as follows: Coordination Committee for Earthquake Prediction (CCEP) under Geospatial Information Authority of Japan, Earthquake Assessment Committee for Areas under Intensified Measures against Earthquake Disaster (EAC) under Japan Meteorological Agency (JMA), Earthquake Research Committee (ERC) under Headquarters of Earthquake Research Promo-

K. Satake (✉)
Earthquake Research Institute, The University of Tokyo, Tokyo, Japan
e-mail: satake@eri.u-tokyo.ac.jp

Y. Himiyama et al. (eds.), *Human Geoscience*, Advances in Geological Science,
https://doi.org/10.1007/978-981-32-9224-6_10

tion (HERP) under Ministry of Education, Culture, Sports, Science and Technology (MEXT), and Central Disaster Management Council (CDMC) under Cabinet Office.

In this section, we first review the history of damaging earthquakes and following actions of the government in Sect. 10.2. Then, the activities of the current governmental committees and their recent products are introduced. The CCEP and EAC have made efforts to assess possible precursors with a purpose of earthquake short-term prediction. The HERP promoted to expand the geophysical observation network (Sect. 10.4.1), made long-term forecast of earthquakes on inland active faults and subductions zones (Sect. 10.4.2), predicted ground motion from such earthquakes (Sect. 10.4.3) and made probabilistic seismic hazard maps (Sect. 10.4.4). The long-term forecast and ground motion prediction made before the 2016 Kumamoto earthquake are shown as an example. The CDMC has estimated the maximum possible earthquakes along the Nankai Trough and around the Tokyo metropolitan area, estimated the building, human and economic losses from such earthquakes, and made policies to reduce such damage (Sect. 10.5).

10.2 History of Earthquake Disasters and Committees

In 1891, the Nobi earthquake (magnitude or M 8.0) caused devastating damage in Gifu and Aichi prefectures in central Honshu, with approximately 7000 casualties and 90,000 collapsed houses. This earthquake was one of the largest inland earthquakes in Japan, with surface rupture of 80 km. After this earthquake, the Imperial Earthquake Investigation Committee (IEIC) was set up in 1892 (Clancey 2006; Tomari 2015). It was the first governmental committee with interdisciplinary members. The purpose of IEIC was promoting research for earthquake prediction and reducing earthquake disasters. Their activities to promote research included publication of reports and bulletins in both Japanese and English, and many seismological research as well as report of earthquakes were published by Professors Omori and Imamura among others.

In 1923, the Kanto earthquake (M 7.9) caused the worst earthquake disaster in Japan. This earthquake was an interplate earthquake on the boundary of subducting Philippine Sea plate and overlying plate along the Sagami Trough. It caused direct damage by strong ground shaking and tsunamis, but most damages are from numerous fires occurred in Tokyo and Yokohama. The total number of casualties was 105,000, of which approximately 90% was due to fire. After the Kanto earthquake, in 1925, the Imperial Earthquake Investigation Committee was dissolved and Earthquake Research Institute was set up in Tokyo Imperial University (now the University of Tokyo).

Following typhoon Vera of 1959 which caused significant storm surge in Ise-bay and caused about 5,000 casualties, Central Disaster Management Council (CDMC) was established in 1962. The CDMC targets not only earthquake disaster but also disasters from meteorological hazards (Cabinet Office 2015).

In 1995, the Kobe earthquake (M 7.3) occurred and caused approximately 6,500 casualties. This earthquake was an inland earthquake on active faults, and most casualties were due to building collapse by strong ground motion. After the Kobe earthquake, the Japanese government set up Headquarters of Earthquake Research Promotion (HERP 2014).

In 2011, the largest earthquake instrumentally recorded in Japan (M 9.0), the Tohoku earthquake, occurred offshore Miyagi prefecture of northern Honshu. The earthquake caused devastating tsunami damage with 18,500 casualties (see Chap. 11). Such a gigantic earthquake was not anticipated in Japan, hence the governmental committees revised their strategies. However, no major reorganizations have been made for the committees, except for the Nuclear Regulation Authority, which was set up following the accident of Fukushima Dai-ichi nuclear power station.

In 2016, an earthquake sequence consisting of a foreshock (M 6.5) and mainshock (M 7.3) occurred on Hinagu and Futagawa faults, respectively, in Kumamoto prefecture in central Kyushu. This earthquake sequence, called the Kumamoto earthquakes, caused 50 immediate casualties (and additional >100 casualties due to secondary effects during evacuation), and 8,300 totally collapsed houses. The ERC had made long-term forecast and ground motion prediction of possible earthquake in the region, but such knowledge was not fully utilized before the earthquake (see Sect. 10.4).

10.3 Committees for Earthquake Prediction

One of the societal expectations to seismological community is to "predict" earthquakes and reduce disasters. While the latter (earthquake disaster reduction) requires holistic approaches not limited to natural science, the former (earthquake prediction) is a pure scientific issue.

Earthquake "prediction" and "forecast" are similar concepts but used in different meaning by different community. An international commission consisting of seismological experts in the world (International Commission on Earthquake Forecasting for Civil Protection, or ICEF) defined that prediction is deterministic while forecast is probabilistic (Jordan et al. 2011). They stated, "A prediction is defined as a deterministic statement that a future earthquake will or will not occur in a particular geographic region, time window, and magnitude range, whereas a forecast gives a probability (greater than zero but less than one) that such an event will occur."

In 1965, the national earthquake prediction program started in Japan. The planning of earthquake prediction program started as early as in 1947, followed by preparatory document (so-called "blue print") published in 1962 (Tomari 2015). The governmental Coordination Committee for Earthquake Prediction (CCEP) was set up in 1969. Since around 1975, a possibility of future large earthquake on the eastern end of the Nankai Trough has been discussed. The anticipated future earthquake was named "Tokai earthquake." In 1977, Assessment Committee for Tokai Areas was formed under CCEP. This committee was renamed as "Earthquake Assessment Committee for Areas under Intensified Measures against Earthquake Disaster" (EAC) following the establishment of Act on Special Measures Concerning Countermeasures for Large-Scale Earthquakes in 1978.

The Coordination Committee for Earthquake Prediction (CCEP) consists of 30 members from the governmental organizations such as Japan Meteorological Agency or Geospatial Information Authorities of Japan and the university researchers. At their regular meetings, held four times a year, the members exchange information on earthquake activity and discuss for earthquake prediction.

The Earthquake Assessment Committee (EAC) for Tokai earthquake was convened by Japan Meteorological Agency and consisted of six seismologists. The purpose of the committee was to examine the precursory phenomena and judge if the anticipated Tokai earthquake was likely to occur soon. In order to evaluate the possibility, the committee regularly met, examined the data and reported the status of seismic and crustal activity in Tokai region. Because of recent judgement that deterministic earthquake prediction is diffucult (see next paragraph), the EAC terminated its activity in 2017.

The earthquake prediction relies on observed precursors such as stress changes of ground, earthquake activities or electromagnetic variations. After reviewing these proposed precursors, the ICEF concluded that earthquake prediction is currently impossible because the observable behavior of earthquake systems is poorly understood (Jordan et al. 2011). The ICEF also evaluated the probabilistic forecast with short-term scales of days and weeks. While the short-term forecasting has operational utility such as probabilistic estimate of aftershocks, which has been adopted in Japan, the absolute value of probability is very low and translating such information for decision-making is a challenge.

10.4 Headquarters of Earthquake Research Promotion

The basic objectives of the HERP are to strengthen earthquake disaster prevention measures, especially to promote earthquake research in order to reduce earthquake damage (HERP 2014). The roles include (1) planning of comprehensive and basic policies; (2) coordination of budgets and other administrative work of relevant governmental organizations; (3) establishment of comprehensive survey and observation plans; (4) collection, arrangement, analyses and comprehensive evaluation of survey results by relevant governmental organizations and universities; and (5) publication based on the above evaluations. The HERP hosts two committees: Policy Committee and Earthquake Research Committee (ERC).

The ERC holds monthly meetings to collect, analyze and evaluate current earthquake activities in Japan. After damaging earthquakes, the ERC holds emergency meetings and publicize the evaluation. For example, after the 2016 Kumamoto earthquake, ERC made two emergency meetings and evaluated that the foreshock and mainshock occurred on Hinagu and Futagawa faults, respectively. The ERC has annually updated long-term forecasts of large earthquakes in and around Japan and published national seismic hazard maps.

10.4.1 Observation Networks

In 1997, the HERP published fundamental earthquake survey and observation plans, including networks of high-sensitivity seismographs, broad-band seismographs, strong-motion seismographs, and GPS (now called GNSS to include other systems) stations, as well as active fault surveys. The high-sensitivity seismograph network (Hi-net), strong-motion seismograph networks (K-NET and KiK-net) and continuous GNSS stations (GEONET) are fundamental observation network for earthquake activity, and consists of ~1,000 stations on Japanese islands (Fig. 10.1). These seismic and geodetic networks were completed by around 1998 and have provided basic observational data for seismic activity and strain accumulation of Japanese Islands.

By using these data, the earthquake activity is monitored. All the observed data on high-sensitivity seismograph network are sent to Japan Meteorological Agency, who monitor the seismic activity in 24 h a day, 7 days a week. Once an earthquake occurs, the JMA reports the recorded seismic intensity at each station in about 2 min, and estimated location (latitude, longitude and depth) and size (magnitude) of the earthquake, as well as possibility of tsunami in 5 min (JMA 2016, see Chap. 11 for

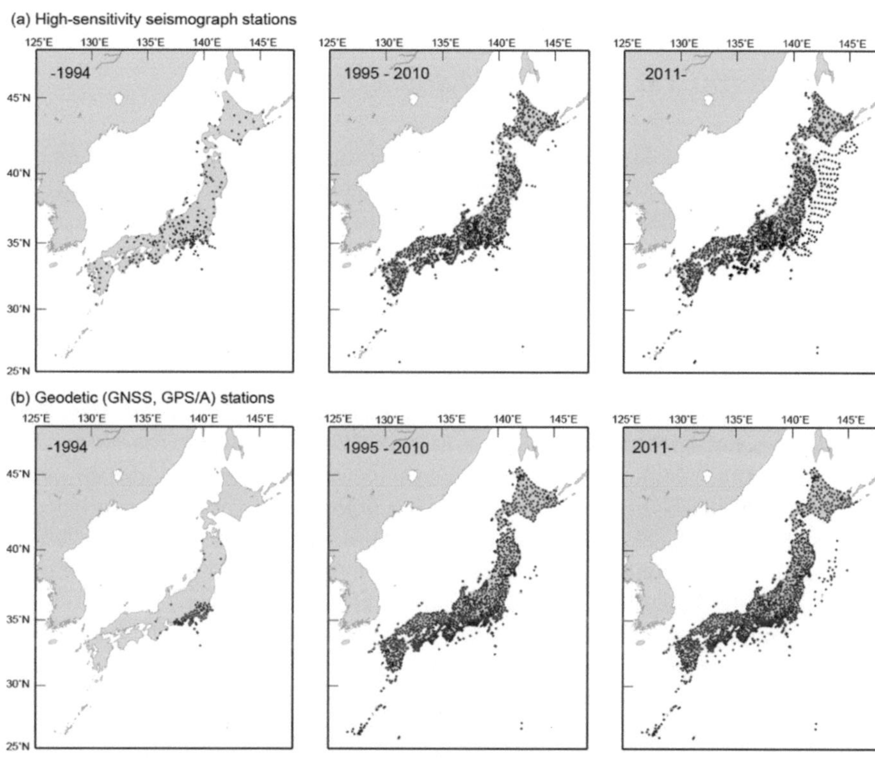

Fig. 10.1 Distribution of **a** high-sensitivity seismograph network and **b** geodetic (GNSS, GPS/A) network stations for the time periods of –1994 (left), 1995–2010 (center) and 2011– (right)

tsunami warning system). The annual number of earthquakes that JMA determines the location and magnitude is more than 100,000; JMA locates all earthquakes with M > 2 on Japanese inland and M > 3 around Japan.

When the seismic intensity of 5 or larger on JMA scale is anticipated, Earthquake Early Warning (EEW) information is issued. The EEW system forecasts ground shaking after the earthquake occurrence but before the arrival of seismic waves, based on quick analysis of seismic data recorded near the earthquake source. The EEW has been implemented and in operation since 2007 in Japan (Kamigaichi et al. 2009). Typical lead time between the announcement and start of large ground shaking is from several to several tens of seconds, yet providing useful information through TV, radio or cell phones.

The submarine geophysical observations played an important role in the 2011 Tohoku earthquake. The tsunami was first detected on offshore bottom pressure and GPS buoy gauges before its arrival on coast (see Chap. 11). The huge (>50 m) seafloor displacement was recorded on bottom pressure and GPS/Acoustic stations. After the 2011 Tohoku earthquake, offshore stations (S-net) consisting of ~150 stations of seismograph and tsunami gauges were installed. Offshore geodetic stations (GPS/A) stations were also installed after 2011.

10.4.2 Long-Term Forecast of Large Earthquakes

In addition to the geophysical measurements, the fundamental observations include surveys for about 100 active faults (out of about 2,000 mapped ones) in Japan. The location and size of active faults were estimated from aerial photographs and digital elevation models, and in situ paleoseismological surveys were conducted to reveal the past activities.

The ERC conducts long-term evaluation of inland earthquakes on active faults and subduction-zone earthquakes, based on active fault surveys and historical and paleoseismological data. The long-term forecasts are made in terms of probabilities of future earthquake occurrence in the next 30 years. If earthquakes of similar size, often called "characteristic earthquakes" recur more or less regularly, the time-dependent probabilities are calculated from the average recurrence interval and date of the most recent event. In this case, the probability is low soon after the occurrence of "characteristic earthquake" but it increases with time toward the next "characteristic earthquake." On the contrary, if earthquakes occur randomly in time, called Poisson process, the time-independent probability is calculated.

In the source area of the 2016 Kumamoto earthquakes, two fault systems, the Futagawa fault, and the Hinagu fault, were mapped and evaluated in 2002 and revised in 2013. Size and occurrence probabilities of future earthquakes on each segment were estimated from the fault mapping and paleoseismological data. On the western section of Futagawa fault (19 km long), where the mainshock (M 7.3) occurred on April 16, an earthquake with M 7.0 and fault slip of 2 m was forecasted with the 30-year probability of 0–0.9% from the evidence of past two earthquakes. While the probability value is small, it belongs to a group of high probability of active faults. On the northern section of Hinagu fault zone (16 km long), where the foreshock (M 6.5) occurred on April 14, an earthquake with M 6.8 and 2 m slip was forecasted but the probability was not estimated because only one past event was dated.

In addition to the evaluation of individual faults, the ERC made regional evaluation of entire Kyushu. The estimated occurrence probabilities of an M 6.8 or larger earthquake as 18–27% in central Kyushu and 7–18% in southern Kyushu during the 30 years from 2016 to 2045. They also estimated the 30-year probability of an earthquake from the current seismicity data; they are 11 and 19% for the central and southern Kyushu, respectively.

The long-term forecasts have been also made for large interplate earthquakes along subduction zones (Fig. 10.2), on the basis of "characteristic earthquake" model. As of 2010, the highest 30 year probability, 99%, was calculated for a large earthquake with M ~ 7.5 in the Miyagi-oki region, just west of the 2011 Tohoku earthquake source. However, the 2011 Tohoku earthquake (M 9.0) was much larger than the long-term forecast. After the 2011 Tohoku earthquake, the ERC is in the process of revising the long-term forecasts. In their revised forecast for the earthquakes along the Nankai Trough and the Sagami Trough, the variability of earthquake size different from "characteristic earthquake" has been considered (Satake 2015).

Because the typical recurrence interval of earthquakes are several decades to 100 years for subduction-zone earthquakes and 1,000 years or longer for active faults, the 30-year probability is usually very small. For the Futagawa fault, the 30-year probability of 0–0.9% was considered as "high," but general public considered that the probability value is low. After the Kumamoto earthquake, the ERC has

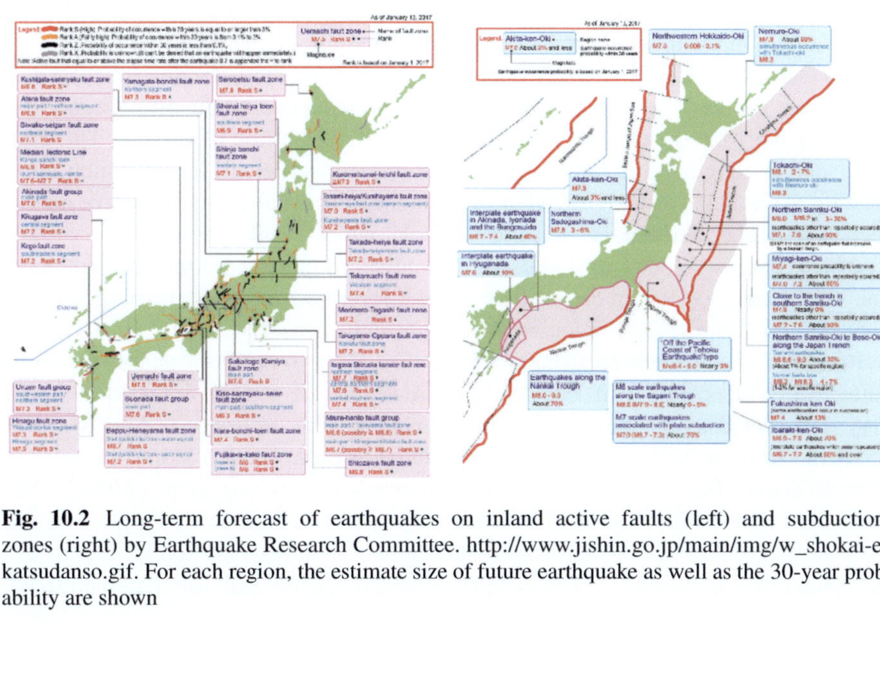

Fig. 10.2 Long-term forecast of earthquakes on inland active faults (left) and subduction-zones (right) by Earthquake Research Committee. http://www.jishin.go.jp/main/img/w_shokai-e_katsudanso.gif. For each region, the estimate size of future earthquake as well as the 30-year probability are shown

recently classified that those active faults with 30-year probability of 3% or larger as S rank, and 0.1–3% as A rank.

10.4.3 Ground Motion Prediction

Once the earthquake source is assigned and the fault parameters are specified, the ground shaking from the earthquake can be computed, and its distribution can be shown as a map. Such a map shows the shaking intensity of the areas around the fault when the earthquake occurs. It is useful to prepare for the ground shaking and possible damage from future earthquakes.

The ERC has proposed a standard method for the strong motion prediction as "recipe for predicting strong ground motion for specified seismic source faults." (Irikura and Miyake 2011). The recipe starts from the source characterization of earthquake fault. The source model is characterized by three kinds of parameters, outer, inner, and extra fault parameters. The outer fault parameters define the entire source area for possible earthquake. The inner fault parameters characterize fault heterogeneity, asperity and background region. The extra fault parameters are related to the propagation pattern of the rupture.

Based on past records of strong ground motion, peak ground acceleration and other intensity measures can be described by ground motion prediction equations (GMPE). For example, an empirical relation between the distance from the earthquake source and the seismic intensity for different magnitude have been used to predict ground motion. If the subsurface seismic velocity structure is known, wave propagation can be simulated by numerical method. A hybrid method of empirical method at short period and numerical simulation at long period are also proposed in the recipe.

The ground-shaking is strongly controlled by shallow subsurface layers. The definitions of "shallow" structure or its "bedrock" may depend on the purpose, and the following two "bedrocks" are defined in ERC's strong motion prediction. The shallower one is "engineering bedrock" usually defined as the surface on which the S velocity of about 0.4 km/s. The engineering bedrock is usually several tens of meters deep. The deeper bedrock is "seismic bedrock" which is often defined as a surface on which S wave velocity of 3 km/s, and usually several to more than ten km deep. The amplification factor in the shallow surface layer above the engineering bedrock is used to compute seismic intensity on the ground surface.

Predicted ground motion from an earthquake (M 7.0) on the western section of Futagawa fault is shown in Fig. 10.3. Narrow zone of seismic intensity 6 upper (6+) is expected along the fault trace, whereas seismic intensity is 5 or lower at about 10 km from the fault. Such a map is called Seismic Hazard Map for Specified Seismic Source Faults or simply Scenario Earthquake Shaking Maps (SESM). The predicted seismic intensity distribution is very similar to the observed ones during the mainshock of the 2016 Kumamoto earthquakes, although the actual fault length was longer than assumed. This indicates that once the active faults are parameterized,

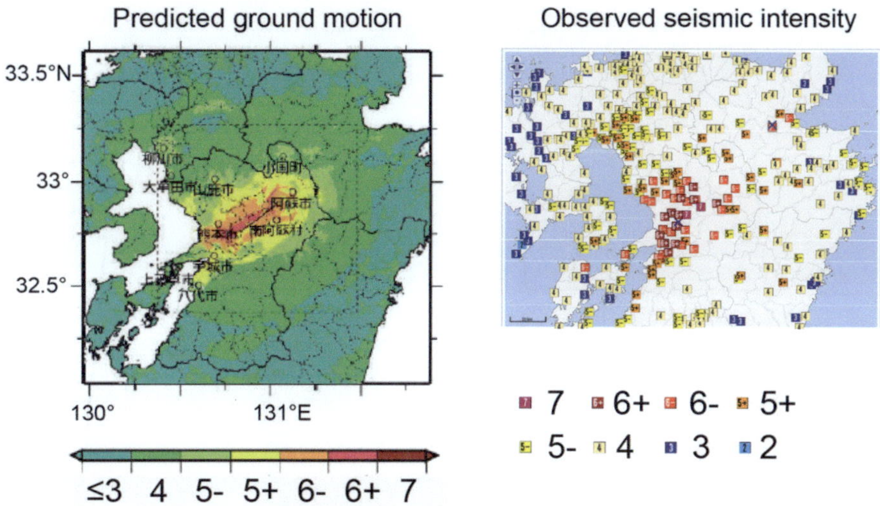

Fig. 10.3 Left: Predicted ground motion for Futagawa section of Futagawa fault (ERC 2014). http://jishin.go.jp/main/chousa/14_yosokuchizu/bessatsu.pdf. Right: Actual observed seismic intensity of the 2016 Kumamoto earthquake (Japan Meteorological Agency)

the ground shaking from an earthquake can be predicted. The SESMs, however, do not involve probability of such ground shaking.

10.4.4 Probabilistic Seismic Hazard Maps

The Probabilistic Seismic Hazard Maps (PSHM) show the possibilities of ground shaking for the entire Japan (Fujiwara et al. 2006). These maps are prepared by combining long-term forecasts and ground motion prediction, and are useful to examine regional differences of ground motion probabilities. In addition to those earthquakes on active faults and subduction zones with the 30-year probability was assigned by the long-term forecast, earthquakes whose source are difficult to identify before the occurrence are also considered. Among the three quantities, i.e., time period, seismic intensity and probability, two of them are fixed and the distribution for the last one is expressed as a map. For example, time period is fixed as 30 years and seismic intensity is fixed as 6 lower (6−), then the distribution of probability is shown in map (Fig. 10.4).

In Fig. 10.4, the 30-year probability is shown in different colors. The probability of 0.1% in 30 years roughly corresponds to a return period of 30,000 years if the Poisson process is assumed. Similarly, 3, 6, and 26% probabilities correspond to return periods of 1,000, 500, and 100 years, respectively. This means that the darkest red color in Fig. 10.4 indicates a region where strong ground shaking (seismic intensity of 6− or larger) is expected in the next 100 years. Such regions are on the Pacific coast of

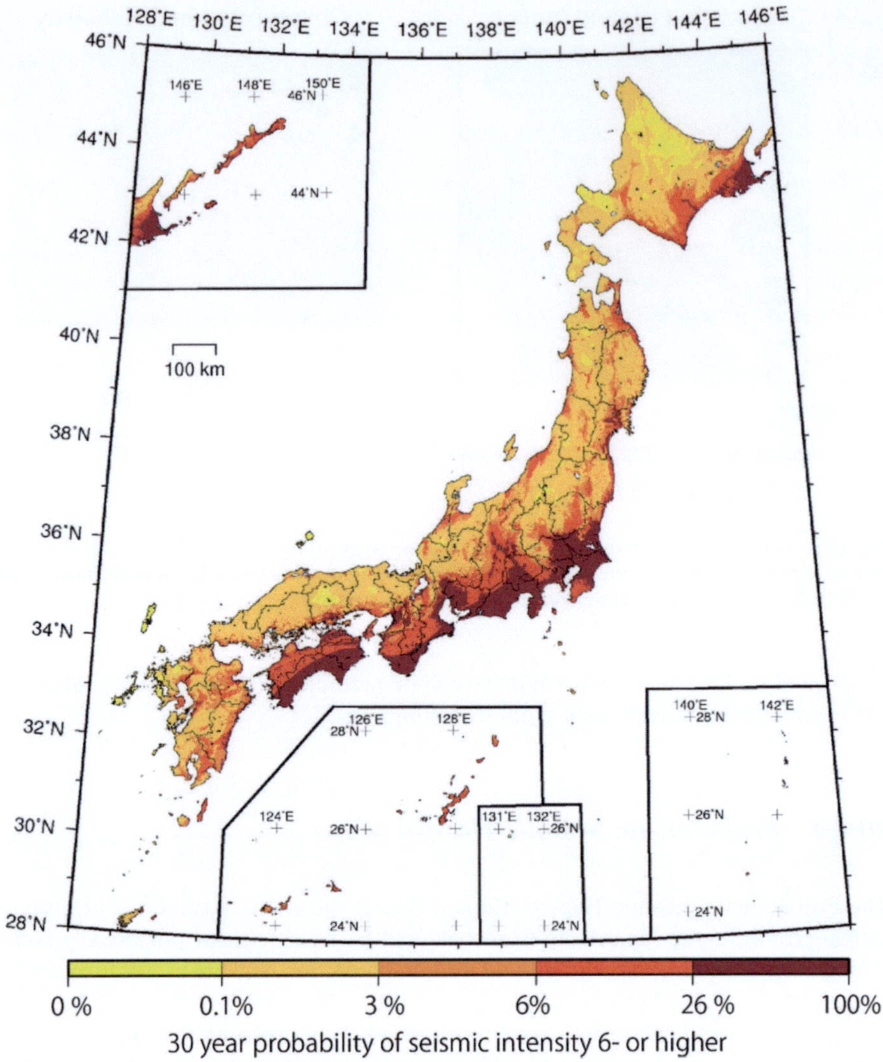

Fig. 10.4 A map showing distribution of probability (the color scale on the bottom) for ground motion of seismic intensity of 6 lower (6−) or larger occurring within 30 years. http://www.jishin.go.jp/main/img/w_shokai-e_hazardmap.jpg

Hokkaido, Kanto to Tokai, Kii peninsula and Shikoku, where strong ground motion is expected from subduction-zone earthquakes. Relatively higher probabilities at inland locations are due to inland earthquakes on active faults with higher probabilities of occurrence.

Scenario Earthquake Shaking Maps (SESM) and Probabilistic Seismic Hazard Maps (PSHM) constitutes National Seismic Hazard Map of Japan. The maps can be accessed through Japan Seismic Hazard Information Station (J-SHIS) provided by National Research Institute for Earth Science and Disaster Resilience (NIED). The J-SHIS users can access various information such as active faults or subduction zones (location, past earthquake history and future earthquake probability), geomorphological classification using 250 m mesh and site amplification factor (between engineering bedrock and ground surface), probabilistic seismic hazard (probability of exceedance for different seismic intensity levels), and exposed population.

10.5 Central Disaster Management Council

The Central Disaster Management Council assessed damage from earthquakes along the Nankai Trough and Kanto region beneath the Tokyo metropolitan area (Cabinet Office 2015). They considered the worst-case scenario for the earthquake source, computed ground shaking, and tsunami heights, and estimated human, building, and economic losses.

For the subduction-zone earthquake along the Nankai Trough, the CDMG assumed an Mw 9 earthquake. They divided the source region into four regions, assumed strong ground-motion generation area (SMGA) for each region, and computed seismic intensity distribution using a method similar to the ERC's recipe. Seismic intensity distribution from one of such models (Fig. 10.5) shows that the

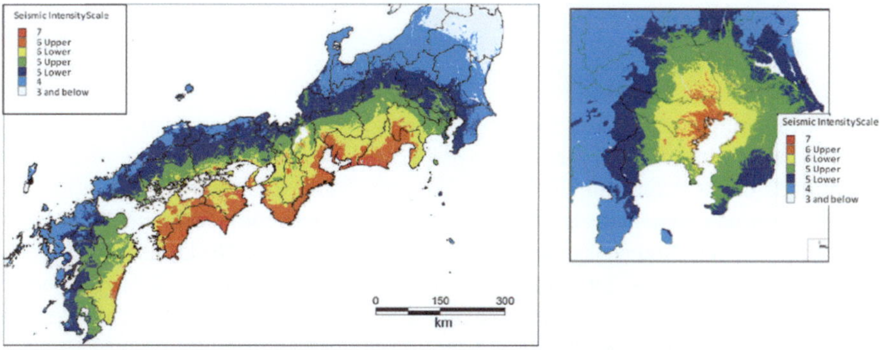

Fig. 10.5 Seismic intensity distribution from an M ~ 9 earthquake along the Nankai Trough (left) and an M ~ 7 earthquake beneath the Tokyo metropolitan area (right). Damage such as human, properties and economic losses are estimated from the above-ground shaking. http://www.bousai.go.jp/jishin/index.html

strong ground shaking is expected in a wide area extending from Tokai to Kyushu regions.

Based on such seismic intensity distribution and tsunami heights, the CDMC estimated damage, both human fatalities, and numbers of damaged buildings. The estimated human fatalities ranged from 32,000 to 323,000, nearly and order of magnitude different, due to uncertainties in assumptions and modeling methods. The number of houses destroyed or damaged by ground shaking, tsunami or fire ranges from 940,000 to 2,386,000. The estimated economic loss is 170 trillion Japanese yen in the affected region, and additional 50 trillion yen in other regions. This estimated economic loss is significant, considering that the annual budget of Japan is about 100 trillion yen. On the basis of such damage estimates, the CDMC and the government promote disaster management by designating areas (cities and towns), and support reinforcement of buildings or seawalls, prepare emergency response plans. By making such preparation for disaster reduction, the government attempts to reduce damage. For example, they estimated that the number of damaged or destroyed houses will be halved by retrofitting buildings.

Similar estimates have been made for the Tokyo metropolitan area (Fig. 10.5). Because only 90 years have passed since the last Kanto earthquake (M 7.9) of 1923, for which the recurrence interval is estimated as 200 to 400 years, the CDMC considered that an M ~ 7 earthquake beneath metropolitan Tokyo is more imminent. Various types of M ~ 7 earthquake were assumed, then the seismic intensity distribution was computed, and resulting damage from both ground shaking and fire was estimated. The estimated number of collapses buildings is about 610,000 and fatalities is 23,000. Among them, 60–70% of both building and human damage are from fire. The estimated economic loss is 95 trillion Japanese yen, similar to the annual budget of Japan. The government attempts to reduce disaster, by various policies such as retrofitting houses or preventing fires. The goal of the next 10 years is to reduce both damage to half of the estimated numbers.

10.6 Conclusions

After devastating earthquakes, the Japanese government has set up various committees to promote research for earthquake prediction and disaster reduction. While two committees target deterministic earthquake prediction, international seismological communities consider that earthquake prediction is currently difficult. Probabilistic estimates of earthquake occurrence, both in short-term or long-term, yield very small probability values for a few days or 30 years, respectively. It is difficult to motivate local government or general public to prepare for future earthquakes by using these probability values. Ground motion prediction from scenario earthquake is a useful tool to assess possible damage from future earthquake, though it does not include any information of occurrence probability. The result of earthquake research conducted by such committees was not fully utilized for recent earthquakes such as the Kumamoto earthquakes in 2016. It is important to better communicate and educate the local government officials and general public using the scientific products produced by the committees.

References

Cabinet Office (2015) White paper on disaster management in Japan 2015, 222 pp. http://www.bousai.go.jp/kaigirep/hakusho/pdf/WP2015_DM_Full_Version.pdf

Clancey G (2006) Earthquake nation. University of California Press, Berkeley and Los Angeles, p 331

Fujiwara H, Kawai S, Aoi S, Morikawa N, Senna S, Kobayashi K, Ishii T, Okumura T, Hayakawa Y (2006) National seismic hazard maps of Japan. Bull Earthq Res Inst Univ Tokyo 81:221–232. http://www.eri.u-tokyo.ac.jp/BERI/pdf/IHO81304.pdf

Headquarters for Earthquake Research Promotion (2014) Understanding earthquakes, 30 pp. http://jishin.go.jp/main/pamphlet/brochures2014en/understanding_earthquakes.pdf

Irikura K, Miyake H (2011) Recipe for predicting strong ground motion from crustal earthquake scenarios. Pure Appl Geophys 168:85–104. https://doi.org/10.1007/s00024-010-0150-9

Japan Meteorological Agency (2016) Earthquakes and Tsunamis—disaster prevention and mitigation efforts. 26 pp. http://www.jma.go.jp/jma/kishou/books/jishintsunami/jishintsunami.pdf

Jordan T, Chen Y-T, Gasparini P, Madariaga R, Main I, Marzocchi W, Papadopoulos G, Sobolev G, Yamaoka K, Zschau J (2011) Operational earthquake forecasting. State of knowledge and guidelines for utilization. Ann Geophys 54. https://doi.org/10.4401/ag-5350. Report by International Commission on Earthquake Forecasting for Civil Protection

Kamigaichi O, Saito M, Doi K, Matsumori T, Tsukada S, Takeda K, Shimoyama T, Nakamura K, Kiyomoto M, Watanabe Y (2009) Earthquake early warning in Japan: warning the general public and future prospects. Seismol Res Lett 80:717–726. https://doi.org/10.1785/gssrl.80.5.717

Satake K (2015) Geological and historical evidence of irregular recurrent earthquakes in Japan. Philos Trans R Soc A 373:2014375. https://doi.org/10.1098/rsta.2014.0375

Tomari J (2015) 130 years history of earthquake prediction research in Japan. University of Tokyo Press, Tokyo, 671 pp (in Japanese)

Chapter 11
Tsunami Science and Disaster Prevention

Kenji Satake

Abstract The March 11, 2011, Tohoku earthquake (M 9.0) caused devastating tsunami disaster on the Pacific coast of East Japan, including 18,500 casualties and accident of Fukushima Dai-ichi nuclear power station. Earthquake tsunamis are caused by seafloor deformation due to faulting, and its propagation speed is basically controlled by water depth. The tsunami generation and propagation can be simulated by computer and the arrival times and water heights on coasts can be predicted if the initial condition and ocean depth topography are known. The 2011 tsunami was first detected by offshore tsunami gauges, before it arrived on the Sanriku coast in about 30 min after the earthquake with a maximum height of nearly 40 m, then inundated Sendai plain with a maximum distance of nearly 5 km from the coast. The tsunami warning was issued 3 min after the earthquake, which saved many lives, but several issues were identified to further improve the tsunami warning system. Similar tsunami disasters occurred in the past, but the experience was not fully utilized to reduce the 2011 disaster. To mitigate future tsunami disaster, coastal infrastructures such as seawalls or breakwaters and soft countermeasures such as tsunami hazard maps or education and awareness are being developed.

Keywords Tsunami · Tohoku earthquake · East Japan earthquake disaster · Tsunami warning system · Hazard map

11.1 The 2011 East Japan Earthquake Disaster

On March 11, 2011, the Japanese coasts were hit by a devastating tsunami caused by a giant earthquake (M 9.0) off the Pacific coast of the Tohoku region. The tsunami also caused partial destruction of the Fukushima Dai-ichi nuclear power station. The earthquake, whose magnitude was the largest as instrumentally recorded one in Japan, was officially named "off the Pacific coast of Tohoku earthquake" by the

---

K. Satake (✉)
Earthquake Research Institute, The University of Tokyo, Tokyo, Japan
e-mail: satake@eri.u-tokyo.ac.jp

Y. Himiyama et al. (eds.), *Human Geoscience*, Advances in Geological Science,
https://doi.org/10.1007/978-981-32-9224-6_11

Japan Meteorological Agency (JMA), but often abbreviated as the Tohoku (-oki) earthquake in seismological literature. The earthquake and tsunami disaster, including the Fukushima accident, was named "the Great East Japan Earthquake Disaster" by the Cabinet Office.

Human, property and economic damages of the 2011 disaster were summarized by the Central Disaster Management Council. As of March 2016, the human casualties are approximately 15,900 deaths and 2600 missing, of which approximately 92% were drowned by the tsunami. More than two-thirds of the casualties were elderly (60 years or older), who experienced difficulties in promptly evacuating. The total inundation (flooded) area of the 2011 tsunami was estimated as 561 km^2 where the population was about 600,000. Had the tsunami arrived in the night time and/or the without tsunami warning, the casualties would have been even larger. The damaged houses, mostly due to the tsunami, consisted of 122,000 totally collapsed, 280,000 partially collapsed and 726,000 partially damaged. The total economic loss, 16,900 billion yen (about 130 billion U.S. dollars), is about 20% of the annual national budget of Japan.

In this chapter, we first review the generation and propagation of tsunamis with the 2011 tsunami as an example. We then focus on issues related to human geosciences, i.e., tsunami warning, history of past tsunamis, tsunami hazard maps, tsunami countermeasures, and education/awareness.

11.2 Tsunami Generation and Propagation

The Tohoku earthquake occurred at 14 h 46 m 18.1 s Japan Standard time (UT+9 h) on March 11, 2011. The earthquake was felt in most of the Japanese islands, with the largest seismic intensity of 7 (the highest) on JMA's scale, which corresponds to 11 or 12 on the Modified Mercalli scale. The earthquake occurred at the boundary between the Pacific plate and the overlying plate (Fig. 11.1). The Pacific plate subducts beneath northern Honshu along the Japan Trench at a rate of about 8 cm per year. At the time of the interplate earthquake, thrust-type fault movement between the subducting and overlying plates causes uplift of seafloor just above the fault plane and subsidence of landward part of seafloor and coast (Fig. 11.2). Such a vertical movement of seafloor causes similar displacement of the sea surface, if the wavelength of displacement is much longer than the water depth. The sea surface displacement then becomes the source of a tsunami.

Tsunami is considered as long-waves or shallow-water waves in hydrodynamics. When the horizontal scale of motion, or the wavelength, is much larger than the water depth, the horizontal motion of water mass is almost uniform from bottom to the surface (Fig. 11.3). The typical wavelength of seafloor (or sea surface) displacement is several tens to hundred km, while typical ocean depth is about 4–5 km. Hence the long-wave approximation is appropriate for most earthquake tsunamis. This is different from ordinary wind-waves, which are called short or surface waves, where

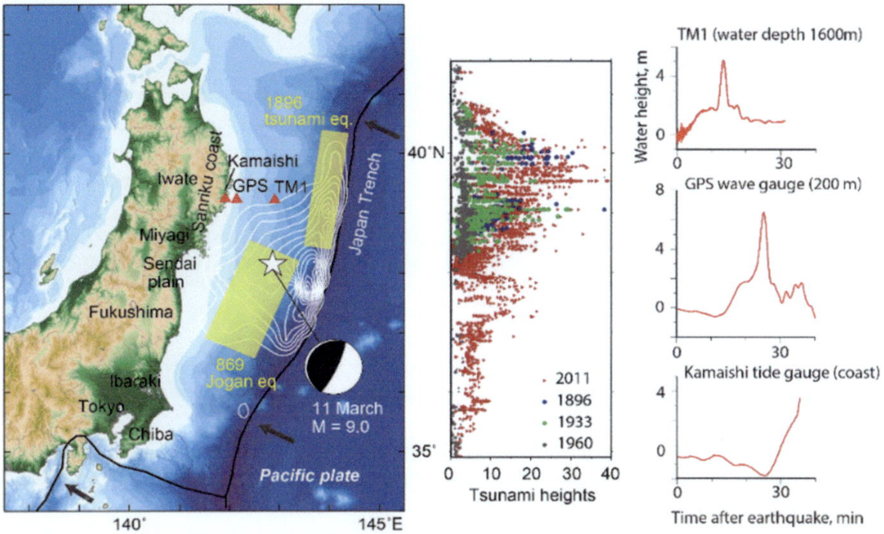

Fig. 11.1 (Left) The source region of the 2011 Tohoku earthquake (the epicentre is shown by the white star). The red triangles show the locations of the bottom pressure, GPS wave and coastal tide gauges that recorded tsunamis waveforms shown in Right. The plate boundaries and the relative motions are shown by thick black curves and arrows, respectively. The proposed fault models (yellow rectangles) of the 1896 Sanriku tsunami earthquake (Tanioka and Satake 1996) and the AD 869 Jogan earthquake (Sawai et al. 2012) are shown. The beach ball shows the focal mechanism solution of the mainshock USGS. The slip distribution estimated from the tsunami waveforms is shown by contours with 4 m intervals (Satake et al. 2013). (Center) The coastal heights of the 2011 Tohoku tsunami compiled by the Joint Survey Group (Mori et al. 2011), the 1896 and 1933 Sanriku tsunamis and the 1960 Chile tsunami. For the historical tsunamis, the heights are compiled by NOAA and Japan Tsunami Trace Database are used. (Right) The tsunami waveforms recorded by the bottom pressure gauge, GPS wave gauge and coastal tide gauge (Fujii et al. 2011)

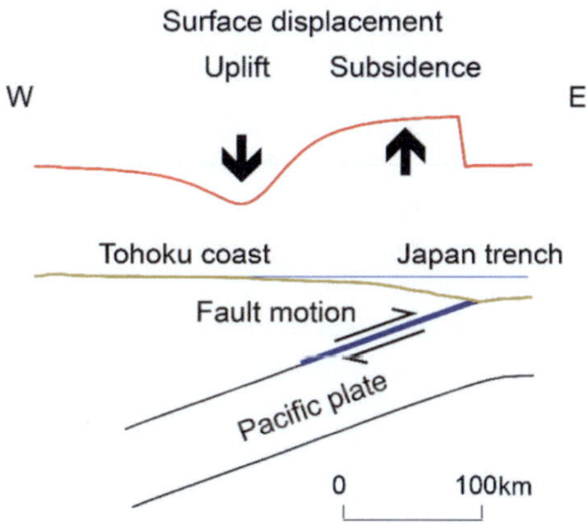

Fig. 11.2 Schematic cross-section across the Japan Trench where the Pacific plate subducts beneath Honshu. When a thrust-fault movement occurs at the plate interface, seafloor (and sea surface) just above the fault plane is uplifted while that of deeper extension is subsided

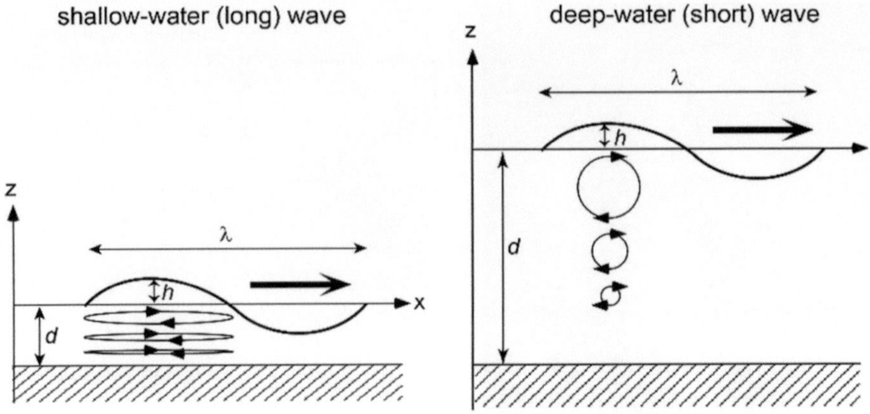

Fig. 11.3 Schematic figure of shallow-water (long) wave (*Left*) and deep-water (short) wave (*Right*). Tsunami generated by great earthquakes belongs to left, while wind-wave belongs to right

the wavelength is comparable to the water depth, and the motion is larger near the water surface but decays with depth (Fig. 11.3).

The speed of long-waves is given as $\sqrt{gd}$ where g is gravitational acceleration (9.8 m/s^2) and d is water depth. This indicates that the deeper the ocean, the faster the tsunami travels. For the deep ocean such as the Pacific Ocean, with 4000 m depth, the tsunami speed is 200 m/s or 720 km/h, similar to that of a jetliner. Near the coast, at the depth of 10 m, the speed becomes 10 m/s or 36 km/h, similar to that of an automobile. However, the amplitude of a tsunami becomes large as the speed becomes slower. Note that the speed of tsunami, even in the deep ocean, is much slower than seismic waves, whose typical speed is 5–8 km/s.

Because the tsunami speed depends only on water depth, tsunami propagation can be computed if seafloor topography, or bathymetry, is given. More specifically, the equation of motion for shallow-water waves and the equation of continuity are solved numerically by using the finite-difference method. The arrival times, as well as amplitudes on coasts, can be predicted by such numerical simulation.

The 2011 tsunami was first recorded on offshore gauges (Fig. 11.1). On ocean bottom pressure gauge TM1, located about 76 km off Sanriku coast at a 1600 m water depth and connected with seafloor cable, ~2 m water rise was recorded immediately after the earthquake, followed by an impulsive wave with additional 3 m rise within 2 min. On a nearshore GPS wave gauge, located about 10 km offshore at a water depth of 200 m, similar tsunami waveforms with larger amplitudes were recorded from 12 min after the earthquake. The tsunami arrival on the coast was recorded on tide gauge at Kamaishi port at about 30 min after the earthquake, but the large tsunami destroyed the system hence the maximum amplitude was not recorded. The coastal heights of the tsunami were measured by various organizations. The surveys show that the tsunami heights varied locally, particularly along the sawtooth-shaped

Sanriku coast, with the maximum heights of almost 40 m (Fig. 11.1). The tsunami reached the Sendai plain about 1 h after the earthquake and inundated regions as far as 5 km from the coast.

11.3 Tsunami Warning System

The velocity difference between the seismic and tsunami waves makes it possible to issue a tsunami warning by detecting seismic waves before the actual tsunami arrival on the coast. In order to promptly issue tsunami warning messages, most near-field tsunami warning systems rely on seismic observations to estimate the location, depth, and size of the earthquake. The epicentre and magnitude M of earthquakes are first estimated within a few minutes of occurrence using the seismographic network.

The tsunami warning system was first introduced in the 1950s by the Japan Meteorological Agency (JMA). At that time, it took about 20 min to issue tsunami warnings. In 1983, when a large earthquake occurred in the Japan Sea, the warning was issued in 12 min after the earthquake, but the tsunami arrived in 7 min and 100 people died. JMA improved the warning system, and when another large earthquake occurred in 1993 also in the Japan Sea, the tsunami warning was issued in 5 min. However, the tsunami arrived in less than 5 min and the 230 people died. JMA further improved the system and now issues tsunami warning within 3–5 min after a large earthquake (Fig. 11.4).

In addition to the seismic observation, the tsunami warning system utilizes tsunami numerical simulation. Tsunami generation and propagation for different earthquake sources, an order of 100,000 cases with various locations and magnitudes, were calculated in advance, and the results have been stored as a database. When a large earthquake occurs, the most appropriate case in terms of location and magnitude is retrieved from the database, and expected arrival times and tsunami heights on coasts are announced. The JMA's tsunami warning system consists of three categories: major tsunami warning (estimated tsunami heights >3 m), tsunami warning (1 m < tsunami height < 3 m), and tsunami advisory (0.2 m < height < 1 m) for the 66 coastal segments of Japan. The tsunami warning messages will be updated based on more comprehensive seismic observation and confirmation of tsunami on sea-level measurements.

At the time of the 2011 Tohoku earthquake, JMA issued the first tsunami warning at 3 min after the earthquake (Ozaki 2011). The estimated tsunami heights were 6 m on Miyagi coast, and 3 m or less for the other coasts. These heights were significantly underestimated compared to the actual tsunami heights because the initial estimate of magnitude ($M = 7.9$) was much smaller than the actual size (M 9.0). Nevertheless, very strong ground shaking and the rapid tsunami warning urged many coastal residents to evacuate to high ground and thus saved their lives.

JMA observed the first tsunami arrival at Ofunato on Sanriku coast with 0.2 m amplitude and announced this information at 13 min after the earthquake. While this

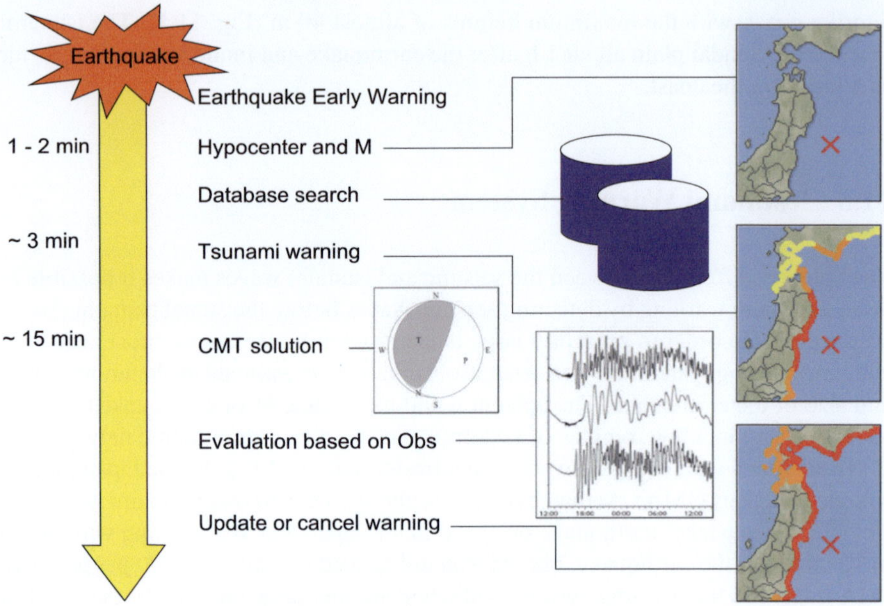

Fig. 11.4 Timeline of Japanese Tsunami Warning System (Japan Meteorological Agency http://www.data.jma.go.jp/svd/eqev/data/en/guide/info.html)

was a precursor to the following large tsunami, some people erroneously interpreted the information that the actual tsunami was small and slowed down their evacuation.

After detecting the large offshore tsunami on the GPS wave gauges (Fig. 11.1), JMA upgraded the tsunami warning messages at 28 min after the earthquake but before the tsunami arrival on the coast to a higher level of estimated tsunami heights. The tsunami warning messages were further updated several times later on. However, the updated tsunami warning message did not reach all the coastal communities, because of electronic power failure and the fact that many people had already started evacuation.

JMA, based on lessons from the 2011 Tohoku earthquake, had made efforts to improve the tsunami warning system. This includes technical developments to estimate the earthquake size in a few mins by using a suite of information, with some being redundant, and deploying and utilizing the offshore tsunami observation systems. On the dissemination aspect, if a danger exists that the earthquake size is potentially underestimated, a warning is issued based on the worst-case scenario. The expected tsunami heights will be announced in simpler expressions considering the uncertainties.

11.4 History of Past Tsunamis

The Pacific coast of Japan has been hit by many tsunamis in the past. In particular, the Sanriku coast had been devastated by previous tsunamis in 1896, 1933 and 1960 (Fig. 11.1). Detailed and quantitative comparison of the 2011 and past tsunamis was made by Tsuji et al. (2014).

The 1896 Sanriku earthquake caused a very large tsunami, with a maximum height of 38 m, similar to the 2011 Tohoku earthquake tsunami. The 1896 tsunami caused about 22,000 casualties, somewhat more than the 2011 tsunami, but mostly limited to Iwate prefecture. In Taro village (now a part of Miyako city) of Iwate prefecture (Fig. 11.5), about 1900 residents, or 80% of the population, lost their lives. One notable feature of this tsunami was that the preceding earthquake shaking was very weak. It was a typical "tsunami earthquake," which produces much larger tsunamis than expected from seismic waves.

The 1933 Sanriku earthquake caused a large tsunami, with heights up to 24 m but mostly between 5 and 10 min the Sanriku coast. While it occurred at night, very strong shaking let many people escape to high ground to survive. The total casualties were about 3000, of which nearly 1000 were in Taro, which corresponds to about 20% of the population at that time. The large tsunami, damage, and casualties were limited to the Sanriku coast.

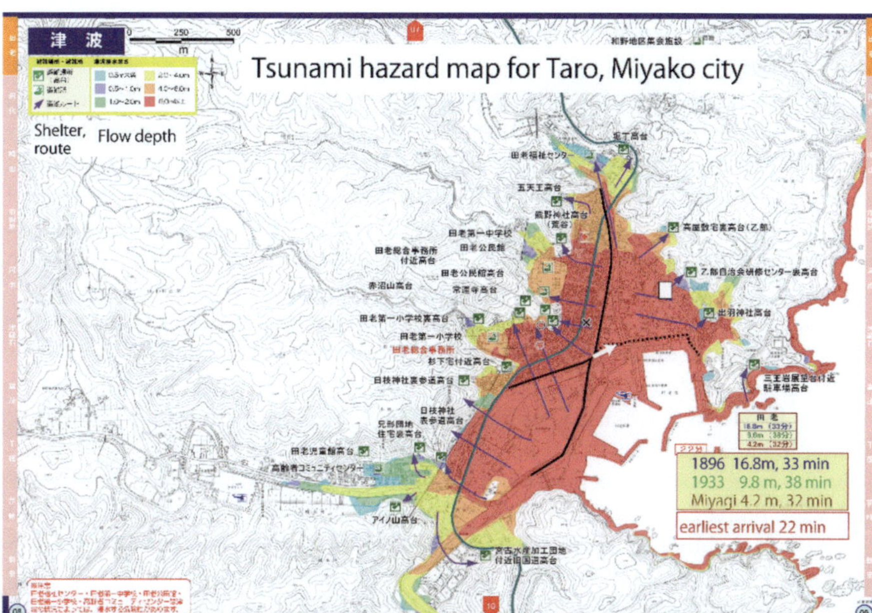

Fig. 11.5 Tsunami hazard map for Taro, Miyako city, Iwate prefecture. Black curves indicate the seawall constructed before the 2011 tsunami. Dashed part was destroyed by the 2011 tsunami. White arrow indicates the view direction and white box is hotel seen in Fig. 11.6

In 1960, the largest ($M = 9.5$) earthquake in the world in the twentieth century occurred offshore southern Chile, and its tsunami arrived on the Japanese coasts about 23 h after the earthquake and caused damage. The tsunami heights on the Sanriku coast were up to 6 m, resulting in the deaths of 62 and 54 people in Iwate and Miyagi prefectures, respectively.

While the above three tsunamis did not inundate the Sendai plain, a similarly large earthquake and tsunami occurred there in AD 869, during the Jogan era on the Japanese calendar. A national history book, "Nihon Sandai Jitsuroku", depicts strong shaking, ground fissures, the collapse of houses, and casualties from the earthquake. In addition, it describes the tsunami as having a sound like thunder, which inundated along the river up to a castle. The flooded area is said to be hundreds of kilometres, with 1000 people drowned. In addition, sand carried by this tsunami has been deposited as geological layers in the coastal lowlands, more than 4 km from the current coast on the Sendai plain (Sawai et al. 2012). Based on the distribution of tsunami deposits, the 869 Jogan earthquake has been modeled as an interplate $M = 8.4$ earthquake. Younger and older tsunami deposits have also been found, corresponding to either 1454 Kyotoku or 1611 Keicho earthquakes recorded in historical documents. These indicate that a recurrence interval of large tsunamis in Sendai plain is 500–800 years (Sawai et al. 2015).

11.5 Tsunami Hazard Maps

The tsunami heights and inundation (flooded) areas from past tsunamis are summarized as tsunami hazard maps. If there is no historical record of past tsunamis, inundation zones from future earthquakes can be estimated by tsunami numerical simulation. Tsunami hazard maps can also include safe places such as tsunami shelters or tall buildings. Once the coastal residents receive a tsunami warning message, they need to know where to evacuate. Tsunami hazard map is an effective tool for the coastal communities to prepare for tsunami hazards.

Tsunami hazard maps were made and distributed to the coastal residents, before the 2011 tsunamis disaster, in most of the affected coastal communities. On the Sanriku coast, most hazard maps show the inundation areas from the past tsunamis, typically the 1896 Sanriku tsunami, as in the case of Taro (Fig. 11.5). The 2011 tsunami inundation area was similar to that shown on the hazard map.

On the contrary, in the Sendai plain, tsunami hazard maps and other countermeasures were prepared for a smaller ($M \sim 8$) earthquake which was estimated to occur with 99% probability in the next 30 years by the Earthquake Research Committee (see Chap. 10, or Satake 2015). The predicted inundation area for the $M \sim 8$ earthquake was mostly within less than 1 km from the coast, and much smaller than the actual tsunami inundation area from the 2011 $M \sim 9$ earthquake. The distribution of the 869 tsunami deposits, however, was similar to the inundation area of the 2011 tsunami.

11.6 Coastal Infrastructures

To reduce social vulnerability, various infrastructures, or hardware, have been implemented since the past tsunamis. Speakers to broadcast tsunami warning messages have been installed. Numerous signboards showing the altitude and route to evacuation places have been installed. Seawalls and breakwaters have been constructed for coastal cities, particularly to protect critical facilities such as nuclear power stations.

In Taro, seawalls with a height of 10 m and a total length of 1.4 km (later extended to 2.4 km) around the town were constructed after the 1933 Sanriku tsunami (Figs. 11.5 and 11.6). The seawalls protected the town from the 1960 Chilean tsunami. However, the 2011 tsunami came over and destroyed a part of seawall, and caused 210 casualties (5% of the population). The tsunami destroyed up to the 3rd floor of a hotel which is located inside the destroyed seawall (Fig. 11.6).

In Kamaishi, a large breakwater was constructed in 2009 at the water depth of 63 m in the entrance to the port with a cost of 120 billion yen (about 1 billion U.S. dollars). It was designed for 5 m high tsunami, similar to the 1896 or 1933 tsunamis. The 2011 tsunami height was estimated as 10 m and the breakwater was partially destructed. While the breakwater delayed the tsunami arrival to Kamaishi port by 5 min and reduced the coastal tsunami heights from 13.7 to 8.7 m (Takahashi et al. 2011), it could not protect the port and town of Kamaishi and caused significant human and property damages.

These demonstrate the limitation of protection by coastal infrastructures. It should be also noted that hardware infrastructures have their lifetimes which may be shorter than the recurrence interval of great tsunamis. After the 2011 Tohoku earthquake and

Fig. 11.6 Seawall survived (left) and destroyed (right) by the 2011 tsunami in Taro. View direction is shown on the map of Fig. 11.5. The hotel in the centre was destroyed to its 3rd floor

tsunami, the Japanese government established a general policy for future tsunami hazards (Central Disaster Management Council 2011). The possible future tsunamis are classified into two levels: L1 and L2. The L2 tsunamis are the largest possible tsunamis with a low frequency of occurrence, but cause devastating disaster once they occur. For such events, saving people's lives is the first priority and soft measures such as tsunami hazard maps, evacuation facilities or disaster education need to be prepared. The L1 tsunamis are expected to occur more frequently, typically once in a few decades, for which hard countermeasures, such as seawall or breakwater must be constructed to protect lives and properties of residents as well as economic and industrial activities.

11.7 Disaster Education and Practices

Gigantic earthquakes and tsunamis are rare phenomena. Only a few $M \sim 9$ earthquakes occurred in the world in the last century (Satake and Atwater 2007). For such infrequent hazard, education is the most important and effective tool to reduce disasters.

The Sanriku coast has experienced tsunami disasters repeatedly. But the average interval between each tsunami is a few decades, which is about the same as the life span of an individual. Thus most people experience one or two tsunamis in their lifetime. A concept, "Tsunami Tendenko," which calls for a quick tsunami evacuation without waiting for others, not even one's parents or children, was taught on the Sanriku coast from the past experience.

On other coasts, for example along Nankai trough in southwestern Japan, the recurrence interval of tsunami disaster is even longer, typically about 100 years for the interplate earthquakes, which means that some generation does not have their own experience. The experience from the past tsunamis needs to be taught to the next generations and to be utilized to reduce the tsunami disasters. A famous story, called "Inamura-no-hi" (fire of rice sheaves), has been used for tsunami education. After a strong earthquake was felt at a coastal village of Kii Peninsula (Wakayama prefecture), the village chief put fire on his just-harvested rice crops to guide villagers to high ground and to save their lives. This is based on the true story during the 1854 Nankai earthquake which occurred on December 24, 1854 (10th day of 11th month on Japanese lunar calendar).

Periodic practice and drills are also important to keep the tsunami warning and mitigation system functional. The Japanese government assigned November 5 as "World tsunami awareness day," and it was endorsed by the United Nations General Assembly on December 2015. The date was chosen based on the above story of "Inamura-no-hi," though it was on the Japanese lunar calendar. Preparedness exercises are conducted in all parts of Japan by various local governments and private companies and organizations.

11.8 Conclusions

Most tsunamis are caused by seafloor deformation due to large submarine earthquakes. The tsunami propagation is controlled by water depth, hence the tsunami arrival times and heights on coasts can be computed if the tsunami source and bathymetry are known. The tsunami source can be estimated from analysis of seismograph data.

The 2011 Tohoku tsunami was first detected by offshore observations, before its arrival on Sanriku coast about 30 min and Sendai plain in about 1 h after the earthquake. The tsunami warning system, based on seismic and offshore tsunami observation, can be further improved to inform coastal residents about the tsunami arrival. After the 2011 Tohoku tsunami, the Japanese government constructed S-net (Seafloor observation network for earthquakes and tsunamis along the Japan Trench), which consists of 150 seismic and tsunami observation stations along with the submarine cables of 5700 km length.

The devastating 2011 tsunami was not the first event of such type. Similar tsunamis have repeatedly hit the Sanriku coast and Sendai plains. History of past tsunami damage as well as assessment of future tsunami can be summarized as tsunami hazard maps.

Coastal infrastructures can reduce the tsunami disaster, but there is a limitation. We should not heavily rely on hard infrastructures. Because tsunami hazard is infrequent, education and awareness are important to reduce future tsunami disaster.

References

Central Disaster Management Council (2011) Report of the committee for technical investigation on countermeasures for earthquakes and tsunamis based on the lessons learned from the "2011 off the Pacific coast of Tohoku earthquake." 46 pp. http://www.bousai.go.jp/kaigirep/chousakai/tohokukyokun/pdf/Report.pdf

Fujii Y, Satake K, Sakai S, Shinohara M, Kanazawa T (2011) Tsunami source of the 2011 off the Pacific coast of Tohoku earthquake. Earth Planets Space 63:815–820. https://doi.org/10.5047/eps.2011.06.010

Mori N, Takahashi T, Yasuda T, Yanagisawa H (2011) Survey of 2011 Tohoku earthquake tsunami inundation and run-up. Geophys Res Lett 38. https://doi.org/10.1029/2011gl049210

Ozaki T (2011) Outline of the 2011 off the Pacific coast of Tohoku earthquake (Mw 9.0) -Tsunami warnings/advisories and observations. Earth Planets Space 63:827–830. https://doi.org/10.5047/eps.2011.06.029

Satake K (2015) Geological and historical evidence of irregular recurrent earthquakes in Japan. Phil Trans R Soc A 373:2014375. https://doi.org/10.1098/rsta.2014.0375

Satake K, Atwater BF (2007) Long-term perspectives on giant earthquakes and tsunamis at subduction zones. Annu Rev Earth Planet Sci 35:274–349. https://doi.org/10.1146/annurev.earth.35.031306.140302

Satake K, Fujii Y, Harada T, Namegaya Y (2013) Time and space distribution of coseismic slip of the 2011 Tohoku earthquake as inferred from tsunami waveform data. Bull Seism Soc Am 103:1473–1492. https://doi.org/10.1785/0120120122

Sawai Y, Namegaya Y, Okamura Y, Satake S, Shishikura M (2012) Challenges of anticipating the 2011 Tohoku earthquake and tsunami using coastal geology. Geophys Res Lett 39:L21309. https://doi.org/10.1029/2012GL053692

Sawai Y, Namegaya Y, Tamura T, Nakashima R, Tanigawa K (2015) Shorter intervals between great earthquakes near Sendai: scour ponds and a sand layer attributable to A.D. 1454 overwash. Geophys Res Lett 42:4795–4800. https://doi.org/10.1002/2015gl064167

Takahasi S et al (2011) Urgent survey of the 2011 Great East Japan earthquake and tsunami disaster in ports and coasts. Technical Note of the Port and Airport Research Institute, No. 1231. 200 pp

Tanioka Y, Satake K (1996) Fault parameters of the 1896 Sanriku tsunami earthquake estimated from tsunami numerical modeling. Geophys Res Lett 23:1549–1552

Tsuji Y, Satake K, Ishibe T, Harada T, Nishiyama A, Kusumoto S (2014) Tsunami heights along the Pacific Coast of Northern Honshu recorded from the 2011 Tohoku and previous great earthquakes. Pure Appl Geophys 171:3183–3215. https://doi.org/10.1007/s00024-014-0779-x

Chapter 12
Volcanic Eruption and Human Geoscience

Setsuya Nakada

Abstract Volcanic hazards are very variable and the impacts are not proportional to the scales of eruptions. It is clear that volcanic activity in Japan has been quiet during these hundred years in comparison with those in other volcanic countries. Very large eruptions may be possible in the near future, by considering the probability and the time that passed after the last large eruptions. The impacts from the large eruptions were discussed at the timing of resumption of the nuclear power plants. Although the method to forecast volcanic eruptions had been developed, we repeatedly failed to forecast onsets of eruptions and to expect the development of eruptions. By experiencing these failures, the eruption scenarios covering possible eruptive phenomena and sequences (eruption event-tree models) became to consider in the recent research project. The Volcanological Society of Japan commented about the issue related to large volcanic eruptions; for example, how the nation and volcanologists should be and how the research results should be used. The research results should be utilized for increasing the resilience of the residents against volcanic hazards.

Keywords Volcanic disasters · Caldera-forming eruption · Eruption scenario

12.1 Introduction

We cannot avoid encountering natural hazards as far as we live in tectonic moving terrain of the geological background such as the Japan archipelago, where the oceanic plates are colliding against the continental plates. Especially, earthquakes, tsunamis, and volcanic eruptions are common phenomena in there. For these decades, the Japanese people have received severe damages from several large earthquakes that occurred in the inland and ocean sides and from resultantly generated tsunamis, and a huge number of casualties has been reported up to the present. Volcanic phenomenon is one of the events repeated from a long time ago since people started to live in this archipelago.

S. Nakada (✉)
National Research Institute for Earth Science and Disaster Resilience, Tsukuba, Japan
e-mail: nakada@bosai.go.jp

Y. Himiyama et al. (eds.), *Human Geoscience*, Advances in Geological Science,
https://doi.org/10.1007/978-981-32-9224-6_12

Volcanic activity produce a large variety of hazards, including ash falling, lava flow, pyroclastic flow (gravitational current of the mixture of volcanic ash, pumice, lava fragment and volcanic gas), lahar (debris or mudflow), sector collapse or landslide, ballistic ejection of lava fragments, volcanic tsunami, seismicity, ground deformation, geothermal effect, and so on. Being different from other natural hazards, those are various in the time and space scales. The most extensive impact comes from falling of volcanic ash, while the most devastating impacts come from the pyroclastic flow and volcanic tsunami (described late in the eruption at Unzen). Furthermore, it is characteristics of volcanic disasters that these volcanic hazards frequently occur in combination. Volcanic events themselves are less frequent than earthquake events. Sites of volcanic eruption and areas affected by volcanic eruptions are generally limited to on and near the volcanoes.

In this chapter, the present state of volcanism in Japan is shown by comparing with other volcanic countries, suggesting elevated risks from large eruptions in Japan. The difficulty of forecasting volcanic eruptions is shown by exampling recent volcanic eruptions, and the necessity of scientific research especially for less-frequent large eruptions is insisted.

12.2 Characteristics of Volcanism in Japan

12.2.1 Scale of Eruptions

The scale of volcanic eruption is expressed by the Volcanic Explosivity Index (VEI), proportional to the volume of tephra (volcanic ashes including lapilli and blocks) from one explosion (Newhall and Self 1982), ranging from 0 (smallest) to 8 (largest), similar to the seismological magnitude. VEI is also represented by the height of the eruption column of volcanic ash standing on the volcano during an eruption, so that the area covered by volcanic ash depends on the severity of eruption (the taller the column, the more extensive the ash covered area).

Non-explosive eruption	VEI 0	$<10^4$ m^3 of tephra
Small eruption	VEI 1	10^4–10^6
Moderate eruption	VEI 2	10^6–10^7
Moderate Large eruption	VEI 3	10^7–10^8
Large eruption	VEI 4	10^8–10^9
Very large eruption	VEI 5 and >5	$>10^9$

The scale of eruption is not proportional to the size of damages from volcanic disasters (Auker et al. 2013). According to their analysis, the fatal volcanic events are much frequent in the eruptions of VEI 2–5 rather than VEI > 5. Of course, a larger fatal event happens in a less-frequent very large eruption. The relationship of VEIs

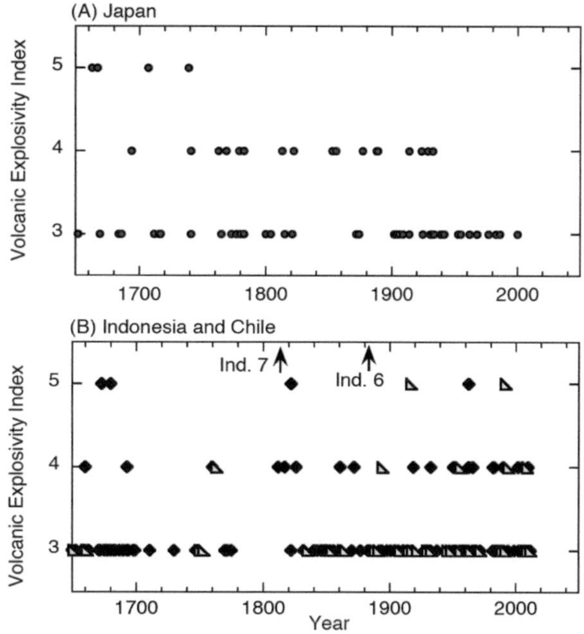

Fig. 12.1 Volcanic Explosivity Index (VEI) of eruptions during A.D. 1650–2010 in Japan **a** compared with those in Indonesia and Chile **b** data from Smithsonian Volcanism Network archives

and years of eruptions in Japan for these three hundred years is shown in Fig. 12.1. The same diagrams are also shown to Indonesia and Chile, both of which possess abundant active volcanoes with the similar geological background to Japan. Evident is that VEI 5 or larger eruptions had never occurred during these three hundred years in Japan. This contrasts to the common happening of the same size eruptions in Indonesia and Chile. One of the two last VEI-5 eruptions is the Hoei eruption at Mount Fuji in AD 1707. Furthermore, VEI-4 eruptions have never occurred during these one hundred years in Japan, being again large contrasts to Indonesia and Chile. As the recent geological background is not different among Japan and these two countries, the recent eruption quiescence in Japan may be accidental. Since it is considered that the magma production rate does not change during these hundred years, such the quiescence may imply progress of steady magma accumulation under the Japanese volcanoes.

12.2.2 Hazards from Very Large Eruptions

The relationship of the magnitude and frequency of volcanic eruptions is very similar to that of earthquakes (Fig. 12.2), famous as the Gutenberg and Richter's (1944) law. Figure 12.2 shows that the probability of VEI 7 and larger eruptions is ~0.01 for 1000 years in Japan. The last VEI 7-eruption occurred at Kikai, south of Kagoshima at 7300 years ago. Therefore, we may be present within the period to consider the

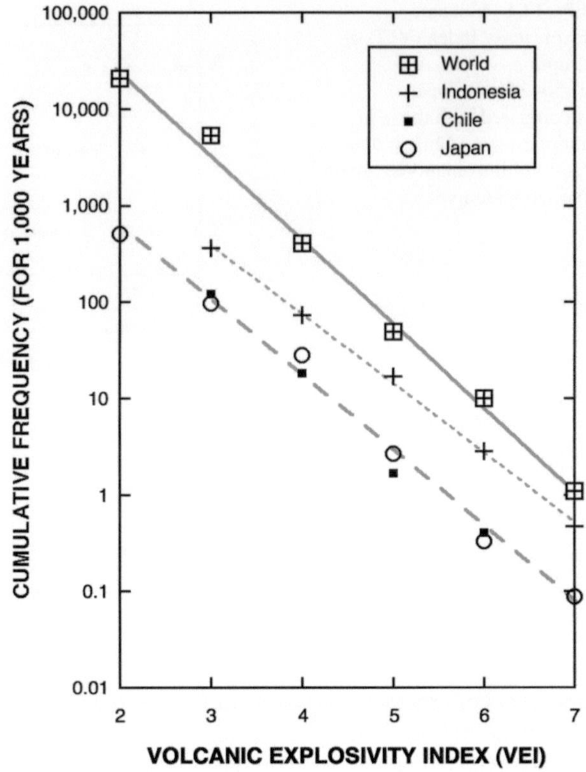

Fig. 12.2 Relationship of the Volcanic Explosivity Index and the frequency in the world, Japan, Indonesia, and Chile. The frequency was given as the values normalized for one thousand years. After Nakada (2015). Indonesia is more active than Japan and Chile. This kind of relationship is also found in the volcanic zones and individual volcanoes

countermeasures against risks from a very large eruption. The danger of a caldera-forming eruption in Japan was already examined by Tatsumi and Suzuki-Kamata (2014) that predicted too serious situation throughout Japan if it occurs in Kyushu. The expected impact from such the eruption is tephra deposits which cover the huge area of the Japan Islands. However, tephra sedimentation itself is not critical for human lives as far as shelters with meals and water are kept. Large-scale pyroclastic flows which are always associated with very large eruptions are more destructive. Their moving distance reached ~150 km from the volcanic centre in the case of eruption at Mount Aso at about 90 thousand years ago (Fig. 12.3). Lives and houses attached by pyroclastic flows are completely burnt, destroyed, and covered with a thick deposit. Ancient culture populated in the southern Kyushu disappeared by the VEI-7 eruption at Kikai at 7 thousand years ago (Machida and Arai 2003).

A serious impact may happen on permanent infrastructures such as nuclear power plants (Fig. 12.3). This is because they cannot take quick actions when they notice the crisis of a very large eruption. That is, fast cooling of the reactor is impossible after its shut down, and the relocation of all used fuels takes more than a few years. Considering the difficulty of forecasting very large eruptions, the time between the recognition of such the event and the happening (lead time) may be insufficient to

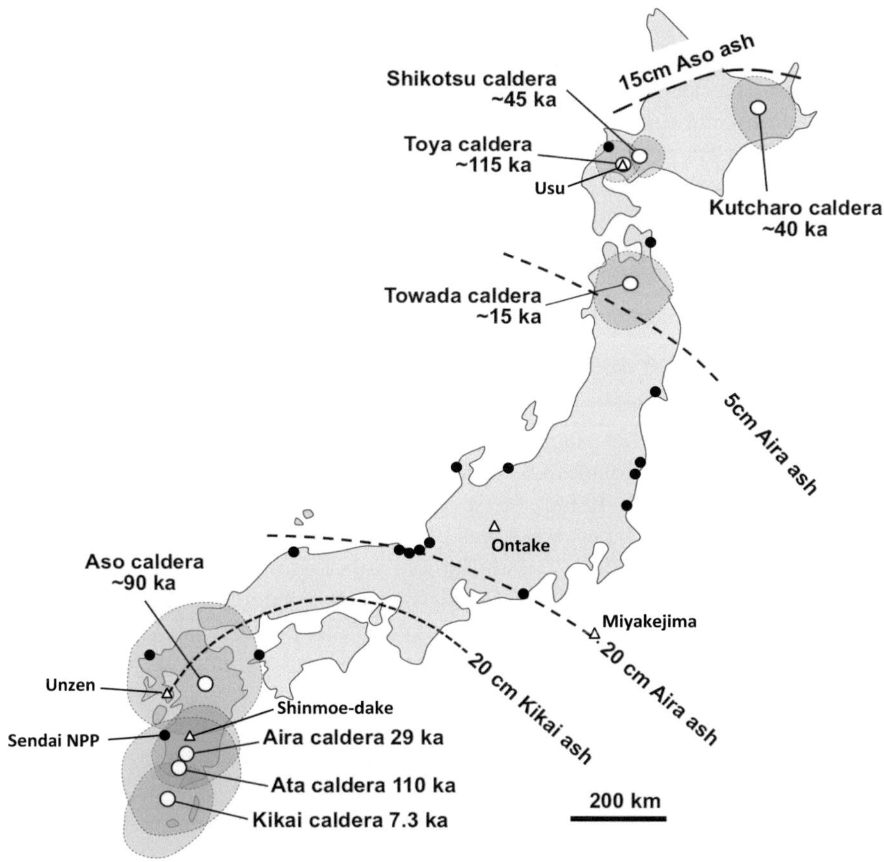

Fig. 12.3 Distribution of pyroclastic-flow deposits and tephra of caldera-forming (VEI 6–8) eruptions with the sites of nuclear power plants (NPP's) in Japan. The area of pyroclastic-flow deposits is shown in gray areas surrounding the caldera centres (open circles). The partial contour lines of tephra are shown for the Aso-4 (~90 thousand years ago), Aira (AT ash, ~30 thousand years ago), and Kikai (Akahoya, ~7300 years ago) caldera-forming eruptions. Solid circles are the NPP sites

avoid the serious accident on nuclear power plants. As the introduction of nuclear power plants was carried out in the period when no large eruptions occurred in Japan as seen in Fig. 12.1, some nuclear power plants seem to have been located within the areas possibly affected by very large volcanic eruptions. The safety standard issue for nuclear power plants had been incomplete for volcanic disasters before 2012 when the IAEA Safety Standards for volcanic hazard was published.

Geologically the precursory phenomena are mentioned for several caldera-forming eruptions in Indonesia and Japan (Takada 2010; Kobayashi et al. 2010). They include temporal change in magma chemistry, an eruption of high-temperature lava flows, smaller but large eruption, multiple magma intrusions under the ground, multiple faulting and landslides in the targeted area, and so on. However, we have no experience to have observed very large eruptions such as the caldera-forming eruption. It is not known whether the time scales of the above geological precursory phenomena can be detectable by a monitoring system with modern technology. Recent petrological studies are focused on estimating the detail time history of crystals involved in the products of very large eruptions, based on theoretical analysis with new analytical technologies (e.g., Druitt et al. 2012). The results may suggest the residence time of magmas of caldera-forming eruptions as short as several hundred to thousand years, which might be detectable in the modern monitoring system.

12.3 Earthquake and Volcanic Eruption

Some scientists believe that a large earthquake may trigger an eruption, and they were worried about what the Tohoku-Oki Earthquake of M9.0 on March 11 2011, would trigger eruptions in Japan. In fact, the seismic activity on about 20 active volcanoes in Tohoku and Central Japan areas was elevated (Fujita et al. 2015). Takada and Fukushima (2011) showed the possibility that subsidence at 5 active volcanoes in Tohoku was triggered by the Tohoku-Oki Earthquake. Tectonic changes derived from a large earthquake might trigger the eruptions of volcanoes. However, from Fig. 12.4, the number of volcanoes erupted did not change before and after the Tohoku-Oki Earthquake in the time window of, at least, 5 years.

There is no evidence clearly showing earthquakes' triggering of volcanic eruptions. The 1707 eruption at Mount Fuji (VEI 5) was considered as one of the examples because the eruption occurred 49 days after the Hoei Earthquake of the M8-class that occurred along the Nankai Trough. Another example is an eruption of Puyehue-

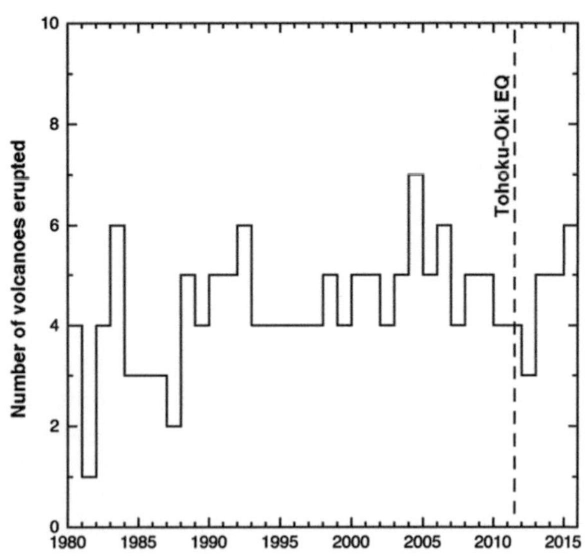

Fig. 12.4 Change in the yearly number of volcanoes erupted

Cordon Caulle volcano (VEI 3) in Central Chile, which occurred 2 days after the largest earthquake (M9.5), offshore of Chile in 1960. There may be an accidental coincidence of timings of two phenomena. On the other hand, the Jogan eruption at Mount Fuji in 864, which is larger in the magma volume than the 1707 eruption, occurred 4 years before the Jogan Earthquake ($M > 8.6$, Namegaya and Satake 2014) in 869 at Tohoku, which triggered a tsunami similarly to that in 2011. About 50 years after the Jogan earthquake, a VEI-5 eruption occurred at Towada Volcano in 915. Earthquakes of M8 and larger in Japan occur ~0.2 times a year, according to the JMA report for 2001–2010. Figure 12.2 shows the frequency of VEI-3 eruption of ~0.1 annually. Therefore, if we accept the time window as long as a few tens of years, it seems difficult to scientifically prove the linkage between the two phenomena. However, the periods with high frequency in both large earthquake and volcanic eruptions probably existed, due to the background where large earthquakes and volcanic eruptions tectonically linked to each other, as summarized by Hotate (2012), based on the study of the historical documents in Japan.

12.4 The Present Situation of Forecasting Volcanic Eruption

12.4.1 Eruptions at Unzen and Miyakejima

The experiences that we failed forecasting recent eruptions are shown below.

12.4.1.1 Unzen

Unzen volcano, located in the Shimabara Peninsula, Kyushu, west Japan, erupted in 1990 and the eruption continued till early 1995 (Nakada et al. 1999). In this volcano, the previous eruptions in 1663 and 1792 were small in scale and lava flows were emitted near the summit of Mount Unzen. Soon after the 1792 eruption, a small old volcano, located behind the populated ancient city of Shimabara, collapsed by a large earthquake sourced just beneath the city. Avalanche of collapsed materials (debris) rushed into the inland sea that the city faced, resulting in the generation of a volcanic tsunami, which killed about 15,000 people in the opposite shore of the inland sea, Kumamoto, and in the shores of the Shimabara Peninsula side. The number of casualties of this event is largest among the volcanic disasters recorded in Japan.

In the last eruption which started in November 1990, a new lava dome was formed on the top of Mount Unzen. Even after the lava appearance, in the beginning, no scientist could forecast the subsequent phenomenon. Partial collapses of the growing lava dome generated pyroclastic flows successively (Fig. 12.5). Forty-three people, including the press, local regidents, and scientists, were killed by pyroclastic flows on

Fig. 12.5 Pyroclastic flows moving in the northwestern part of Shimabara City, taken on June 24, 1993, from a GSDF helicopter. Left behind is the upperslope of Unzen Volcano

June 3, 1991. Although geologists had noticed pyroclastic-flow deposit on the slope of this mountain, they could not connect on-going eruption with their experience; they could not imagine the formation mechanism of pyroclastic flow deposits they observed. Therefore, timely forecasting of pyroclastic flows was impossible when new lava appeared on the mountain tip. The eruption with successive generation of pyroclastic flows continued till early 1995.

12.4.1.2 Miyakejima

The second example is an eruption in 2000 at a volcanic island, Miyakejima, located about 180 km south of Tokyo (Nakada et al. 2005). In this eruption, forecasting by scientists was impossible when they faced new phenomena. However, this eruption was a reproduction of the old eruption about 2500 years ago in this volcano, which issued about 130 million m^3 of tephra and was associated with the formation of a depression at the summit (Tsukui et al. 2005). During these 300 years, however, eruptions repeated every about 20 years with a simple emission of volcanic ash and lava in smaller magnitudes. In addition, the duration of these eruptions was as short as a few days.

The submarine eruption first in the 2000 eruption occurred offshore on the late June, soon after the seismic swarm within the island. Scientists considered that this time eruption was over as they thought the scenario same as previous eruptions. However, another eruption started 10 days after the first eruption, being accompanied by partial subsidence of the summit area (Fig. 12.6). The size of the subsidence area increased with time up to about 1.8 km across, accompanying occasional small explosions until the middle of August. The eruption column of volcanic ash rose about 16 km above the summit in the middle of August. Furthermore, 10 days after this event, a mass of volcanic dust rising within the summit crater overflew the rims, hanging over the residential area in the northern coast of the island. Fortunately, nobody was injured nor killed by this volcanic dust flow. Immediately after this event, a huge amount of poisonous volcanic gas (sulfur dioxides) had been issued from the summit crater continuously, which forced all islanders to evacuate outside the island for more than four years.

12.4.2 Issues of Forecasting Eruptions

Like the above, forecasting the volcanic eruptions is very difficult when we are based on our limited experiences, especially those of the last eruptions. The Miyakejima case was the example in which it had frequently erupted during these hundred years

Fig. 12.6 Photograph showing the subsidence of the summit area at the Miyakejima island, taken on the early morning of July 9, 2000, from a jet plane of the Japan Coast Guard

and the recent events had been monitored with the geophysical method. Even in such the volcano, the phenomenological development could not be forecasted. An example that forecasting the onset of eruption was impossible is the eruption at Shinmoe-dake in 2011. Although some phenomena suggesting magma's accumulation under the ground in geodetically and seismologically had been observed before about one year, the timing and type of eruption could not be forecasted; as a result, the volcanic alert was issued after the onset of a climactic eruption (Kato and Yamasato 2013). The worse example was the eruption at Ontake volcano on September 27, 2014. The eruption itself was moderate in scale, but introduced the worst result under an unfortunate situation; this becomes the largest volcanic disaster in Japan after the World War II (Chap. 13). Soon after this event, the volcanic monitoring system was reexamined and new instruments were installed near the craters of eruption-credible volcanoes including Ontake volcano. Even if dense monitoring near the crater is carried out, however, this does not mean that difficulty of forecasting was solved. The eruption of Usu volcano in 2000 was occasionally introduced as an example of successful forecasting. Precursory seismicity of this eruption occurred similarly to a previous eruption in 1910 in this volcano, and the onset of eruption was after the peak of seismicity passed. The timing of eruption onset was announced luckily. However, the magnitude and type of eruption were different from ones expected by scientists and the site of eruption was an unexpected new place.

12.4.3 Eruption Scenarios

In the "Observation Research Project for Forecasting Earthquake and Volcanic Eruption," which was carried out as a national project during 2009–2013, we began to prepare the eruption scenarios. In order to seek more advanced technology for forecasting volcanic eruptions, it is strongly requested to prepare event-trees consisting of possible eruption phenomena, the precursory and eruption sequences. This was a lesson learnt from many failures in forecasting eruptions. The eruption scenarios can be produced by utilizing the database of eruption phenomena, eruption records studied, and old documents. Re-examining research results and the additional study for the past eruptions were conducted to prepare the eruption scenarios. Thus, we can evaluate eruption potentials for active volcanoes by showing possible phenomena with probability, considering the present observation data on the eruption scenarios.

The probabilistic concept for volcanic eruptions has been already discussed since Newhall and Hobbit (2002). Event trees had been shown for several active volcanoes in the world (e.g. Newhall and Pallister 2015). The method to calculate the probabilities of branches in event-trees are different from researchers to researchers. In Fig. 12.7, the eruption event-tree with the probability was shown for Usu Volcano, which was preliminarily prepared in the above project (Nakada 2016).

12.5 Scientific Research and Increasing the Resilience of Residents

The Sendai Framework for Disaster Risk Reduction (2015–2030) (website) adopted on March 18, 2015, calls for focused action within and across sectors by States at local/national/regional/global levels in the following 4 priority areas. That is, (1) understanding disaster risk, (2) strengthening disaster risk governance to manage disaster risk, (3) investing in disaster risk reduction for resilience, and (4) enhancing disaster preparedness for effective response and to "building back better" in recovery, rehabilitation and reconstruction. Although one of seven global targets in this framework is "substantially reduce the number of affected people globally by 2030," volcanic disaster should be considered in longer time scale. Large eruptions may not occur within the following 15 years. In addition, installing the early warning system on volcanic eruption written in this framework may be impossible by 2030. Physically forecasting volcanic eruption especially for both small and very large eruptions may be different from the concept of this framework.

Forecasting and monitoring of large volcanic eruptions became the important social issue. All Japanese nuclear power plants were stopped soon after the disaster from the Tohoku-Oki earthquake and the associated tsunami in March 2011. Toward resuming the operation of the Sendai NPP in southern Kyushu in 2015, the safety against the volcanic hazards was argued during 2013–2015. Effects from the future very large eruptions became the theme of the argument because the Sendai NPP is located within the area which had possibly devastated by the VEI-8 eruption of 30 thousand years ago (Fig. 12.3). Many of volcanologists were involved in official or unofficial arguments on the safety issue from very large eruptions during 2013–2014; that is, related to the recommendation by the Japanese Cabinet Office on its measures

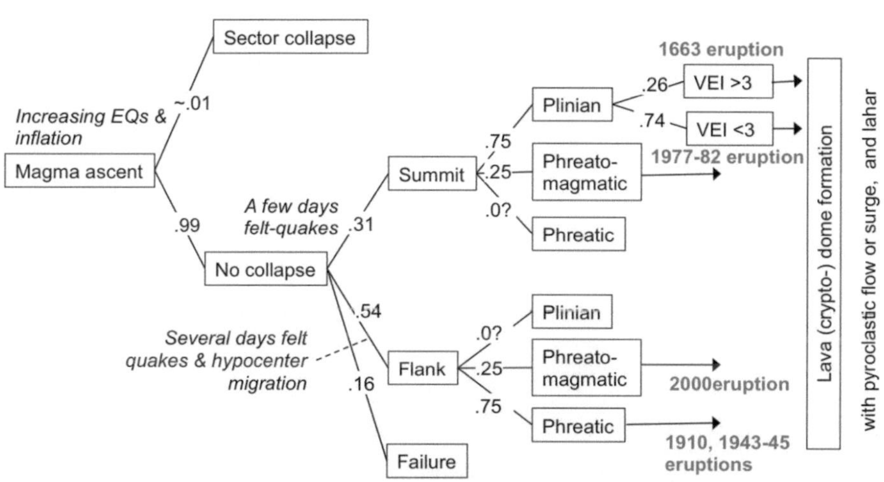

Fig. 12.7 Eruption scenario (event tree) for Usu Volcano

against the large-scale eruption (May 2013), and the guideline for the nuclear power plants to evaluate volcanic effect (June 2013).

The committee related to nuclear power problems of the Volcanological Society of Japan issued the comments on forecasting and monitoring of large eruptions in November 2014, as follows;

1. The subject of how forecasting and monitoring of large eruption (VEI-6 and more) is important to be treated by the Japan Volcanological Society, although the countermeasures against them should be done by the nation. The consultation on this issue by related ministries and agencies should be made. The investigation result should be utilized for raising the safety measures of the nuclear power plants.
2. Research and investigation for forecasting large eruptions should be promoted in the viewpoints of both basic research and application. The results should be utilized to evaluation standards of the eruption alerts with increasing its accuracy.
3. The monitoring system with the eruption alerts should be examined, maintained, and developed systematically by the nation, considering several problems observed in recent eruptions and future development in volcanological research and observation investigation, including view points of technology and human resources. Understanding the possibility of forecasting eruptions with its limit and ambiguity is indispensable to work the eruption alert effectively. Regulations and standards such as the guidelines for evaluating the eruption impacts should be examined carefully by reflecting the above characteristics.

We should challenge interdisciplinary forecasting of less-frequent large volcanic hazards, and understanding their generation mechanisms is in the first priority. A volcanic eruption of Mount Fuji that has rested for more than 300 years and caldera-forming eruptions are good examples. Our scientific understanding of those volcanic eruptions and observation-monitoring with new technology should be linked to the preparedness of stakeholders for those hazards, and to increasing the resilience of inhabitants. People who are not living in the prone areas of the volcanic disasters are also involved because they may have a high chance to encounter volcanic hazards during traveling and may get jobs related to those areas in future.

References

Auker MR, Sparks RSJ, Siebert L, Crosweller HS, Ewert J (2013) A statistical analysis of the global historical volcanic fatalities record. J Appl Volcanol 2. http://www.appliedvolc.com/content/2/1/2 (confirmed on August 31, 2016)

Druitt TH, Costa F, Deloule E, Dungan M, Scaillet B (2012) Decadal to monthly timescales of magma transfer and reservoir growth at a caldera volcano. Nature 482:77–82. https://doi.org/10.1038/nature10706

Fujita E, Kozono T, Ueda H, Kohno Y, Yoshioka S, Toda N, Kikuchi A, Ida Y (2013) Stress field change around the Mount Fuji volcano magma system caused by the Tohoku megathrust earthquake, Japan. Bull Volcanol 75:679. https://doi.org/10.1007/s00445-012-0679-9

Gutenberg B, Richter CF (1944) Frequency of earthquakes in California. Bull Seis Soc Am 34:185–188
Hotate M (2012) Earth disturbance in the history–earthquake and the emperor of Nara and Heian, vol 1381. Iwanami Shinsho, 241 p. (in Japanese)
Kato K, Yamasato H (2013) The 2011 eruptive activity of Shinmoedake volcano, Kirishimayama, Kyushu Japan-Overview of activity and volcanic alert level of the Japan Meteorological Agency. Earth Planets Space 65:489–504
Kobayashi T, Okuno M, Nagaoka S, Miyabuchi I, Iguchi M, Miki D (2010) Precursory events of large-scale caldera-forming eruption—examples at Kikai and Aira calderas. Ann Disas Prev Res Inst Kyoto Univ 53B:269–275. (in Japanese)
Machida H, Arai F (2003) Atlas of Tephra in and around Japan (Revised edn). University Tokyo Press, Tokyo, 336p. (in Japanese)
Nakada S (2015) Regularity of volcanic eruptions in terms of Volcanic Explosivity Index (VEI). Bull Volcanol Soc Jpn 60:143–150. (in Japanese)
Nakada S (2016) Eruption scenarios and probability forecasting. Bull Volcanol Soc Jpn 61:199–209 (in Japanese)
Nakada S, Nagai M, Kaneko T, Nozawa A, Suzuki-Kamata K (2005) Chronology and products of the 2000 eruption of Miyakejima Volcano. Jpn Bull Volcanol 67(3):205–218
Nakada S, Shimizu H, Ohta K (1999) Overview of 1990–1995 eruptions at Unzen Volcano. J Volcanol Geothermal Res 89:1–22
Namegaya Y, Satake K (2014) Reexamination of the A.D. 869 Jogan earthquake size from tsunami deposit distribution, simulated flow depth, and velocity. Geophys Res Lett 41:2297–2303
Newhall C, Hoblitt R (2002) Constructing event trees for volcanic crises. Bull Volcanol 64:3–20
Newhall CG, Pallister JS (2015) Using multiple data sets to populate probabilistic volcanic event trees. In: Papele P (ed) Volcanic hazards, risks, and disasters. Elsevier, pp 203–232. https://doi.org/10.1016/b978-0-12-396453-3.00008-3
Newhall CG, Self S (1982) The volcanic explosivity index (VEI): an estimate of explosive magnitude for historical eruptions. J Geophys Res 87:1231–1238
Takada A (2010) Caldera-forming eruptions and characteristics of caldera volcanoes in the Sunda Arc, Indonesia. J Geol Soc Jpn 116:473–483. (in Japanese)
Takada Y, Fukushima Y (2013) Volcanic subsidence triggered by the 2011 Tohoku earthquake in Japan. Nature Geosci 6:637–641
Tatsumi Y, Suzuki-Kamata K (2014) Cause and risk of catastrophic eruptions in the Japanese Archipelago. Proc Jpn Acad Ser B 90:347–352
Tsukui M, Kawanabe Y, Niihori K (2005) Geological map of Miyake jima Volcano, 1:25,000. Volcano Geological Map Series 12, AIST Geological Survey of Japan

Chapter 13
The 2014 Eruption of Mt. Ontake Volcano and Its Disaster

Koshun Yamaoka

Abstract The eruption of Mt. Ontake, central Japan, in 2014 caused a serious casualty in spite of its eruption strength. Although the main reason for the casualty arose from the time the eruption occurred, holiday noon with good weather condition in beautiful autumn season, we could have had some chances to prevent the tragedy. Appropriate information dissemination, improvement of public literacy of volcanic activity as well as better understanding of hydrothermal volcanic eruptions will reduce the possible tragedy in future.

Keywords Phreatic eruption · Hydrothermal system · Volcano information · Volcano alert level

13.1 Introduction

Mt. Ontake volcano erupted at 11:52 on September 27, 2014 claiming the lives of more than 58. The good weather condition in the midday of a weekend gathered more than 300 hikers in the summit area at the time of the eruption, contributed to the large death toll. The eruption started with ejection of large amount of volcanic ash from newly opened craters. The ash with volcanic gas turned into a pyroclastic density current (PDC) spreading rapidly from the craters. The hikers were covered with the dense PDC being threatened by falling rocks in the darkness. The volcanic alert level of Ontake volcano at the time of the eruption was 1, which is the lowest grade, but was raised to 3 after the eruption started.

Mt. Ontake is an active volcano with the altitude of 3067 m, which is the second tallest among the volcanoes in Japan. The volcano is also popular as one of the Japan's 100 branded mountains ("hyakumeizan" in Japanese). As an isolated mountain locating in the central part of Japan in good accessibility together with a boom in

K. Yamaoka (✉)
Graduate School of Environmental Studies, Nagoya University, Furo-cho, Chikusa-ku, Nagoya 464-8601, Japan
e-mail: kyamaoka@seis.nagoya-u.ac.jp

Y. Himiyama et al. (eds.), *Human Geoscience*, Advances in Geological Science,
https://doi.org/10.1007/978-981-32-9224-6_13

mountain climbing prevailing all the generations, more and more hikers climbed it in recent years. It used to be a mountain of religion and people climbed it for religious purpose only in summer season. It is only in recently that many hikers climb it in September.

The history of the volcanic activity is divided into two period (Takeuchi et al. 1998). The older activity, which is called as old Ontake, ranges from 750,000 to 420,000 with main activity of andesitic eruptions. After 300,000-year dormancy, Mt. Ontake resumed its volcanic activity, which is called as new Ontake, about 100,000 years ago within the range of the old Ontake volcanic body. The new Ontake volcano has grown by repeating magmatic activity. Phreatic eruptions with every several hundred years were also recorded as ash layers in volcanic products. The last magmatic eruption is just 6000 years ago (Oikawa et al. 2015), and Ontake volcano has been classified as an active volcano to be monitored.

The first eruption of Mt. Ontake in written history occurred in 1979. The eruption was a phreatic one with new craters opened in the southern slope near the summit. Volcanic ash of the order of 200,000 tons were ejected in the eruption, but neither injuries nor deaths were reported. After this eruption, two small-scale eruptions were identified in 1991 and 2007 before the 2014 eruption. In both eruptions, very small amount of volcanic ash was ejected from a vent that was created in 1979 eruption. The 2007 eruption was preceded by long-period earthquake and observable crustal deformation, which can be interpreted as an intrusion of magma in the deep part of the vent (Nakamichi et al. 2009).

The Eruption in 2014

The 2014 eruption was preceded by earthquake swarm that started about one month before (Kato et al. 2015). Peak seismic activity was observed on September 10–11, when earthquakes counts more than 50 per day. The seismic activity soon decayed and kept low until the eruption. Low-frequency earthquakes, which can be interpreted as an activation of hydrothermal activity, were observed but their activity was also kept low until the eruption. No crustal deformation was observed in contrast to the small-scale eruption in 2007. This is the main reason why JMA did not raise the volcanic alert level.

Unusual activity that required an emergency action started about 11:45 of September 27, which is only 7 min before the eruption. Volcanic tremor as well as rapid crustal deformation, which clearly indicated an impending eruption, was detected. JMA noticed the activity and promptly began preparation for issuing a warning, but the eruption started before the warning was issued. The hikers at the summit encountered the eruption unexpectedly.

The eruption started with ejection of steam and ash from newly opened vents. The outpouring ash-and-gas mixture with high density spread along the topographic surface of the mountain as a pyroclastic density current (PDC), partly flowed down the slope of a valley to the southeast (Fig. 13.1). The temperature of the PDC was at most slightly higher than boiling temperature as evidenced by damages to the cloths of the hikers and was not fatal to the hikers in contrast to the PDC in magmatic eruption. Instead, fragments of rocks driven by the gas emission from the vents were

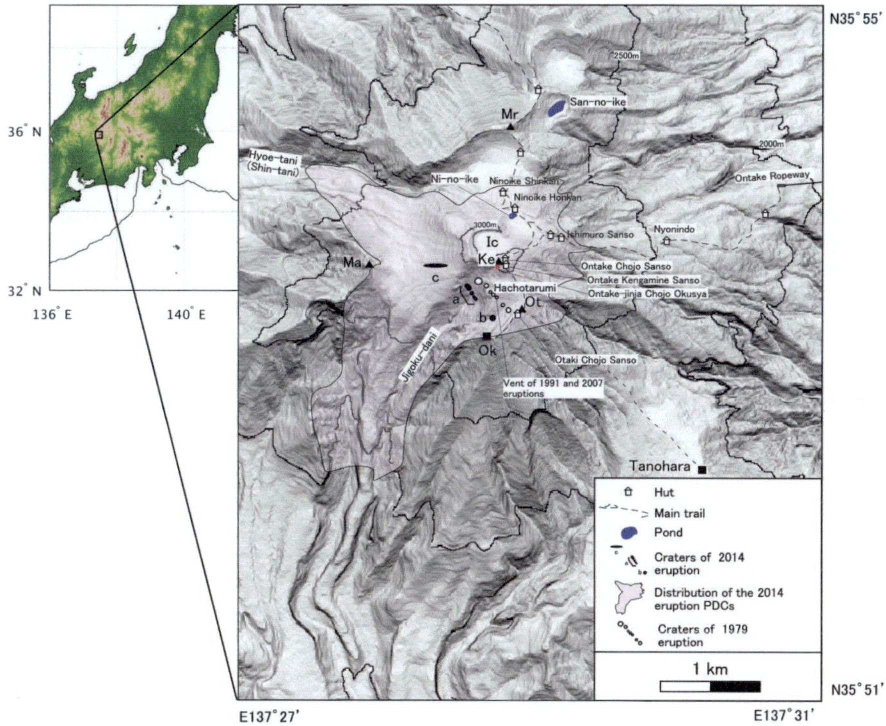

Fig. 13.1 Location map of Mt. Ontake, from Fig. 1 of Yamaoka et al. (2016) with original figure of Mt. Ontake from Oikawa et al. (2016)

fatal to the hikers. Ash was raised from the vents as an eruption column after PDC and fallen as wet ash, which is a characteristic nature of phreatic eruptions. These eruption sequence is described in some papers based on interviews to eyewitnesses and field observation on deposit of volcanic material (Maeno et al. 2016; Oikawa et al. 2016). The total volume of fallout deposit is estimated as about 100,000 tons (Takarada et al. 2016), which is about half of that of the 1979 eruption.

13.2 Short Precursor of Phreatic Eruption

The volcanic tremor and the crustal deformation immediately before the eruption indicate that water vapor ascended from the hydrothermal reservoir by opening cracks in the country rock. The uprising water vapor fractured and opened the country rock to create the path from the reservoir to the surface. This resulted in the new craters aligned in N–S direction which is perpendicular to the maximum slope direction near the summit. The direction is caused by gravity driven stress to generate maximum extension in the direction of the slope. This is similar to magmatic eruption in the way that magma breaks into country rock in many volcanoes. In contrast to the eruption of Ontake volcano in 2014, it takes hours to days for magmatic eruption from the beginning of tremor and crustal deformation to the eruption. The difference can be interpreted as the depth of reservoirs from which fluid is supplied.

The hydrothermal reservoir beneath the crater of Mt. Ontake is estimated to be located at a depth of 500–1000 m based on resistivity structure of Mt. Ontake (Sabry and Mogi 2016), chemical composition of volcanic ash (Minami et al. 2016) and analysis of volcanic gas (Mori et al. 2016). The depth is extremely shallower than that of magma reservoirs in many volcanoes, which is generally 5–10 km. This means that the time length for the fluid to reach the surface in opening cracks is much shorter for the phreatic eruption than usual magmatic eruption. In fact, the tremor and crustal deformation started two hours before the fissure eruption of Izu-Oshima volcano in 1986 (Yamaoka et al. 1988; Shimada et al. 1988). Earthquake swarm started at least five hours before the onset of the fissure eruption of Miyakejima volcano in 1983 (Aramaki et al. 1986). The eruption of Usu volcano in Hokkaido, Japan, in 2000 was also preceded by several day's activity of earthquake swarm and crustal deformation (Jousset et al. 2003). In this eruption, dacitic magma rose to the surface and cause phreatomagmatic eruptions.

13.3 Volcano Alert Level

The main issue that was discussed after the tragedy is the reason why the volcanic alert level was kept 1 until the eruption. Noticeable earthquake swarm beneath the crater occurred about two weeks before the eruption, which was regarded as a sign of eruption. Actually, JMA announced the possibility of small eruption with minor ash ejection from an existing vent in respond to the occurrence of the earthquake swarm. This information, however, was not paid attention to by local municipal office as well as hikers. They only paid attention to the volcanic alert level.

The volcanic alert level was deployed to the Japanese volcanos in 2007. The level is classified into five, and each level is connected to the disaster mitigation action to be taken in a target area. The level 4 and 5 are issued when volcanic activity can affect the residential area around the volcano. The level 2 and 3 are issued when volcanic activity raises but has little possibility to affect to the residential area. The level 1 means that the volcano is in normal activity affecting only inside the crater. The advantage of the volcanic alert level is the message clearness in the disaster mitigation measures. The alert level works as a trigger for disaster mitigation action by the local residents as well as local municipal office. This advantageous nature of the volcanic alert level is simultaneously a disadvantage of the level. For example, the target area of the level 3 is bounded by the minimum distance to the residential area from the crater, although the distance differs from volcanoes to volcanoes. For example, no residential area exists within 4 km distance from the craters at Mt. Ontake, but many people live just beyond 1 km distance from craters at Hakone volcano, Japan, at which a phreatic eruption occurred in 2015. This means that volcanic activity should be classified to fit the human activity at each volcano. The variability of eruption often bother JMA whether the level should be raised or not. The forerunning activity before the 2014 eruption of Mt. Ontake volcano was lower than that of 2007 eruption, in which only minor ash ejection occurred. They observed seismicity that is comparable to that of 2007 eruption but less crustal movement. This is the main reason why they did not raise the volcanic alert level before the 2014 eruption. Only small fluctuation of decision made large difference on the results in the 2014 eruption of Mt. Ontake volcano. This is inevitable nature of volcanic alert level.

13.4 Emergency Action After the Eruption

Being reported possible casualty to the hikers in the eruption of Mt. Ontake, rescue operation was organized. Operation teams were organized by firefighters, police and Japan Self-Defense Forces (JSDF). Headquarters of disaster mitigation measures was organized in the national government at Kasumigaseki, Tokyo. Local headquarters was also organized at the prefectural office in Nagano city. I was asked to be a member of the headquarters as a volcano specialist. A satellite conference system was deployed in Nagoya University where I worked at, connecting the headquarters in Tokyo and local headquarters in Nagano.

Rescue teams were dispatched in the next day of the eruption though the volcano was still very active. JMA watched the volcanic activity with seismometers and tiltmeters and kept ready to inform the rescue team directly with cell phone in case of sudden change of volcanic activity. My task was to give comments in the headquarters meeting from a scientific point of view on the state of volcanic activity. The task needed to be changed when the volcano activated in the evening of September 29, two days after the eruption, when the amplitude of volcanic tremor suddenly increased. As the volcanic tremor is the vibration by the gas flow at the vent, the sudden change indicates the unstableness of the volcanic activity. After this day, I got tasked to have a meeting on telephone with JMA to evaluate the volcano monitoring data. Following the meeting JMA gave an advice to the rescue team every morning before they climb the mountain. The amplitude of the volcanic tremor stayed high in the morning of September 30, JMA suggested the rescue team not to execute the rescue operation on that day. The 'early-morning meeting' continued until the rescue operation ended on October 16.

13.5 Lessons from the Disaster

There are some lessons from the disaster of Mt. Ontake. Three problems arose from the disaster; insufficient knowledge on the behavior of volcanic hydrothermal activity, lack of communication between science and local governmental administration, and blind belief on the volcanic alert level.

Nature of hydrothermal activity of Mt. Ontake should have been studied more carefully. The long-term activity of Ontake volcano that appeared in observation data had been looked decaying since 1979 eruption until 2014 eruption. For example, the fumarole height was decaying and little fumarole was observed in a few-year period before the 2014 eruption (Fig. 13.2, Yamaoka et al. 2016). In magmatic eruptions, this kind of fumarole decay is interpreted as a decay of volcanic activity because it reflects the temperature decrease in the shallow part of the volcano. Magma drains back or gets frozen after the magmatic eruption ended, resulting in cooling in the shallow part of the volcano including hydrothermal activity. But in the case of phreatic eruptions another interpretation is possible. The decay in fumarole activity can be interpreted as a decay of permeability of layers by which water vapor escapes from the hydrothermal reservoir. Precipitation of mineral on crack surface and hydrothermal-alteration process decrease the permeability. The permeability decay causes the pressure and temperature increase in the hydrothermal reservoir, eventually breached the covering layer to create vapor paths to the surface. Upward migration of hypocenters immediately before the eruption of Mt. Ontake in 2014 suggests that water vapor ascent from the hydrothermal reservoir through the fractures of covering layers (Kato et al. 2015). Migration and spreading of hypocenters preceding the eruption indicate crack propagation from the hydrothermal reservoir in the deeper part to the surface to create efficient paths for water vapor. Analysis of long-period earthquakes indicates the crack opening before the eruption (Maeda et al. 2015). Once paths between the surface and the hydrothermal reservoir are established, the pressure in the hydrothermal reservoir drop sufficiently to accelerate the boiling within it. This may give the reason why the eruptions in 1979 and 2014 overwhelm those in 1991 and 2007. The vapor-transportation efficiency is much better by using open cracks than existing vents. For example, the eruption of Izu-Oshima volcano in 1986 produced comparable lava in the eruptions from an existing vent and newly opened fissure while eruption duration from the vent is ten times longer (e.g. Nagaoka 1988). Considering the little difference in the forerunning observation between 2007 and 2014 eruptions, the size of the eruption is controlled by the accidental rupturing of overlying layers. Small-scale eruption such as 2007 may prevent the final rupture by gradual pressure decrease.

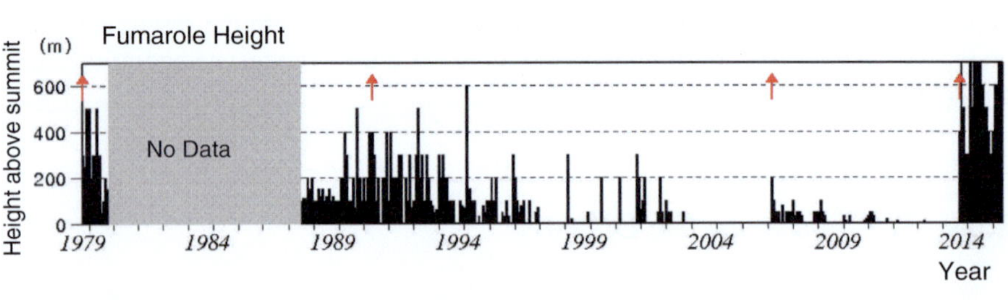

Fig. 13.2 Results of long-term monitoring of fumarole height by the Japan Meteorological Agency (JMA) from Fig. 2 of Yamaoka et al. (2016)

Lack of communication between scientists and administrator of local municipal office reduced chances of information dissemination of volcanic activity to the hikers and local residents. Though a volcano disaster mitigation commission for Mt. Ontake was established recommended by the Central Disaster Mitigation Council of the national government, no volcano specialists participated in the commission. Though data from observation station deployed in Mt. Ontake area were transmitted to research institutions such as Nagoya University as well as JMA and results of volcano activity assessment was informed to the local municipals from JMA, no advice from a scientific point of view was informed to the local municipal office. In response to the earthquake swarm on 10–11 on September 2014, JMA issued volcano information including a possibility of small scale eruption that is similar to that of 2007. The local municipal officials did not understand the meaning of the information, but no communication was made between municipal administrator and scientists. No action was taken to inform the hikers on the status of the volcanic activity to raise their caution. Most of the hikers climbed the mountain without any concern on possible eruption.

Hikers as well as local municipal office did not take any action principally because the volcanic alert level was kept 1 even though a possibility of eruption was announced. The volcanic alert level was deployed in 2007 as an upgrade of volcanic activity level that was previously used. The volcanic activity level was an information only on the state of the volcanic activity issued by JMA. The volcanic alert level was introduced to disseminate simple and clear information on the action to be taken in response to volcanic unrests. The difficulty of issuing volcanic alert level became evident in many of the eruption that occurred after its deployment. In the recent eruption of Aso volcano in Kyushu, Japan in 2016, volcanic alert level was raised from 2 to 3 after an explosive eruption that scattered large ballistic blocks more than 1 km in distance, which is outside of the restricted area of the level 2. It was fortunate that the eruption occurred in midnight and nobody injured in the eruption. In Kuchino-Erabu volcano to the south of Kyushu Island, the volcanic alert level was raised from 1 to 3 after an explosive eruption in 2014. A local elementary school planned an excursion to the summit of the volcano, but did not carry out it because of bad weather condition. If it had been clear, children would have encountered the eruption near the crater. In Kirishima volcano in Kyushu island, the volcanic alert level was raised from 2 to 3 after a sub-Plinian eruption in 2011. Nobody injured because the crater area was restricted under the volcanic alert level of 2 that continued from a small eruption in the previous year. In the eruption of Ontake volcano in 2014, the volcanic alert level was raised from 1 to 3 after the eruption. This means that the large casualty is not unique to the eruption of Ontake volcano but could have happened in other volcanoes. This also shows that the volcanic alert level is still far from the perfect measures to prevent volcanic disasters. Visitor's concern about volcanic activity is the ultimate measure to protect themselves from volcanic disaster. It is necessary to provide volcano information both in crisis and usual time to help the public have intuition on volcanic activity. Raising volcano literacy in the public as well as the improvement volcano monitoring is a necessary measure to reduce the casualty in volcanic eruptions.

References

Aramaki S, Hayakawa Y, Fujii T, Nakamura K, Fukuoka T (1986) The October 1983 eruption of Miyakejima volcano. J Volcanol Geotherm Res 29:203–229

Jousset P, Mori H, Okada H (2003) Elastic models for the magma intrusion associated with the 2000 eruption of Usu volcano, Hokkaido, Japan. J Volcanol Geotherm Res 125:81–106

Kato A, Terakawa T, Yamanaka Y, Maeda Y, Horikawa S, Matsuhiro K, Okuda T (2015) Preparatory and precursory processes leading up to the 2014 phreatic eruption of Mount Ontake, Japan. Earth Planets Space 67:111. https://doi.org/10.1186/s40623-015-0288-x

Maeda Y, Kato A, Terakawa T, Yamanaka Y, Horikawa S, Matsuhiro K, Okuda T (2015) Source mechanism of a VLP event immediately before the 2014 eruption of Mt. Ontake, Japan. Earth Planets Space 67:187. https://doi.org/10.1186/s40623-015-0358-0

Maeno F, Nakada S, Oikawa T, Yoshimoto M, Komori J, Ishizuka Y, Takeshita Y, Shimato T, Kaneko T, Nagai M (2016) Reconstruction of a phreatic eruption on 27 September 2014 at Ontake volcano, central Japan, based on proximal pyroclastic density current and fallout deposits. Earth Planets Space 68:82. https://doi.org/10.1186/s40623-016-0449-6

Minami Y, Imura T, Hayashi S, Ohba T (2016) Mineralogical study on volcanic ash of the eruption on September 27, 2014 at Ontake volcano, central Japan: correlation with porphyry copper systems. Earth Planets Space 68:67. https://doi.org/10.1186/s40623-016-0440-2

Mori T, Hashimoto T, Terada A, Yoshimoto M, Kazahaya R, Shinohara H, Tanaka R (2016) Volcanic plume measurements using a UAV for the 2014 Mt. Ontake eruption. Earth Planets Space 68:49. https://doi.org/10.1186/s40623-016-0418-0

Nakamichi H, Kumagai H, Nakano M, Okubo M, Kimata F, Ito Y, Obara K (2009) Source mechanism of a very-long-period event at Mt Ontake, central Japan: response of a hydrothermal system to magma intrusion beneath the summit. J Volcanol Geotherm Res 187:167–177

Nagaoka M (1988) Geomorphological characteristics of volcanic products of the 1986 eruption of Izu-Oshima volcano and its volume measured by photogrammetry. Bull Volcanol Soc Jpn 33:S7–S15

Oikawa T, Suzuki Y, Chiba T, Kishimoto H, Okuno M, Ishizuka O (2015) Holocene eruption history of the Ontake volcano, central Japan. Abstracts, Fall Meeting of the Volcanological Society of Japan, 102. (in Japanese)

Oikawa T, Yoshimoto M, Nakada S, Maeno F, Komori J, Shimano T, Takeshita Y, Ishizuka Y, Ishimine Y (2016) Reconstruction of the 2014 eruption sequence of Ontake volcano from recorded images and interviews. Earth Planets Space 68:79. https://doi.org/10.1186/s40623-016-0458-5

Sabry AA, Mogi T (2016) Three-dimensional resistivity modeling of GREATEM survey data from Ontake volcano, northwest Japan. Earth Planets Space 68:76. https://doi.org/10.1186/s40623-016-0433-z

Shimada S, Watanabe H, Fukui K, Fukuyama E (1988) Crustal tilt observation associated with the 1986 eruption of Izu-Oshima volcano. Bull Volcanol Soc Jpn 33:S161–S169

Takarada A, Oikawa T, Furukawa R, Hoshizumi H, Itoh J, Geshi N, Miyagi I (2016) Estimation of total discharged mass from the phreatic eruption of Ontake volcano, central Japan, on September 27, 2014. Earth Planets Space 68:138. https://doi.org/10.1186/s40623-016-0511-4

Takeuchi M, Nakano S, Harayama S, Ootuka T (1998) Geology of the Kiso-Fukushima district. In: Geological Survey of Japan, 94p. (in Japanese)

Yamaoka K, Geshi N, Hashimoto T, Ingebritsen SE, Oikawa T (2016) Special issue "The phreatic eruption of Mt. Ontake volcano in 2014". Earth Planets Space 68:175. https://doi.org/10.1186/s40623-016-0548-4

Yamaoka K, Watanabe H, Sakashita S (1988) Seismicity during the 1986 eruption of Izu-Oshima volcano. Bull Volcanol Soc Jpn 33:S91–S101

Chapter 14
Mud Volcanoes and Human Geoscience

Kazuhiro Tanaka, Keiji Asano, Tomokazu Ishihara, Masatoshi Watanabe, Dai Komatsubara, Natsuko Kasirajima, Shingo Tokuyasu, Toshikazu Shinya and Koichi Suzuki

Abstract Mud volcanoes are topographic mounts generated by eruption materials that have risen from deep underground because of abnormal groundwater pressure resulting from the dehydration of clay minerals. The activity of the Lusi mud volcano in Indonesia in 2006 caused serious damage to infrastructure and loss of life. Therefore, understanding the activity of mud volcanoes and the relationship between the geological conditions and the location, timing, and mode of mud volcanism is

K. Tanaka (✉)
Yamaguchi University, Yamaguchi, Japan
e-mail: ka-tanak@yamaguchi-u.ac.jp

K. Asano
CTI Engineering Co., Ltd., Chuo-ku, Japan
e-mail: kj-asano@ctie.co.jp

T. Ishihara
Yamaguchi, Japan
e-mail: ishihara.tomokazu@pref.yamaguchi.jg.jp

M. Watanabe
Soja, Japan
e-mail: soumu@city.soja.okayama.jp

D. Komatsubara
Konoike Construction Co., Ltd., Ishinomaki, Japan
e-mail: komatsubara_di@konoike.co.jp

N. Kasirajima
PKUTECH Co., Ltd., Tokyo, Japan
e-mail: kasirajima.natuko@pkutec.co.jp

S. Tokuyasu
NC Geophysical Survey Co., Ltd., Chiyoda-ku, Japan
e-mail: tokuyasu@chikatansa.co.jp

T. Shinya
Dia Consultant Co., Ltd., Fukuoka, Japan
e-mail: t.shinya@diaconsult.co.jp

K. Suzuki
Hokkaido University, Sapporo, Japan
e-mail: koichi.suzuki@eng.hokudai.ac.jp

Y. Himiyama et al. (eds.), *Human Geoscience*, Advances in Geological Science,
https://doi.org/10.1007/978-981-32-9224-6_14

important. Geological, geochemical, and geophysical studies on mud volcanoes in Japan and Taiwan have shown that the location of mud volcanism, the mode of activity, and the rise of highly saline groundwater are strongly affected by geological structures. In this chapter, we describe onshore mud volcanoes in Japan and Taiwan, and give an overview of mud volcanoes and mud volcanism. We then discuss the effects of mud volcanism on human life and the natural environment and note that mud volcanoes and their relationship to underground geological structures should be considered when siting and assessing critical infrastructure.

Keywords Mud volcano · Groundwater · Natural gas · Geological structure · Natural hazard

14.1 Introduction

Mud volcanoes are geometric high conical mounts composed of erupted mud that originates from deep underground (Bates and Jackson 1987), and they are induced by abnormal pore water pressure. These volcanoes are widely distributed in sedimentary basins onshore and offshore globally (Fig. 14.1; Higgins and Sanders 1974; Kopf

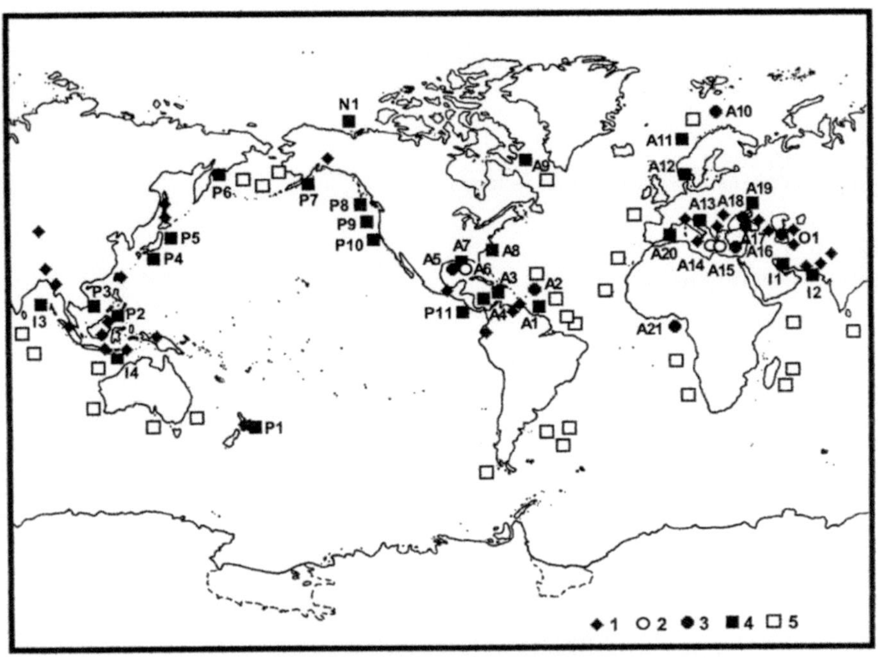

◆: onshore, ○: hydrate, ●: no hydrate, ■: offshore, □: diapir

Fig. 14.1 Map showing the worldwide location of onshore, known, and inferred submarine mud volcanoes (after Mirkov 2000)

2002; Milkov 2000).

Onshore mud volcanoes are found at plate boundaries in places such as Azerbaijan (Lerche and Bagirov 1999), Indonesia (Barber et al. 1986), and Trinidad (Brown 1990; Deville et al. 2009; Dia et al. 1999). Mud volcanoes are also distributed in East Asia, including in Taiwan (Hamada et al. 2009; You et al. 2004) and Japan (Chigira and Tanaka 1997; Ishihara and Tanaka 2009). Generally, onshore mud volcanoes gently erupt groundwater, mud, and gas. However, some volcanoes erupt violently, forming topographic depressions. Therefore, the activity of onshore mud volcanoes may pose a threat to people.

For example, the Lusi mud volcano in Indonesia has erupted vast amounts of groundwater, mud, and gas since 2006, causing serious problems with the ground surface and affecting the lives of local residents (Davies et al. 2007, 2008). Highly saline, pressurized groundwater rising from deep underground in mud volcanoes destroys manmade underground structures by hydro-fracturing and forms mud breccia, which adversely affects tunnel construction through swelling.

Rising groundwater increases the pore water pressure along the landslide plane and can trigger landslides. In addition, regional groundwater flow caused by rising pressurized groundwater may result in long-term groundwater stability. The migration of highly saline groundwater causes rapid weakening and erosion of sedimentary rock via the interaction of clay minerals and groundwater. The volume of methane, which is a potent greenhouse gas, released during eruptions of mud volcanoes is thought to be large enough to affect global warming. Because onshore areas support considerable human activity and the activity of the mud volcanoes can cause many problems, investigating geological and hydrological phenomena related to mud volcanoes is important (Tanaka 2009).

To reduce the risk to human life and the environment, it is necessary to understand where, when, why, and how mud volcanism happens. Here, we describe onshore mud volcanoes in Japan and Taiwan and give an overview of mud volcanoes and mud volcanism. Finally, the effects of mud volcanism on human life and the natural environment are discussed.

14.2 Description of Representative Onshore Mud Volcanoes in Japan and Taiwan

Figure 14.2 shows the distribution of onshore mud volcanoes in Japan and Taiwan. Photos of mud volcanoes are shown in Fig. 14.3 (1)–(7). Four representative mud volcanoes in Japan and Taiwan are described based on geological, geochemical, and geophysical studies.

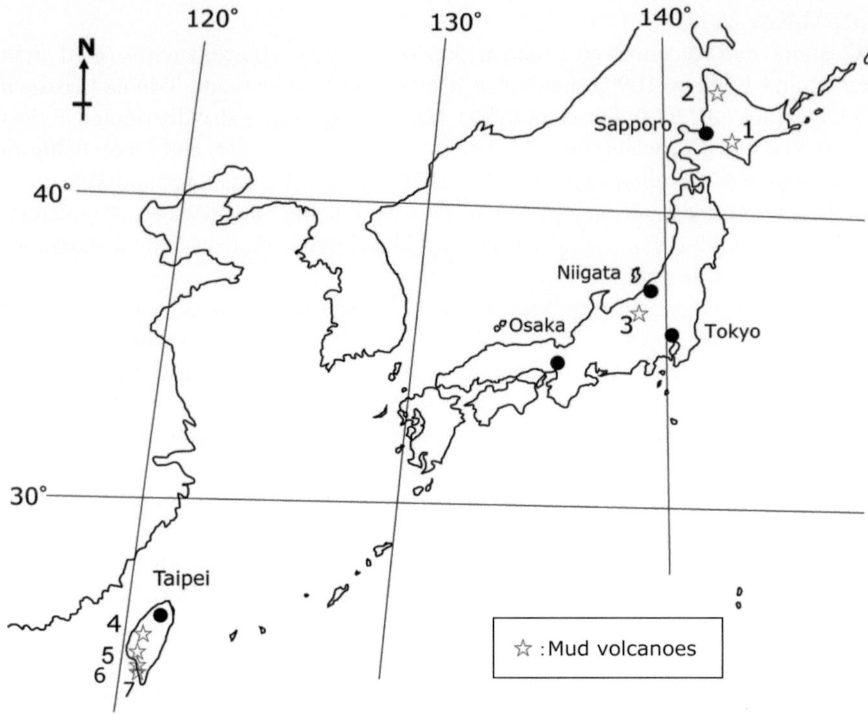

Fig. 14.2 Map showing the location of mud volcanoes in Japan and Taiwan. Japan: 1. Niikappu; 2. Kamihoronobe; 3. Niigata; Taiwan: 4. Denryao; 5. Yanchao; 6. Hsin-yang-nyu-hu; 7. Mantan

14.2.1 Niikappu Mud Volcano

The Niikappu mud volcano is in Niikappu, Hokkaido, north Japan, and it is composed of nine eruption craters along the NW-SE anticline axis (Fig. 14.4; Chigira and Tanaka 1997) in the Neogene sedimentary rocks. There are pool craters in the north of the area and cone craters in the south of the area. The largest crater (N1) is 50 m high, with three terraces on its slope in a subsided topographic depression 1 km wide and 20 m deep in the central part of the volcano, suggesting that N1 has undergone recurrent volcanism [Fig. 14.3(1)]. At present, N1 erupts groundwater and mud at the top of the eruption cone, which is several meters high, during earthquakes with a seismic intensity of more than 5 (Chigira and Tanaka 1997).

14.2.2 Niigata Mud Volcano [Fig. 14.3(3-1, 3-2, 3-3)]

The Niigata mud volcano is situated along the anticline axis in the Neogene sedimentary rocks in the Kamou and Murono areas, Tokamachi City, Niigata Prefecture,

(1) Niikappu mud volcano, Niikappu Town, Hokkaido, Japan (N1)

(2) Kamihoronobe mud volcano, Horonobe Town, Hokkaido, Japan. Red dotted line shows the depression

(3-1) Niigata mud volcano (active), Matsudai area, Tokamachi City, Niigata Prefecture, Japan

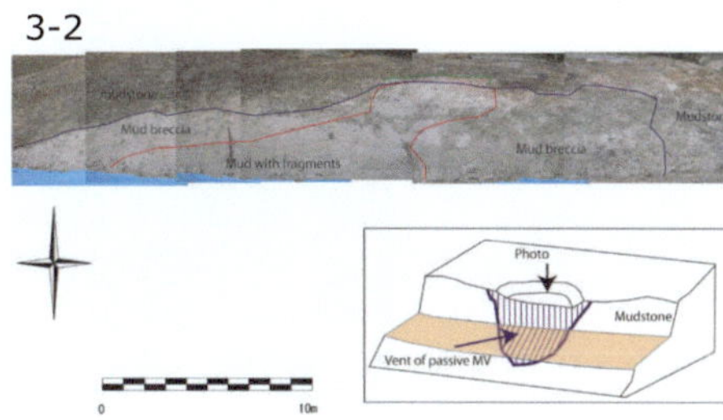

(3-2) Niigata mud volcano (passive), Matsudai area, Tokamachi City, Niigata Prefecture, Japan

(3-3) Niigata mud volcano, Murono area, Tokamachi City, Niigata Prefecture, Japan

(4) Moon World mud volcano, Denryao area Taiwan

Fig. 14.3 Photos of mud volcanoes in Japan and Taiwan. 1. Kamihoronobe mud volcano; 2. Niikappu mud volcano; 3-1. Niigata mud volcano Kamou area (active); 3-2. Niigata mud volcano, Kamou area (passive); 3-3: Niigata mud volcano, Murono area; 4. Moon World mud volcano; 5. Wushanding mud volcano; 6. Hsin-yang-nyu-humud volcano; 7. Mantan mud volcano

(5) Wushanding mud volcano, Yanchao area, Taiwan.

(6) Hsin-yang-nyu-hu mud volcano, Yanchao area, Taiwan

(7) Mantan mud volcano (Fissure eruption in 2014), Pingtung area, Taiwan

Fig. 14.3 (continued)

central Japan (Fig. 14.5; Shinya and Tanaka 2005). In the Kamou area, there are two active pool eruption craters [Fig. 14.3(3-1)] and a passive crater at the rim of a subsided caldera 200 m in diameter and 30 m deep. A borehole 120 m deep was drilled in the center of caldera, and it was concluded that mud breccias consisting of mudstone fragments and clay filled with highly saline groundwater is present deeper than 50 m (Fig. 14.6; Ishihara and Tanaka 2009). The passive crater is exposed at the rim of the subsided caldera in a small depression 30 m wide and 4 m deep, and the volcanic vent is filled with clay and mud breccias [Fig. 14.3(3-2); Shinya and Tanaka 2005].

In the Murono area, there are two small pool craters 50 cm wide on the topographic plain, gently erupting mud, groundwater, and gas [Fig. 14.3(3-3)]. Natural gas erupts in the rice fields around the crater. Geophysical exploration was carried out to identify the shallow underground geological structure (Onishi et al. 2009; Suzuki et al. 2009) and the tectonics during earthquakes (Kusumoto et al. 2015).

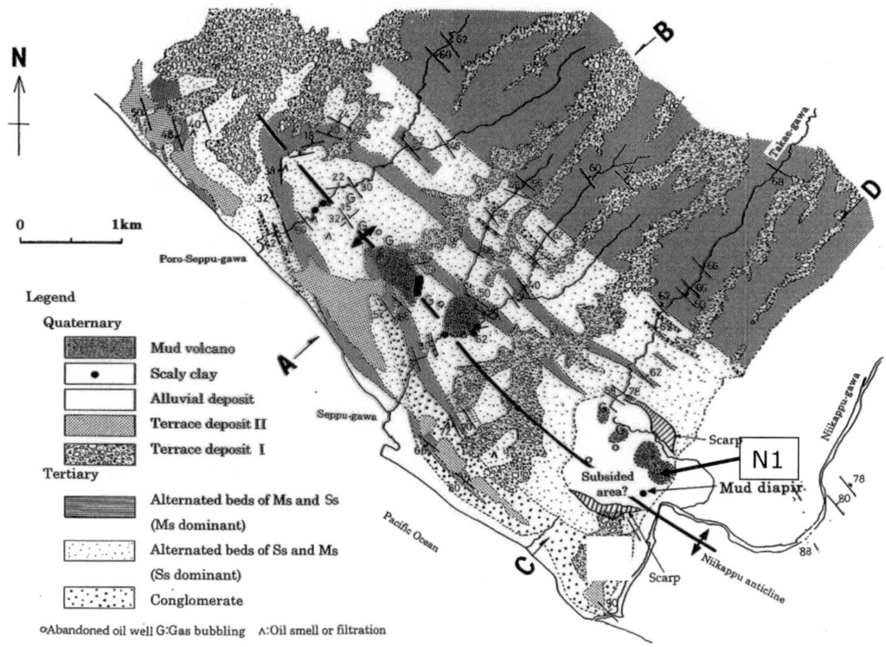

Fig. 14.4 Geological map of the Niikappu mud volcano in Niikappu Town, Hokkaido, Japan (after Chigira and Tanaka 1997)

14.2.3 Moon World Mud Volcano

Many mud volcanoes are distributed in southwestern Taiwan (Fig. 14.7). The Moon World mud volcano is distributed in the Neogene sedimentary rocks in the Denryao area of the northern part of Kaohsiung City, Taiwan (Hamada et al. 2009; You et al. 2004). Six erupting pudding and pool craters are distributed on small hills in a circle 200 m in diameter along the Gutingkeng anticline axis (Fig. 14.8). The craters, which are several meters in diameter, erupt groundwater, mud, and gas gently [Fig. 14.3(4)]. The Moon World badlands are distributed widely 500 m north of the Moon World mud volcano and are devoid of vegetation. Mud breccias and mud intrusions are exposed in the river bottom near the mud volcano.

14.2.4 Wushanting Mud Volcano

The Wushanting mud volcano is in the Gutingkeng formation composed of Neogene sedimentary rocks along the Chishan Fault in the Yanchao area, northern Kaohsuing City, Taiwan. There are five erupting craters on the flat foothill that are mainly cone craters [Figs. 14.9 and 14.3(5)]. The highest mount is up to 7 m high. The craters

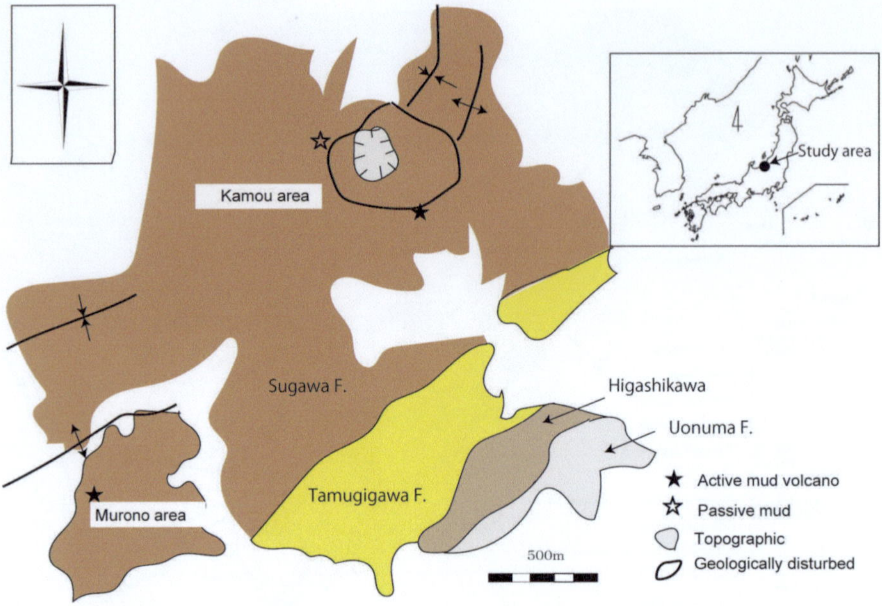

Fig. 14.5 Geological map of the Niigata mud volcano, Kamou and Murono areas, Tokamachi City, Niigata Prefecture, Japan (after Shinya and Tanaka 2009). ☆: Active mud volcano, ★: passive mud volcano

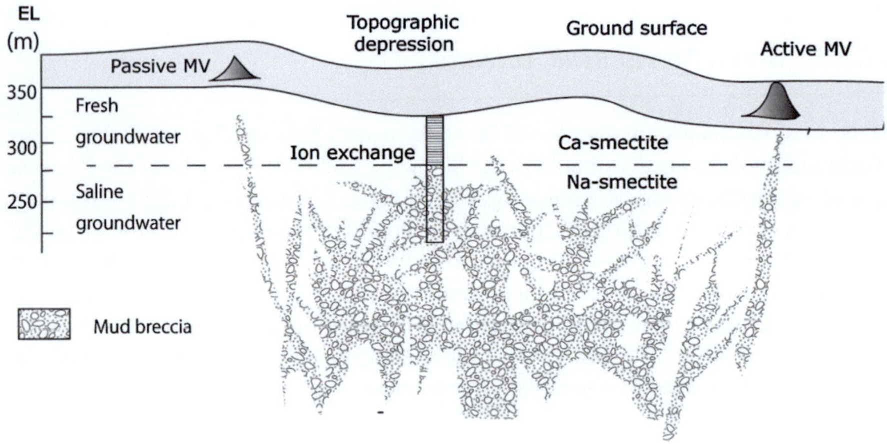

Fig. 14.6 Conceptual model of the underground geological structure of the Matsudai area, Niigata mud volcano, Niigata Prefecture, Japan. The dashed line shows the boundary between fresh and saline groundwater (after Ishihara and Tanaka 2009)

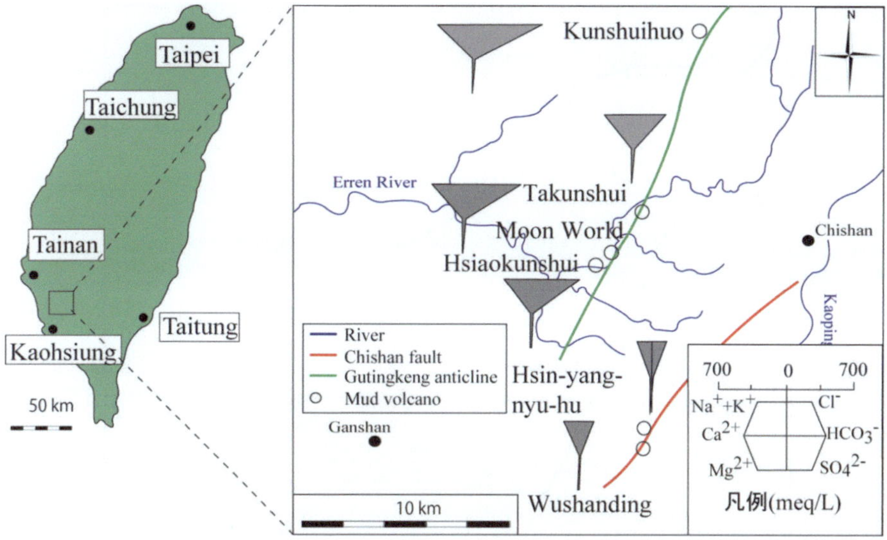

Fig. 14.7 Distribution of mud volcanoes and groundwater quality in southwestern Taiwan. Groundwater samples were collected from mud volcanoes distributed along the Gutinkeng anticline and the Chishan Fault

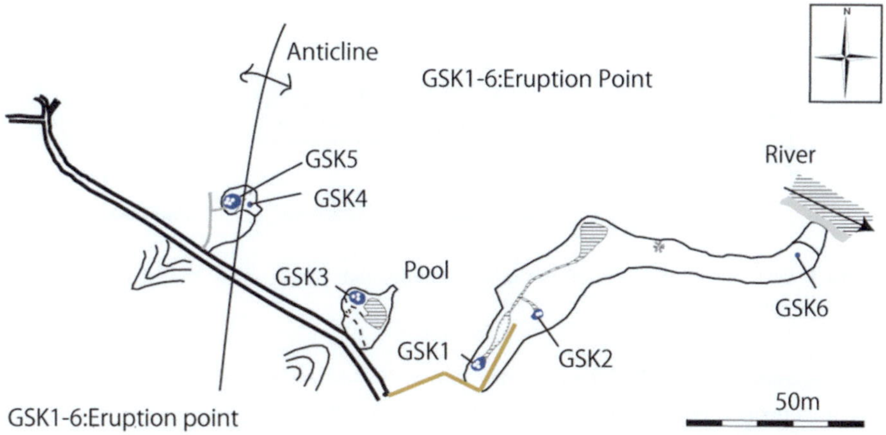

Fig. 14.8 Distribution of the craters of the Moon World mud volcano in the Denryao area, Taiwan. GSK 1–6 indicatethe craters

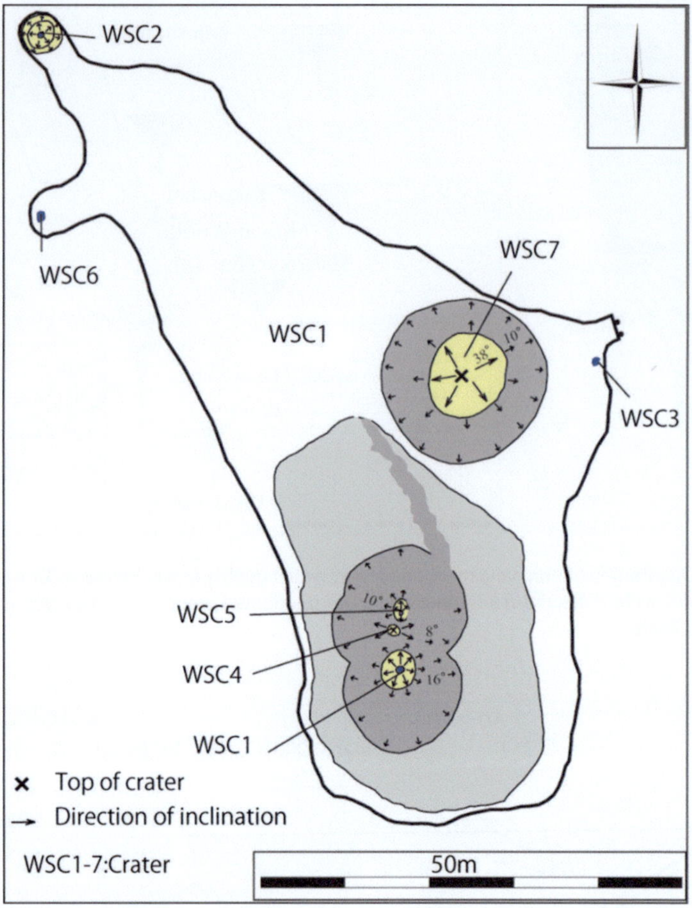

Fig. 14.9 Distribution of the craters of the Wushanting mud volcano in the Yanchao area, Taiwan. WSC 1–7 indicate the craters

erupt groundwater, mud, and gas gently, and the fluid from the crater is highly viscous with a water content lower than 50% (Asano et al. 2015).

14.3 Characteristics of Mud Volcanoes and Mud Volcanism

14.3.1 Topographic Features of Mud Volcanoes and Their Surrounding Areas

Figure 14.10 shows the relationship between the height and diameter of a crater and the water content of erupting fluid (Asano et al. 2014). The higher and wider the

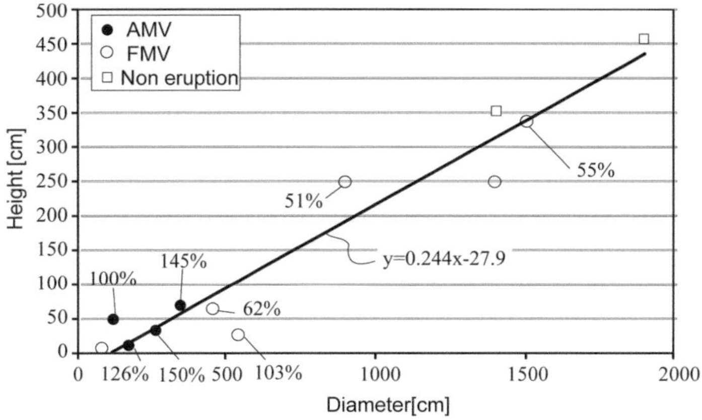

Fig. 14.10 Relationship between the diameter and height of craters of mud volcanoes in Taiwan (after Asano et al. 2014).●: AMV, ○: FMV. The water content of the erupted mud is shown as a percentage

crater, the lower the water content. The distribution of mud volcanoes is controlled by underground geological structures such as anticlines and faults. Mud volcanoes located along faults (fault-related mud volcano; FMV) are characterized by cones with a low water content due to the clayey fault gouge [e.g., Fig. 14.3(5)]. However, mud volcanoes located along anticline axes (anticline-related mud volcano; AMV) are characterized by pools with a high water content [e.g., Fig. 14.3(4)]. The slope angle of mud volcanoes is about 14°.

Topographic depressions occur near the AMV. For example, there is a topographic depression 1 km in diameter and 30 m deep in the central part of the Niikappu mud volcano near the highest cone (N1, Fig. 14.4; Chigira and Tanaka 1997). There is also a topographic depression 200 m in diameter and 20 m deep in the Kamou area of the Niigata mud volcano (Suzuki et al. 2009). These topographic features of AMVs suggest violent mud volcanism with an eruption of a huge volume of groundwater, mud, and gas.

14.4 Erupted Materials

14.4.1 Groundwater

Groundwater that erupts from mud volcanoes is highly saline and shows a higher $\delta\ ^{18}O$ and lower δ D than that of standard mean ocean water (Craig 1961) (Fig. 14.11). Therefore, abnormal pore water pressure may originate from the dehydration of clay minerals (Dahlmann and de Lange 2003). In addition, $\delta\ ^{18}O$ of groundwater erupted from FMV is higher than that of AMV (Fig. 14.11). This suggests that the fluid

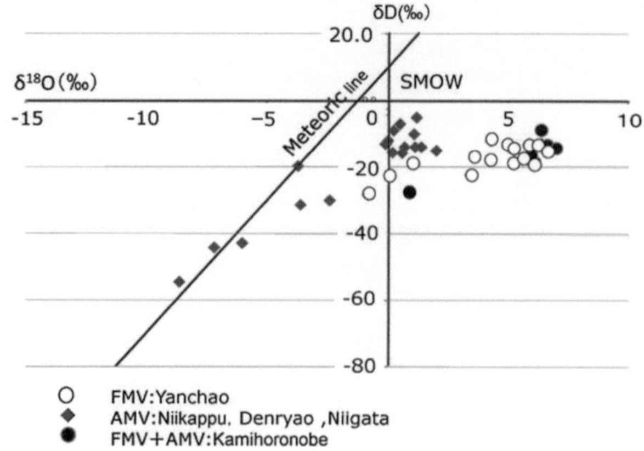

Fig. 14.11 Delta diagram of oxygen and hydrogen stable isotopes in groundwater from mud volcanoes in Japan and Taiwan. FMV: fault-related mud volcano; AMV: anticline-related mud volcano

erupted from FMD originates from deeper underground at a higher temperature than that from AMV because there has been sufficient isotope exchange between the fluid and minerals (Brown 1990; You et al. 2004).

14.4.2 Mud

Mud erupted from mud volcanoes is composed of clay minerals consisting of illite, smectite, illite-smectite mixed-layer minerals, and chlorite (Hamada et al. 2009; Shinya and Tanaka 2005). During the diagenetic transition of 1 mol of smectite to illite, 10 mol of interlayer water is released; thus, this hydration may cause the abnormal pore water (Yamagata and Ogawa 1989). The transition from smectite to illite happens from 60 to 150 °C (Dahlmann and de Lange 2003); thus, the clay originates from a depth of about 3000 m under a geothermal gradient of 3.5 °C/100 m (Aoyagi and Kazama 1980).

14.4.3 Coal Fragments

Coal fragments are intercalated in the Tertiary sedimentary rocks and erupted mud from craters. The burial depth of the coal fragments can be obtained by the vitrinite reflectance (R_o). Figure 14.12 shows the R_o of coal fragments separated from mud erupted from the Niigata mud volcano, and Tertiary sedimentary rock and a boring core from near the volcano (Shinya and Tanaka 2005). The R_o spectrum of the coal fragments showed two peaks at 0.3–1.2% (P1) and 1.5–1.8% (P2), corresponding to the sedimentary rocks and to the boring cores at a depth of 4090 m, respectively, suggesting that coal fragments of erupted mud are derived from a depth of about 4000 m.

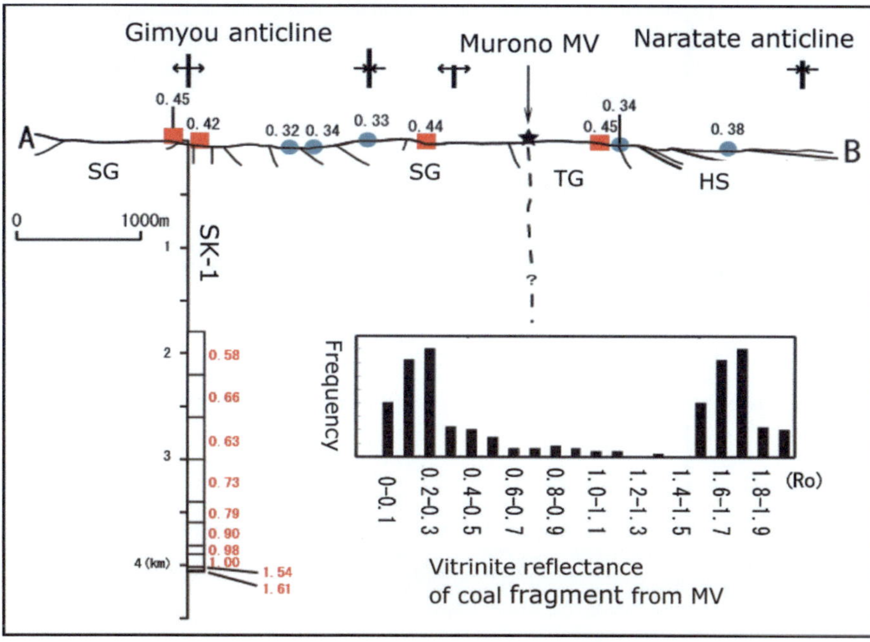

Fig. 14.12 Relationship between the geological structure and vitrinite reflectance (R_o) of coal fragments separated from erupted mud, boring cores and mudstone in an outcrop in the Murono area, Niigata mud volcano, central Japan (after Shinya and Tanaka 2009)

14.4.4 Gas

The gas from mud volcanoes is composed of methane and carbon dioxide (Kopf 2002). The carbon isotope composition of mud volcanoes in Taiwan was analyzed (Fig. 14.13; Asano et al. 2014), and showed that gas from FMV is a thermogenic gas generated deep underground and decomposed by microorganisms. In contrast, the gas from AMV is a mixture of thermogenic and biogenic gas generated under the impermeable layer along the anticline axis.

14.5 Geological Structure Under Mud Volcanoes

Geological and geophysical investigations were carried out to clarify the geological structure under mud volcanoes in Japan and Taiwan. A boring core 120 m long was taken from the center of the topographic depression of the Kamou area of the Niigata mud volcano, Japan. Mud breccia (Clari et al. 2004) composed of mudstone fragments and clay was found from 50 to 120 m deep, corresponding to a diapiritic mélange (Barber et al. 1986). Boring core logging and X-ray CT scanning obser-

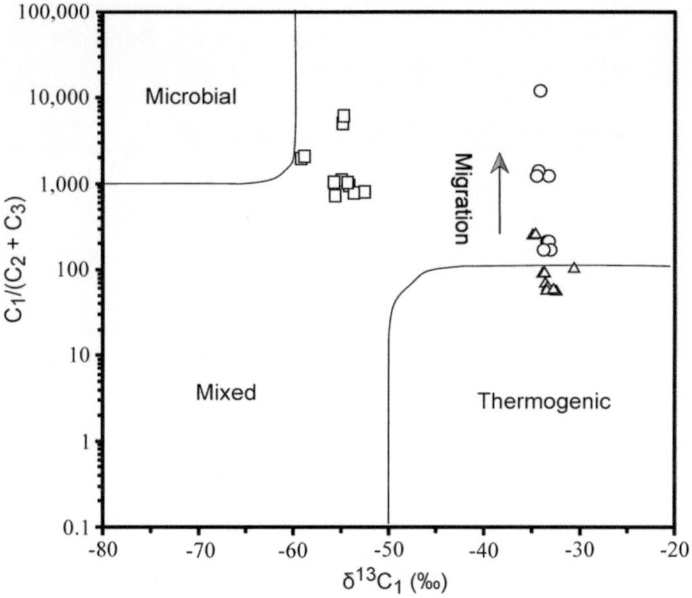

Fig. 14.13 Relationship between $\delta\ ^{13}C_1$ (‰) and the $C_1/(C_2 + C_3)$ ratio. The origins of gases are taken from Waseda et al. (2002). □: Denryao area MVs, ○: Wushanding MV, △: Hsin-yang-nyu-huMV (after Asano et al. 2014)

vations showed that some of the mudstone fragments were generated in situ and others were generated deep underground by hydro-fracturing by gas pressure caused by methane degassing at shallow depths (Fig. 14.14). Mud breccia from the ground surface to 50 m deep was filled with meteoric water, and that deeper than 50 m deep was filled with highly saline groundwater that had risen from underground (Ishihara and Tanaka 2009).

Controlled-source audio magnetotellurics (CSAMT) has been used to determine the underground structure from the surface to a depth of 500 m in the Kamou area of the Niigata mud volcano, Japan, and in the Yanchao area of the Wushanting mud volcano, Kaohsuing City, Taiwan.

In the Kamou area, an extremely low-resistivity zone of less than 1 Ω m at a depth of 400 m within a diameter of 500 m was detected below the topographic depression. It was assumed that the low-resistivity zone corresponded to a mud chamber filled with highly saline groundwater and mud based on laboratory tests measuring the electric conductivity of rock specimens obtained from the survey area. A low-resistivity zone between 1.5 and 2.5 Ω m was present and could be traced to the mud volcano (Fig. 14.15; Suzuki et al. 2009).

In the Yanchao area, a low-resistivity zone of several ohm meters thought to be filled with highly saline groundwater was detected along the Chishan fault and its branch fault (Asano et al. 2015). The mud volcanoes are above the low resistivity zone (Fig. 14.16).

Fig. 14.14 CT scanning photos showing mud breccia from boring cores from the Kamou area, Niigata mud volcano, Japan, and from the outcrop near the Moon World mud volcanoes, Taiwan

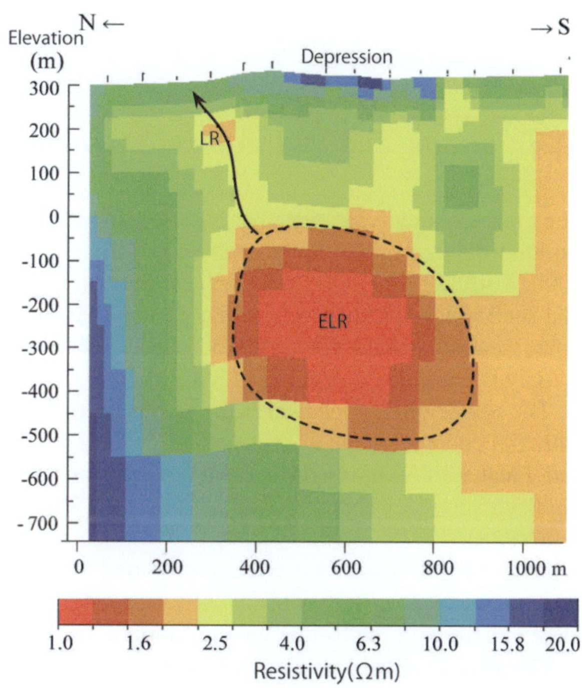

Fig. 14.15 Resistivity profile from the CSAMT survey in the Kamou area, Niigata mud volcano, Niigata Prefecture, Japan (after Suzuki et al. 2009). The extremely low resistivity (ELR) zone at a depth of 300–700 m with a resistivity of less than 1.6 Ω m is present in a geologically disturbed area. A resistivity zone (LR) with a resistivity between 1.6 and 2.5 Ω m is also present and can be traced to the surface

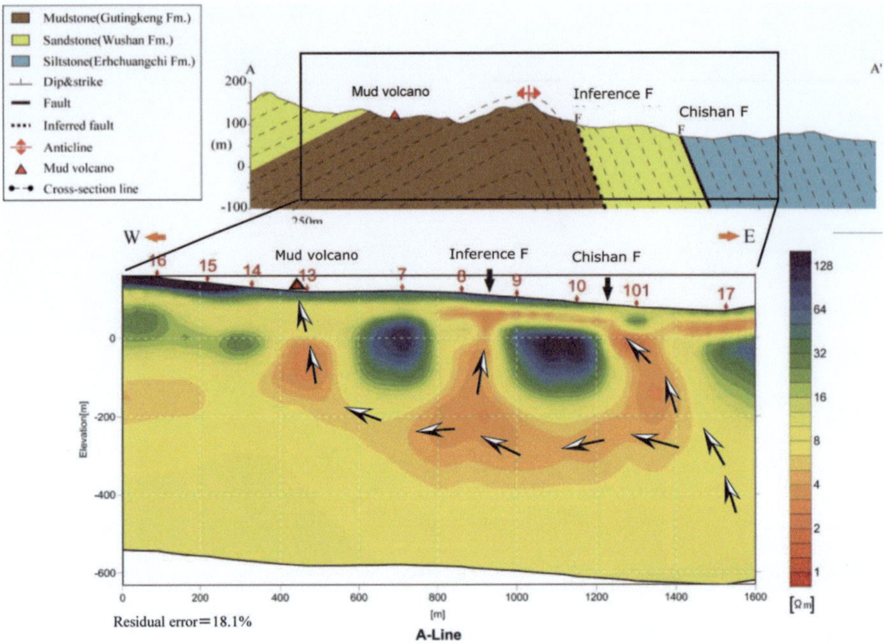

Fig. 14.16 Resistivity profile from the CSAMT survey around the Wushanting mud volcano, Taiwan. A low resistivity zone along the Chishan Fault and Inference Fault with resistivity between 2.0 and 5.0 Ω m is present. Arrows show the expected direction of highly saline groundwater flow (after Asano et al. 2015)

14.6 Timing and Migration of Mud Volcanism

The timing of mud volcanism has been discussed based on the results of geophysical exploration and studies of the erupted materials from mud volcanoes (Fowler et al. 2000; Kopf 2002). Kopf (2002) reviewed the timing of mud volcanism and showed that mud volcanism happens suddenly and that intense activity occurs for a short time. Generally, mud volcanism continues for several hundred years and the first activity is the most violent (Kopf 2002).

The age of mud volcanism in Japan has been determined by radiocarbon isotope dating of charcoal in the Kamou area of the Niigata mud volcano (Fig. 14.6; Ishihara and Tanaka 2009; Shinya and Tanaka 2005) and the marker tephra of the Niikappu mud volcano intercalated in erupted materials. The activity of the Niigata mud volcano started 20,000–30,000 years ago, forming a topographic depression with violent activity. The Niikappu mud volcanism started after oxygen isotope stage 5e and four bouts of violent activity subsequently occurred. The first bout of activity was the most violent with the formation of a topographic depression (Fig. 14.4).

Detailed records of mud volcanism over the last 40 years have been obtained for the Mangtan mud volcano, Bintong Prefecture, Southern Taiwan. The Mangtan mud

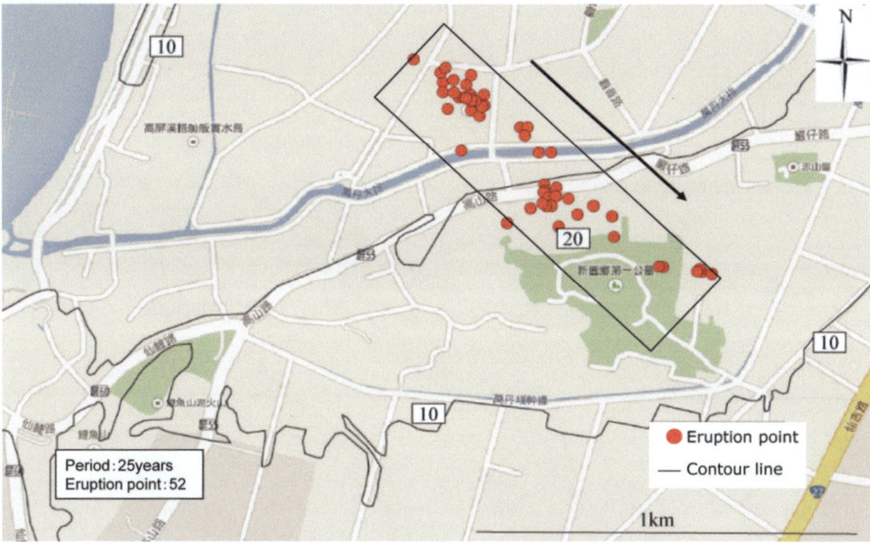

Fig. 14.17 Distribution of craters of the Mantan mud volcano, southern Taiwan over 25 years. Mud volcanism has migrated from the northwest to the southeast in a narrow belt

volcano is in an alluvium plain composed of unconsolidated sediments [Fig. 14.3(7)]. There are more than 20 erupting cones in a narrow linear area in a northwest-southeast direction and they erupt more than once a year. The eruption points have migrated from northwest to southeast over time (Fig. 14.17), and the interval time between episodes of mud volcanism in the area corresponds to the volume of erupted materials from a previous eruption. The larger the former eruption, the longer the interval time is. This suggests that the fluid that has risen remains under the impermeable mud layer, and then rises again by breaking the layer because of the gas pressure caused by degassing.

14.7 Mud Volcanism and Risk to Human Life

14.7.1 Violent Eruption

The violent activity of mud volcanoes, such as the Lusi mud volcano in Indonesia, poses a serious risk to human life (Davies et al. 2007). The relationship between the geological setting and the mode of eruption of highly saline groundwater and mud from deep underground in Tokamachi City, Niigata Prefecture was investigated. A thick impermeable cap rock was found, consisting of Neogene mudstone (Sugawa formation), under the Niigata mud volcano. Therefore, the rising fluid is trapped under the impermeable cap rock, forming a mud chamber (Suzuki et al. 2009). The

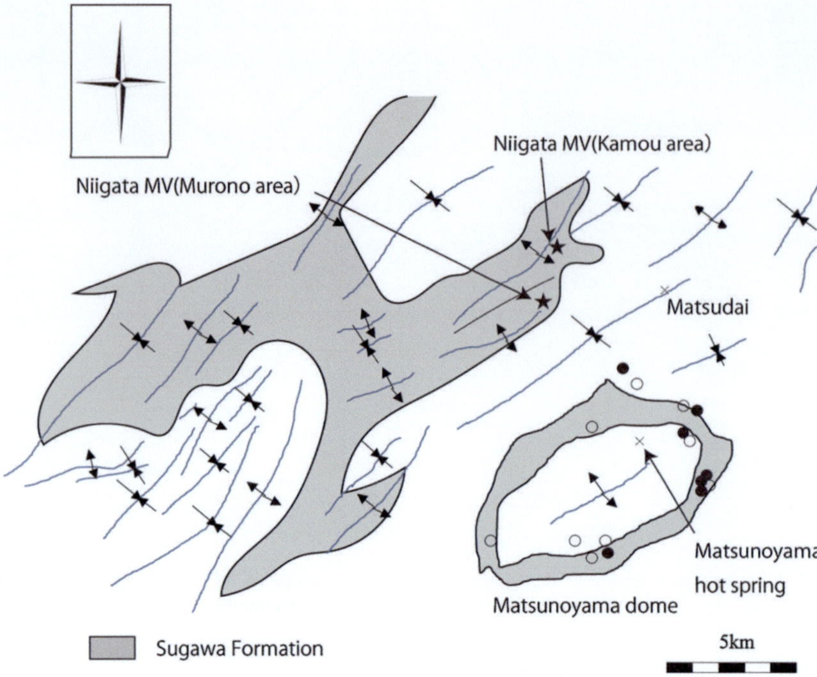

Fig. 14.18 Geotectonic map of Tokamachi Town, Niigata Prefecture, Japan showing the distribution of the Sugawa Formation, consisting of mudstone (modified from Takeuchi et al. 2000). Cl^-: content of groundwater (○: Cl^- >50 mg/L, ●: Cl^- >500 mg/L)

fluid rises by hydro-fracturing owing to the high gas pressure caused by methane degassing. Finally, a huge volume of fluid erupts violently at the mud volcano. After the violent eruption, the ground surface subsides forming a topographic depression (Suzuki et al. 2009).

However, highly saline groundwater seeps gently between the impermeable cap rock and the underlying permeable rock in the Matsunoyama area, Tokamachi City, Niigata Prefecture, 7 km south of the Niigata mud volcano with no violent eruptions (Watanabe and Tanaka 2011), where impermeable cap rock does not cover the ground surface (Fig. 14.18). Hence, the presence of an impermeable cap determines whether violent mud volcanism occurs (Fig. 14.19).

14.7.2 Rock Swelling

The Nabetachiyama tunnel was excavated under the Niigata mud volcano in the Kamou area, Tokamachi City, Niigata Prefecture, Japan, and serious difficulties caused by swelling mudstone were encountered during excavation. A borehole was

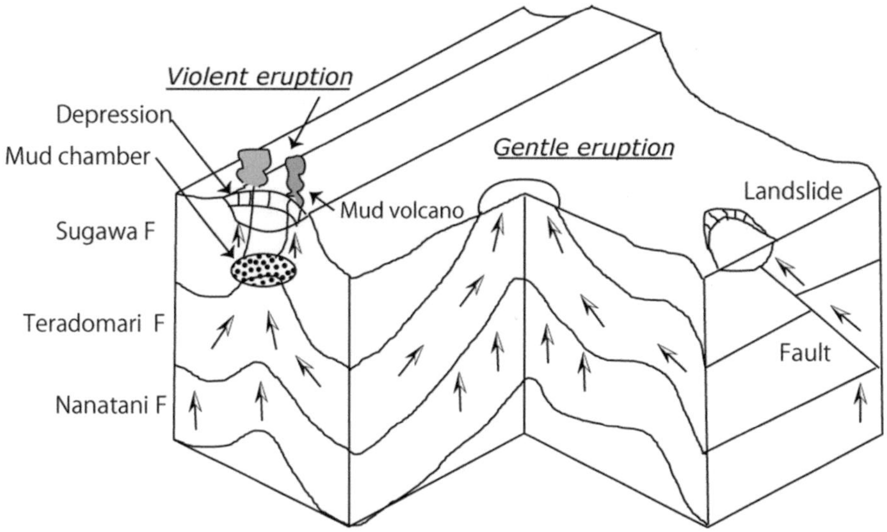

Fig. 14.19 General groundwater flow in Tokamachi City, Niigata Prefecture, Japan. The Niigata mud volcano at the anticline axis with thick impermeable mudstone is characterized by violent eruptions and topographic subsidence. The Matsunoyama dome area with no impermeable layer is characterized by gentle seepage of saline groundwater

excavated to the elevation of the tunnel and the cores were mostly mud breccias consisting of mudstone fragments and a clayey matrix generated by the hydro-fracturing of mudstone. Mud breccias from deeper than 50 m contained large amounts of Na-smectite formed in the highly saline pore water that rose from deep underground. Mud breccias filled with saline groundwater are characterized by quick slaking. The swelling mudstone in the problem section of the excavation was composed of mud breccia filled with saline groundwater and was weakened by quick slaking and hydro-fracturing (Tanaka and Ishihara 2009).

14.7.3 Landslides

Boreholes were excavated in a landslide area composed of Neogene sedimentary rocks in the Matsunoyama area, Tokamachi City, Niigata Prefecture, Japan. The groundwater sampled from the borehole had high salinity and came from dehydration of clay minerals deep underground. This fluid may correspond to that which erupts from mud volcanoes. Because there is no cap rock along the anticline and dome, and the pressurized groundwater erupts gently highly saline groundwater is injected into the landslide plane increasing the pore water pressure and degrading the landslide plane that is weakened by rapid slaking. Figure 14.20 shows the landslide induced by rising saline groundwater in the Matsunoyama area (Watanabe et al. 2009). Therefore,

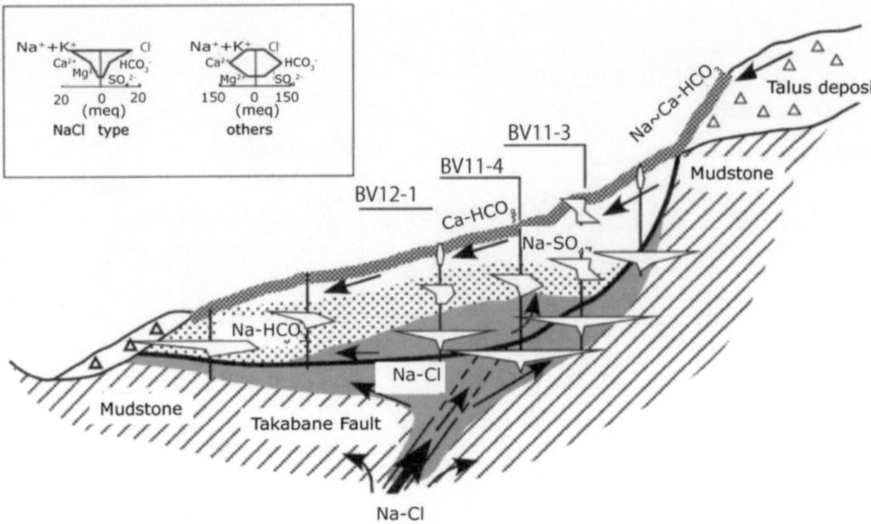

Fig. 14.20 Schematic hydrochemical profile of groundwater in the Utsunomata landslide, Niigata, central Japan (after Watanabe et al. 2009). Highly saline groundwater ascends through the fault and is distributed along the landslide plane

some landslides in sedimentary rocks may be caused by the injection of rising highly saline groundwater.

14.7.4 Greenhouse Effect

Methane is a natural gas that strongly affects global warming. Gas flux has been measured at the mud volcanoes in Taiwan and Japan (You et al. 2004; Etiope et al. 2011). Etiope (2005) estimated the total global flux from inland mud volcanoes as 6–9 Mt/year and proposed that the gas flux from mud volcanoes should be considered in global estimates of greenhouse gas flux.

14.8 Conclusion

Mud volcanism is a common geological phenomenon in areas of sedimentary rocks. The relationship among the location of activity, ground surface topography, underground geological structures, and geochemical properties has been studied at onshore mud volcanoes in East Asia, Japan, and Taiwan. Violent eruption and rock weakening that causes landslide activity and problems with tunnels depend on the underground geological structures. Therefore, the distribution of mud volcanoes and their relationship to underground geological structures should be considered when siting and assessing critical infrastructure on the surface and underground in sedimentary basins.

Acknowledgements I am very appreciative of Prof. Masahiro Chigira of Kyoto University, Prof. of Emeritus Yuichirou Miyata of Yamaguchi University, Dr. Yuji Kanaori and Prof. Tomochika Tokunaga of the University of Tokyo, Prof. of Emeritus Jun Shimada of Kumamoto University, Sumito Morita of the National Institute of Advanced Industrial Science and Technology, and Dr. Amane Waseda of Japan Petroleum Exploration Co., Ltd. for their valuable advice and many discussions. I would like to give my sincere thanks to Prof. Shu-Huei Liu of National Kaohsiung Normal University and Dr. Wen-Chi Lai of National Cheng Kung University for their help and kind discussion about geological surveys in Taiwan.

References

Aoyagi K, Kazama T (1980) Transformation changes of clay minerals, zeolite and silica minerals during diagenesis. Sedimentology 27:179–188
Asano K, Tanaka K, Waseda A (2014) Ascending process of fluids from mud volcanoes in southwestern part of Taiwan. In: Proceedings of the 2014 meeting of the Japan Society of Engineering Geology, Kyoto, pp 201–202. (in Japanese)
Asano K, Tanaka K, Suzuki K (2015) Studies of underground geologic structure of Wushanding mud volcanoes in southwest Taiwan by electromagnetic exploration. In: Proceedings of the 2015 meeting of the Japan Society of Engineering Geology, Kyoto, pp 47–48. (in Japanese)
Barber AJ, Tjokrosapoetro S, Charlton TR (1986) Mud volcanoes, shale diapirs, wrench faults, and mélanges in accretionary complexes, Eastern Indonesia. AAPGG Bull 70:1729–1741
Bates R, Jackson J (1987) Glossary of geology. Virginia, American Geological Institute, p 788
Brown KM (1990) The nature and hydrogeologic significance of mud diapirs and diatremes for accretionary system. J Geophys Res 95:8952–8969
Chigira M, Tanaka K (1997) Structural features and the history of mud volcanoes in southern Hokkaido, southern Hokkaido, northern Japan. J Geol Soc Jpn 103:781–791. (in Japanese with English abstract)
Clari P, Cavagna S, Martire L, Hunziker J (2004) A Miocene mud volcano and its plumbing system: a chaotic complex revised. (Monferrato, NW Italy). J Sediment Res 74:662–676
Craig H (1961) Standard for reporting concentration of deuterium and oxygen-18 in natural waters. Science 133:1833–1834
Dahlmann A, de Lange GJ (2003) Fluid-sediment interactions at Eastern Mediterranean mud volcanoes: a stable isotope study from ODP Leg 160. Earth Planet Sci Lett 212:377–391
Davies JR, Swarbrick ER, Evans JR, Huuse M (2007) Birth of a mud volcano: East Java, 29 May 2006. GSA Today 17:4–9
Davies JR, Brumm M, Manga M, Rubiandini R, Swarbrick R, Tingay M (2008) The East Java mud volcano (2006 to present): an earthquake or drilling trigger? Earth Planet Sci Lett 272:627–638
Deville E, GuerlaisDia SH (2009) Cyclic activity of mud volcanoes: evidences from Trinidad (SE Caribbean). Mar Petrol Geol 26:1681–1691
Dia AN, Castrec-Rouelle M, Boulegue J, Comeau P (1999) Trinidad mud volcanoes: where does the expelled fluid come from? Geochimicaet Cosmochimica Acta 63:1023–1038
Etiope G (2005) Methane emissions from mud volcanoes: towards a global estimate. In: Martinelli G, Panahi B (eds) Mud volcanoes, geodynamics and seismicity. NATO Science Series, vol 51. Springer, Dordrecht, pp 141–146

Etiope G, Nakada R, Tanaka K, Yoshida N (2011) Gas seepage from Tokamachi mud volcanoes, onshore Niigata Basin (Japan): origin, post-genetic alterations and CH_4–CO_2 fluxes. Appl Geochem 26:348–359
Fowler SR, Mildenhall J, Zaloba S, Riley G, Elsley G, Desplanques A, Guliyev F (2000) Mud volcanoes and structural development on Shah Deniz. J Petrol Sci Eng 28:189–206
Hamada Y, Tanaka K, Miyata Y (2009) Geologic structure and geochemistry of Taiwanese mud volcanoes. J Geogr 118:408–423. (in Japanese with English abstract)
Higgins GE, Saunders JB (1974) Mud volcanoes-their nature and origin. Verh Naturforsch Ges Basel 84:101–152
Ishihara T, Tanaka K (2009) Geological structure and groundwater geochemistry around mud volcanoes in the Kamou area, Tokamachi City, Niigata Prefecture. J Geogr 118:350–372. (in Japanese with English abstract)
Kopf JA (2002) Significance of mud volcanism. Rev Geophys 40:1–51
Kusumoto S, Sudo K, Kawabata M, Uda T, Fukuda Y (2015) Vertical movements of the Murono mud volcano in Japan caused by the Naganoken Kamishiro Fault Earthquake in 2014. Earth Planet Space 67:53. https://doi.org/10.1186/s40623-015-0223-1
Lerche I, Bagirov E (1999) Impact of natural hazard on oil and gas extraction in the south Caspian Basin. Kluwer Academic/Plenum Publishers, New York, p 353
Milkov AV (2000) Worldwide distribution of submarine mud volcanoes and associated gas hydrates. Mar Geol 167:29–42
Onishi K, Sanada Y, Yokota T, Tokunaga T, Mogi K, Safani J, O'Neil A (2009) Investigation of subsurface s-wave velocity structures beneath a mud volcano in the Matsudai-Murono district by surface wave method. J Geogr 118:390–407. (in Japanese with English abstract)
Shinya T, Tanaka K (2005) Mud volcanoes in Matsudai, Tokamachi City, Niigata Prefecture. Nat Disaster Sci 24:48–58. (in Japanese with English abstract)
Shinya T, Tanaka K (2009) Origin of materials erupting from mud volcanoes in Tokamachi City, Niigata Prefecture, Central Japan. J Geogr 118:340–349. (in Japanese with English abstract)
Suzuki K, Tokuyasu S, Tanaka K (2009) Underground structure of mud volcanoes in Tokamachi City, Niigata Prefecture determined by electromagnetic exploration, and geophysical and geological surveys. J Geogr 118:373–389. (in Japanese with English abstract)
Takeuchi K, Yosikawa T, Kamai T. (2000) Geology of the Matsunoyama Onsenarea. Japan Geological Survey, Tsukuba, p 76. (in Japanese with English abstract)
Tanaka K (2009) Significance of mud volcanoes and related geological phenomena with special reference to applied geoscience. J Geogr 118:578–586. (in Japanese with English abstract)
Tanaka K, Ishihara T (2009) Mud volcanism near the Nabetachiyama Tunnel and the formation mechanism of swelling rock mass. J Geogr 118:499–510. (in Japanese with English abstract)
Waseda S, Iwano H, Takeda N (2002) Geochemical study on origin and maturity of natural gases. J Jpn Assoc Petrol Technol 67:3–15. (in Japanese with English abstract)
Watanabe M, Tanaka K (2011) Distribution of groundwater containing high concentration of Na−Cl ions in the Matsunoyama area Niigata Prefecture. In: Proceedings of the 2011 meeting of the Japan Society of Engineering Geology, Kyoto, pp 87–88. (in Japanese)
Watanabe N, Sato H, Furuya G (2009) Large scale landslide and overpressured system in the Niigata area, central Japan. J Geogr 118:543–563. (in Japanese with English abstract)
Yamagata T, Ogawa Y (1989) Role of mud diapirism for the formation of chaeotic rocks. J Geol Soc Jpn 95:297–310. (in Japanese with English abstract)
You CF, Gieskes JM, Lee T, Yui TF, Chen HW (2004) Geochemistry of mud volcano fluids in the Taiwan accretionary prism. Appl Geochem 19:695–707

Chapter 15
Land-Use Change and Flood Risk Transformation in Urban Area

Shigeko Haruyama

Abstract Regarding to geomorphologic view, the fluvial plain is composed of vulnerable zones against flooding because of different fluvial process and fluvial landforms are categorized for vulnerability risk mapping on the geomorphologic land classification map. Under the land-use change in urban area, transformation of each landform with cutting and filling connotes to be shifted to be a more vulnerable stage in the fluvial plain. In hand with land-use change, social structure for disaster prevention and mitigation has been transfigured to be individualism with nonintervention in local community and indifference in others. The traditional citizen's participation for mitigation and lifesaving has been bereaved in local community under the natural disaster, however, flood experience and remembering the past evacuation from flooding are precious teaching to suspect or imagine to be lifesaving method and to prepare harmonize community opinions for regional risk communication and risk reduction direction in the local community. It is important for consideration of two dimensions between human dimension and physical dimension to future flood mitigation and prevention.

Keywords Land-use change · Flood risk · Urbanization · The Shonai River Basin

15.1 Introduction

Asia has global issues for flood mitigation and prevention and there are several factors inducing huge disaster damages, such as geomorphologic condition, climate condition and primary cities location in Asia. Monsoon Asia has been experienced flooding every year and the central plain of Thailand has been attacked by torrential rainfall and 2011 is one of the memorial floods in Thailand. The important lesson is geomorphologic factors for damage and land-use change gave the second factors for damage in this severe flooding. The worldwide supply chains were impacted by local Thailand flood because of the industrial production located on global position

S. Haruyama (✉)
Mie University, 1577 Kurimachiyacho, Tsu, Mie, Japan
e-mail: haruyama@bio.mie-u.ac.jp

Y. Himiyama et al. (eds.), *Human Geoscience*, Advances in Geological Science,
https://doi.org/10.1007/978-981-32-9224-6_15

in recently. These mean both human dimension and natural dimension related to damage (Haruyama 1993; Haruyama and Simking 2012).

Regarding to Japan, the Ise Bay typhoon that is memorial flood hit the Nobi Plain in 1959, and one of the rare positive results of the disastrous flooding that followed was an unprecedented opportunity to understand the landform combination of coastal plain and flooding features in Japan (Oya 1956; Nakano and Oya 1960). This fluvial plain was attacked again by a huge typhoon and torrential rainfall in September 1979, and in 2000 the city of Nagoya and the surrounding area were once more inundated by torrential rainfall. The flooding in 2000 is memorable for its urban nature, with large numbers of displaced persons returning and flood-bound traffic in urban areas. The Nobi Plain has suffered several periods of tremendous flood damage under the urban sprawl of Nagoya and rapid land-use change. In 2011, heavy rainfall associated with yet another typhoon hit the area around Nagoya along the urbanized Shonai River, which flows out onto the Nobi Plain. The area also faces new, emerging flood risks of the low-probability high-consequence (LPHC) type caused by dike failure (Haruyama 2002; Yoshida 2013). Because Japan is located in the eastern part of monsoon Asia, which receives torrential rain in the typhoon seasons, the *baiu* seasons(the East Asian rainy season), and the fall rainfall season, it has a long history of experience of flooding and flood-control engineering works, including river management—both structural and non-structural measures—while its social fabric has changed with economic growth. In the last five decades, flood damage in Japan has been drastically reduced, and this has been accompanied by a decrease in associated loss of life. However, flood risk encompasses complex phenomena associated with both human and physical dimensions (Haruyama eds 2013).

Community-based disaster-fighting activities need to be strengthened so as to save lives during natural disasters and address the new disaster risks that have emerged in recent years. In general, monsoon Asia suffers greatly from disastrous floods because of its high rainfall, its low-relief coastal and fluvial plains, and its high population densities (Haruyama and Mizuno 2007; Haruyama and Tujimura 2009; Haruyama and Simking 2012; Haruyama and Hayashi 2013).

The amount of flood damage fluctuates annually, and even though the total rate of economic damage has diminished, there is still concern about the total areas inundated and the total flood damage under LPHC flooding during natural disasters. For this reason, the construction of hard structural measures and maintenance of management infrastructure along rivers continue. Cause-specific economic losses from flooding have continued for the last two decades, and damage from inland flooding occurs every year. Huge amounts of damage from flooding caused by dike failure occur intermittently during torrential rainfalls. Japan has the potential to share its flooding and flood-reduction experience with countries with growing economies. Ways of achieving sustainable reductions in flood disaster risk are discussed in light of the positive and negative consequences of the Japanese experience.

Reducing the damage caused by huge-scale flooding requires comprehensive measures, including those that address the human dimension. Such measures include hazard management, ensuring the resilience of local communities to flooding, disaster exposure management, and damage management. Generally speaking, these measures are achieved by suitable disaster science education so that people understand the mechanisms of natural disasters; sustainable and suitable on-site land-use planning to avoid or reduce disaster damage; appropriate watershed management in harmony with nature without increasing the threat of disaster; and visible local planning between inhabitants and government planners (Haruyama 2009). In on-site investigations of the security of human life in the natural environment, natural environmental factors are important for managing watersheds, but the direction of development also affects our convenience of living. We sometimes forget the importance of natural environmental factors, and the possibility of natural disasters occurring can slip our minds.

Fluvial landforms provide evidence of past flooding events and show us the areas likely to be affected by future disasters, as well as the associated risks. The concept of geomorphologic land classification mapping started in 1959 when the memorial year is Ise Bay Typhoon attacked to Nobi plain including the lower Shonai River Basin was published in 1979 by Oya and Koike. This thematic map was revised, with the addition of information on new development areas, in 2004. An early geomorphologic land classification map showing flood-affected areas on the Nobi Plain was produced by Oya in 1956; this thematic map demonstrated that each landform unit had important roles in terms of flooding type, inundation period, and flood direction on the Nobi Plain. After the Ise Bay typhoon, the geomorphologic land classification map-production project was extended in important river basins under government management in Japan for the prioritization of flood-prevention works. Local river works offices have thus prepared geomorphologic land classification maps along the rivers under their management. From a civil engineering perspective, river control works based on flood simulations over long periods, as well as early warning systems based on rainfall distribution analysis using satellite data and planning for the construction of retarding areas, are important flood-prevention activities.

From one point of view of human dimension of flooding, the important factors for damage are the followings; land-use change and urban sprawls, weakening of social bond for association for giving support each other in local community under the emergency period, loss of local community activities and person who performs the role of leader for local management, dispersing of remembrance and experience of real flooding, a small number of children and aging population in the urban area, solitary life in urban area, lack of understanding of physical condition of residence etc. The coastal and fluvial plain is vulnerable for natural disaster and the large number of flood is related with human and physical dimension of urban area (Hara et al. 2007; Haruyama and Murai 2004; Oya and Haruyama 1987; Taresawa and Haruyama 2013; Haruyama 1997, 1996a, b).

15.2 Flooding in Urban Area

15.2.1 Urban Rivers and Landform Reclamation

Local community flood-fighting and evacuation activities under the emergency period should be strengthened and the appropriate land-use planning for avoiding flood damage and mitigation should be discussed in depth by local governments together with local communities and other stakeholders in flood-prevention activities. In regional planning and land-use planning projects, it is best to consider ground conditions in evaluations over the long term. However, the flood-stricken area will be extensive in the case of a disaster that exceeds the occurrence probability, because in such cases land-use-planning projects will have proceeded without the necessary risk analysis of some areas (Haruyama and Mizuno 2007; Haruyama and Tujimura 2009; Yoshida 2013).

The basins of some urban rivers such as the Shonai have a core of urban expansion—in this case, in Nagoya—and have experienced a number of urban floods and their associated urban problems. The population and population density on the flood plain are large in this urban river basin. Nagoya is located to Central Japan and is one of the country's most important cities in terms of economic growth. Urban sprawl carries the stressor of disruption of land-use balance in the absence of flood-control infrastructure in the river basin or the presence of only low levels of flood-avoidance infrastructure. The watershed of the Shonai is 1010 km^2 in area, with a flood plain of 296 km^2, including Nagoya. The river is only 96 km long, but the river basin includes the important Nagoya–Chukyo Sphere Economic Bloc. The flood plain is handled as a combination of urban and rural zones. In geomorphologic features, the lower Shonai River basin includes a delta, natural levees with back swamps, sand dunes, valley plains, and fluvial terraces (Oya 1956; Haruyama and Oya 1987).

In the upper part of the watershed, three mountainous basins separated by gorges play important roles as flood buffer zones. The beds of the main river and its branches carry characteristic deposits, forming a flying river, because large volumes of sediment have been transported from the upper reaches from the mining of ceramic materials, a decrease in forested area since mediaeval times, and erosion of the granitic rocks in the mountains. After the Edo period, river entraining work was conducted along this river for flood protection of Nagoya castle and the castle town; river dikes and drainage water conduits were constructed, and retention areas were set up. This work has continued in the modern period (Haruyama 2002). Landforms of the lower Shonai River basin (Fig. 15.1), such as natural levees, back swamps, alluvial fans, fluvial terraces, valley plains, former river courses, and deltas have been formed through repeated flooding, storm surges, sea-level changes, and artificial development with cutting and filling. The above-mentioned landforms reveal that the flooding and inundation history of the river basin was closely associated with the fluvial process. In the river basin, there are remarkable huge flood Ise Bay

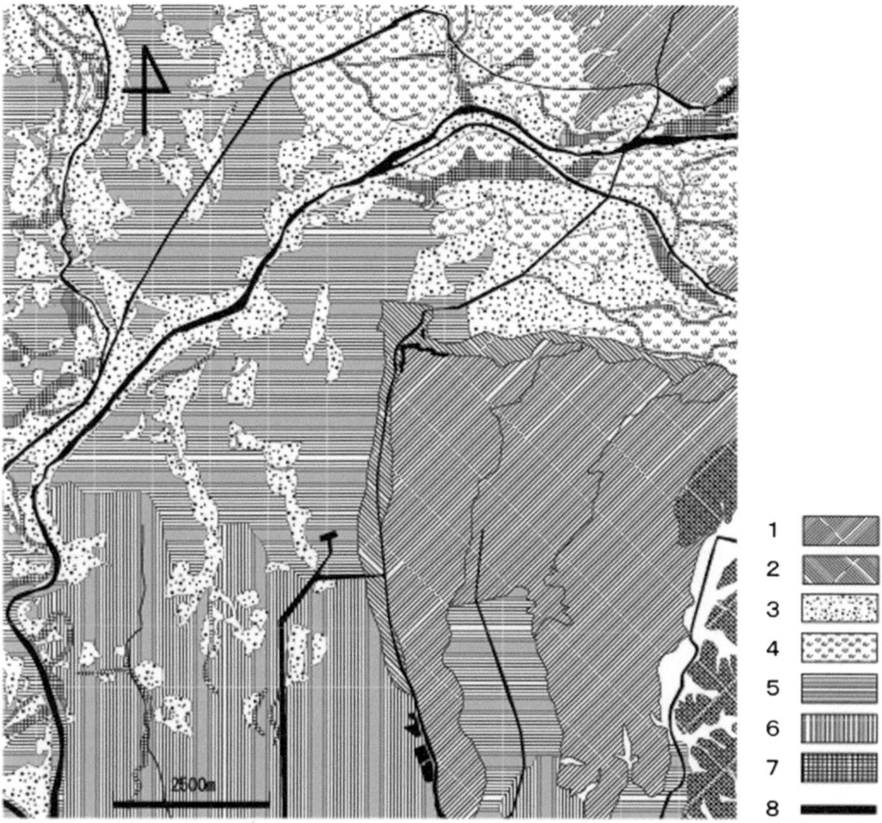

Fig. 15.1 Geomorphologic land classification map of the lower Shonai including greater Nagoya (explanatory note; 1 hill, 2 terrace, 3 natural levee, 4 back swamp, 5 upper delta, 6 lower delta, 7 fill and cutting, 8 water surface)

Typhoon and the flood history (Table 15.1). The urban area located on lower part of flood plain has been attacked by flooding in a second half of 1950.

15.2.2 Land-Use Change and Vulnerability of Flooding

Regarding human dimension of flooding in the study area, mesh map of land-use change is used as a tool for understanding. In light of the recent land-use changes, mesh maps (250 × 250-m mesh) of the lower Shonai River basin for the last 100 years were prepared for discussions of vulnerability to flooding aimed at determining the

Table 15.1 Major flood of the Shonai River Basin

Flood year	Rainfall (upper part of Biwajima observation station) [mm]	Discharge at Biwajima observation station (m^3/s)	Flood damage
1959	103		140,569 houses (Aichi), 6227 houses (Gifu)
1961	137	1530	Flood area 7347 ha (Aichi), 379 ha (Gifu), 39,604 houses (Aichi), 171 houses (Gifu)
1971	117	1600	Flood area 229 ha (Aichi), 344 ha (Gifu), 832 houses (Aichi), 1515 houses (Gifu)
1975	135	1600	Flood area 3091 ha (Aichi), 96 ha (Gifu), 10,315 houses (Aichi), 107 houses (Gifu)
1976	154	1300	Flood area 376 ha (Aichi), 8713 houses (Aichi)
1983	154	1900	Flood area 526 ha (Aichi), 8 ha (Gifu)
1988	106	1600	Flood area 317 ha (Aichi), 62 ha (Gifu), 1896 houses (Aichi), 91 houses (Gifu)
1991	156	2200	Flood area 965 ha (Aichi),4 ha (Gifu), 6440 houses (Aichi), 16 houses (Gifu)
2000	334	3500	Flood area 10,476 ha (Aichi), 10 ha (Gifu), 3441 houses (Aichi), 8 houses (Gifu)

Note Shonai river construction office (1957)

important flood risk zones with physical condition of the study area. The land-use change of the river basin is one of the flooding transformation urban areas.

Mesh maps show land-use mesh maps from 1911 to 2008 of the lower Shonai River Basin, including Nagoya (Figs. 15.2, 15.3, 15.4 and 15.5). Calculated each land-use category in each mesh map, the categories in this area are classified by rural area with rural and urban land uses, taking into account the priority of urbanization. In Fig. 15.3 (1911) the mainland use was paddy field (43%), and there were small areas of upland crops. In this period, the traditional land uses on the landforms were

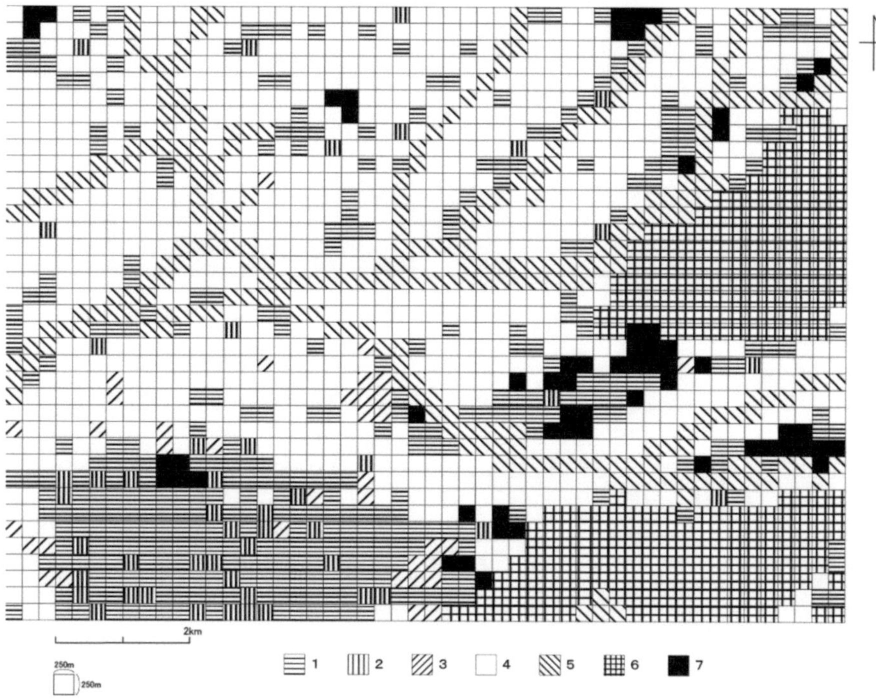

Fig. 15.2 Land-use mesh map in 1911 in Nagoya and surroundings (explanatory note: 1 Residential area, 2 Public space, 3 Industry space, 4 Paddy field, 5 Upland crop field, 6 Forest cover, 7 The other)

nature-oriented: paddy fields were placed in deltas and back swamps, and residential areas were sited on natural levees and terraces. Forest coverage was 14%, and the slopes of hills, the cliffs of terrace, and the valley plains were covered by thick woods and grasslands. In the 18th and 19th centuries, Nagoya castle town and the surrounding urban areas were located on river terraces along the Shonai because these areas were flood-free.

In 1934 (Fig. 15.3), paddy fields still constituted 41% of the land, and the landscape continued to be rural, with farmlands surrounding an urban core. However, the forest cover had decreased to 10% on the hillslopes and terraces. Marked land-use changes began in the early 1950s after the Second World War. Urban expansion and rapid land-use change with urban sprawl began to occur around Nagoya and brought into same condition of land-use change pattern at the boundary zone between urban and rural. In 1911 land-use patterns were traditional and people lived in harmony with their physical surroundings. There were a few areas of artificial landfill or slope cutting to prepare for urbanization. In contrast, by 1973, the area of paddy fields had decreased to 13% and that of upland crops had decreased to 3% because of urban sprawl from Nagoya. The residential district and other densely populated areas had increased to 48% and the industrial area to 7%. Urban land uses, including new

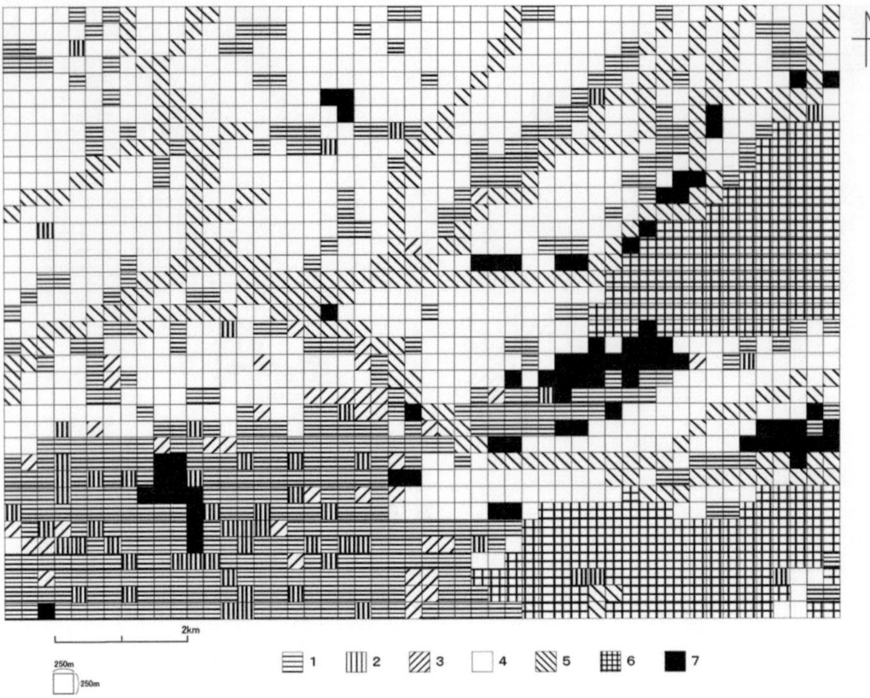

Fig. 15.3 Land-use mesh map in 1934 in Nagoya and surrounding (explanatory note is same as Fig. 15.3.)

schools, public community facilities, and apartments, had advanced onto the lower alluvial plain in the absence of flood mitigation infrastructure during this period. This rapid urban sprawl decreased the size of the buffer zones that had unintentionally acted as natural flood-retarding areas. Areas with no infiltration increased with the land-use changes because of an increase in concrete surface coverage by roads and housing, and the forest cover decreased.

In the face of modern city construction and Nagoya's urban expansion, the percentage of paddy fields continued gradually to decrease, and, conversely, that of residential areas rapidly increased, reaching 78% by 2008. Industrial areas also increased, from 2% in 1911 to 7% in 2008; the industrial zone has been decorated suburban area of Nagoya. Agricultural land use is now very limited: the farmlands are located on a few valley plains with wet conditions or irrigation systems. The recent urban land-use changes have led to the building of underground shopping malls, subway networks, and sophisticated underground utilities near the railway stations. Public facilities, including primary schools, which have been built quite recently, are located near the river on low-lying areas that once followed the river course, on valley plains, and in back swamps and deltas. This building has been conducted with a lack of awareness of the original landforms and landform processes associated with repeated flooding of the alluvial plain.

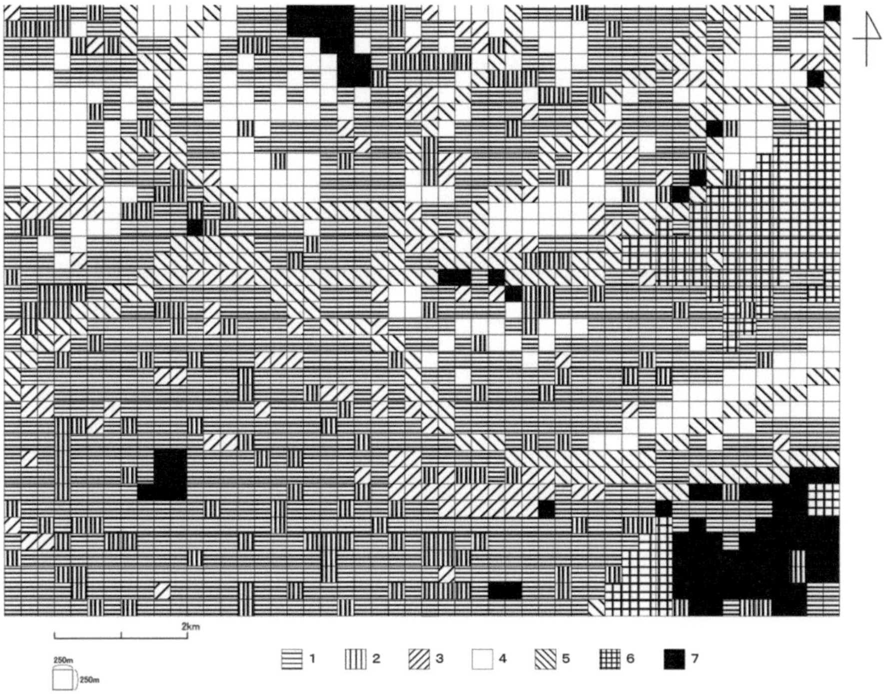

Fig. 15.4 Land-use mesh map in 1973 in Nagoya and surrounding (explanatory note is same as Fig. 15.3)

As a result of the land-use change, the residential zone has expanded onto the flood plain on the former course of the Shonai and into the back swamps behind the natural levees. These two landform units are usually affected by inundation and by dike failure during flooding. Comparison of land elevation maps with land-use changes reveals that the lower zones are the most common targets of residential development projects. Also, highly elevated areas on the fluvial terraces and hillsides are targets of artificial land cutting for new housing supply.

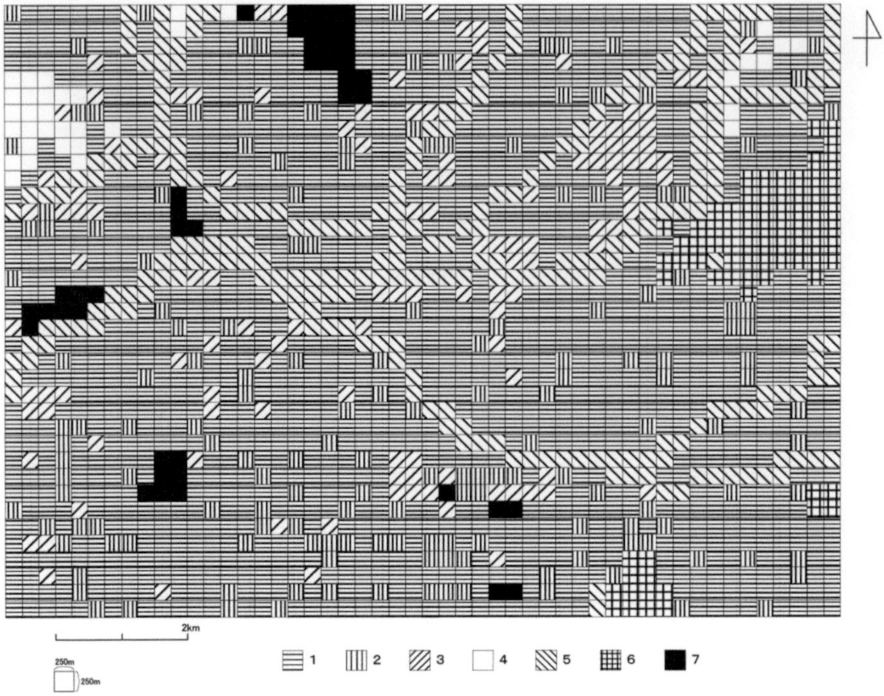

Fig. 15.5 Land-use mesh map in 2008 in Nagoya and surrounding (explanatory note is same as Fig. 15.2.)

15.3 Inundation Risk and Vulnerability on Lowlands

15.3.1 Land Form and Flooding Risk

In terms of ground conditions and elevation, the valley plain includes an alluvial fan formed by the river's branches; a continuous natural levee with a back swamp along the main Shonai River in the east, where the river is narrowed by high fluvial terraces; and a wide delta plain including small natural levees with huge back swamps and former river courses on the western lower plain, which merges with the Kiso River and Nagara River flood plains (Haruyama and Oya 1987). The marked landform contrast between the eastern and western parts is related to crustal movement associated with active faults and the tectonic subsidence of the southern part of the Nobi Plain. Assessment of the flood risk in the lower Shonai River Basin reveals that the delta plain is at risk of flooding with tsunami intrusion and high tides with typhoons; this zone is likely the most vulnerable to residential destruction by typhoons. The region with natural levees and back swamps is spreading widely in the Shonai lower Basin. This area suffers different types of flooding: the natural levees are higher than the back swamps, and when flooding occurs, the natural levees flood for only a short period. The floodwaters run off to fill the back swamps, which are then inundated for a long time. The alluvial fan accounts for only a small part of the study area because not the main river but the smaller branches form the alluvial fan (Haruyama 1996a, b). However, because of this, when severe torrential rainfall occurs, the alluvial fan suffers rapid flooding. The valley plains occur along the main river and its branches; these areas usually suffer relatively moderate flooding, depending on river levels.

The flooding and inundation risks based on geomorphologic land classification mapping emphasize the potential vulnerability of the area to flooding and long-term inundation. However, future flooding risks are likely related to recent land-use changes without appropriate land-use policies. Designated evacuation places are several of these facilities located near river courses and are thus vulnerable during flooding. The degree of vulnerability of these locations depends not only on environmental factors but also on the level of infrastructure for flood mitigation; the amount of river management preparation conducted; the presence of an early warning system; and the level of preparedness, for example, by conducting local community evacuation and resuscitation activities. The other important factor in mitigation is knowledge of natural disaster science and education at several levels of natural disasters. Regardless, because these flood evacuation places are located in areas of potential risk, flood evacuation planning in relation to flood potential needs to be evaluated in future. In relation to recent land-use changes, the flood vulnerability has been changed in terms of landform units.

15.3.2 Local Community in Changing Land Use and Resilience in the Social Structure Change

In hand with land-use change, the social structure for disaster prevention and mitigation has been transfigured to be individualism with nonintervention in local community and indifference in others in the urban area and surroundings. The traditional citizen's participation for flood mitigation and lifesaving with evacuation has been bereaved in the several local communities under the natural disaster, however, the flood experience and remembering the past evacuation from flooding are precious teaching to suspect or imagine to be lifesaving method in the local community and to prepare harmonize community opinions for regional risk communication and risk reduction direction in the local community action plan.

Local fire brigade (member of local fire brigade) is one of the key role of evacuation under emergency period and there are well-informed persons for disaster management. However, disaster information and communication have been decreasing in the local community, and the board of self-government of towns and villages existed has been week in urban area. Citizen's participation and mitigation planning which involves the ordinary citizen participation should be important for life-saving under emergency. Encourage citizens to participate should be in disaster mitigation planning in each region and saving against emergency is important for sustainable social activities. Regarding local community for flood mitigation in future, drilling for evacuation and lifesaving training are instructions of the guidance for experience of caring natural disaster. The independence prevention and mitigation structure in local community has been started after 1995 huge earthquake in this study area but there are different styles of local community activities in normally and non-normally period. Regarding 2011 flood in this area, disaster communication and well-worked disaster management communities were good practice for evacuation and rehabilitation after disaster. Also, the flooding is including the north and south problem in local area and the well knowledge of disaster and appropriated action for disaster are related with the training and learning from the past experiences (Haruyama 1996a, b).

After Yoshida (2013), 2011 flood of the study area is showing one of the lessons as above argument and the following activities as flood evacuation drill, sharing the information of flood experience with risk communication, consideration or caring each other in the community for mutual cooperation under the emergency period, saving for emergency, early prediction and warning, mitigation planning in the local government and continuous support system in local community make use of the results of emergency conquest.

15.4 Conclusion

The flooding types and features in this area are basically associated with the geomorphologic conditions and micro-landform series on the fluvial plain. However, urban sprawl with rapid land-use change has brought other natural disaster-related flood vulnerabilities linked to the human dimension. Zoning of flood levels needs to be discussed more in terms of land-use change and geomorphology. The flood risk in the area has changed and been transformed with expansion of urban area. Regarding loss of buffer zone for flood retarding under the land-use change, the original landform should be considered for local land-use planning for strengthening flooding. Loss of supporting system for flood mitigation in local community and mitigation activities has resulted in the vulnerable social structure. Overcoming weakness in physical and human dimension of flooding should include social resilience. In the future, the local government policy on flood mitigation should include appropriate land-use planning based on the natural environment because of the shortage of flood and inundation buffer zones that has arisen since the start of urban expansion. The community resilience needs to be supported by sophisticated land-use planning in future.

References

Hara Y, Haruyama S, Okhubo S and Takeuchi K (2007) Landfill development causes changes in distribution of flood-prone areas: case study in Laguna Plain, Metro Manila The Philippines. In: Hurumai H et al (eds) Southern Asian water environment. IWA Publishing, UK, pp 11–18

Haruyama S (1993) Geomorphology of the central plain of Thailand and its relationship with recent flood conditions. Geo J 26(12):327–334

Haruyama S (1996a) Zoning of flood inundated areas utilizing the geomorphologic land classification map. In: Proceeding of the United State and Japan of Science and Technology workshop on the technology of disaster prevention against local severe storms, pp 185–187

Haruyama S (1996b) The North-South problem in the Flooding in Asia. J Geogr 41(9):71–72 (J). (J) means Japanese

Haruyama S (1997) Geomorphology and flooding in the lower part of U-Tapao River in Southern Thailand. J Gakujyutu Kenkyu (Waseda Univ) 45:1–9 (J). (J) means Japanese

Haruyama S (2002) Tokai torrential rainfall and geomorphological land classification map of the Shonai River Basin. The study, creation and utilization of a geomorphologic land classification map. Kokonshoin in Tokyo, pp 110–114 (J). (J) means Japanese

Haruyama S (2009) Living with flood in Mekong, Kokon-Shoin in Japan (J). (J) means Japanese

Haruyama S eds (2013) Coastal geomorphology and vulnerability of disaster towards disaster risk reduction. TERRAPUB Publisher in Tokyo

Haruyama S, Hayashi K (2013) Tsunami 2004 disaster impacted by recent land use change in Phuket, Thailand. Coastal geomorphology and vulnerability of disaster towards disaster risk reduction. TERRAPUB Publisher in Tokyo, pp 69–86

Haruyama S, Murai A (2004) Geomorphology and flood plain management for flooding mitigation—the lesson from the Kumozu River. J Suiri Kagaku 278:21–32 (J). (J) means Japanese

Haruyama S, Mizuno S (2007) Consideration of disaster feature and local prevention strength under 2004 Fukui flood. J Nat Disaster 26(3):307–322 (J). (J) means Japanese

Haruyama S, Oya M (1987) Comparative study of fluvial plain of Shonai and Yahagi River basins based on geomorphologic land classification. Geog Rev 59(10):571–588 (J). (J) means Japanese

Haruyama S, Simking L (2012) Geomorphology and flooding of the central plain in Thailand at 2011 flooding. J Nat Disaster 30(4):463–476 (J). (J) means Japanese

Haruyama S, Tujimura A (2009) The local prevention activities of the smallest community unit-lesson from 2004 Toyooka flood. E-J GEO 4(1):1–20 (J). (J) means Japanese

Nakano T, Oya M (1960) Relationship between high tide, flooding and landform in Ise bay Typhoon. Geographical Survey Office (J). (J) means Japanese

Oya M, Koike K (1976) Nobi plain fluvial geomorphologic map, upper Kiso river construction office (Ministry of Construction)

Oya M (1956) Geomorphologic land classification map showing flood affected area of the Kiso River Basin_Nobi plain (J). (J) means Japanese

Oya M, Haruyama S (1987) Flooding and urbanization in the lowlands of Tokyo and Vicinity. J Nat Disaster Sci 9(2):1–12

Shonai river construction office (1957) https://www.cbr.mlit.go.jp/shonai/origawa/familia/saigaishi/index.html (J). (J) means Japanese

Taresawa Y, Haruyama S (2013) Coastal landform in the southern part of Miyagi Prefecture and 2011 Tsunami disaster. Coastal geomorphology and vulnerability of disaster towards disaster risk reduction. TERRAPUB Publisher in Tokyo, pp 87–96

Yoshida A (2013) The evaluation of the flooding in the Shonai River Basin and future measures. The Graduate thesis of Mie University (J). (J) means Japanese

Chapter 16
Landslides and Human Geoscience

Masahiro Chigira

Abstract Landslides are important natural phenomena and have the potential to generate natural disasters. I summarize here the importance of landslides in human society, landslide classification, and mobility. I then describe the features of earthquake- and rain-induced landslides. Some landslides are highly mobile and can form a landslide dam, which can then be breached and flood as far as 100 km downstream. Landslides are thus not only a hazard in mountainous regions. Slow-moving landslides may damage infrastructure, but can be managed unless they fail catastrophically. The prediction of potential sites of catastrophic landslides is becoming feasible, but global climate change may affect the frequency of landslides due to changes in rainfall patterns. Although intense rainfall induces numerous shallow landslides, large amounts of rainfall result in deep-seated catastrophic landslides.

Keywords Landslide · Earthquake · Rainstorm · Prediction

16.1 Landslides and Human Society

The word "landslide" is commonly used as a comprehensive term to include all types of mass movements that occur on hillslopes and valleys. It is a phenomenon that causes severe disruption to human society, as well as being an important natural process that degrades the Earth's surface (Cruden and Varnes 1996). Landslides are one of the most common geohazards and have a significant impact on society. They damage agricultural fields, buildings, and infrastructure. When landslides move rapidly there is also the potential for fatalities. Petley (2012) compiled global data on aseismic landslides and reported that 2620 fatal landslides were recorded between 2004 and 2010, which caused a total of 32,322 fatalities. The majority of loss of life occurs in Asia, particularly along the Himalaya and in China, which is also seismically active areas. Earthquakes can also induce landslides that lead to many fatalities, with the most recent large event being the 2009 Wenchuan Earthquake in China, when >20,000 lives were lost by landslides (Huang and Fan 2013). However, the fatalities

M. Chigira (✉)
Disaster Prevention Research Institute, Kyoto University, Gokasho, Uji 611-0011, Japan
e-mail: chigira@slope.dpri.kyoto-u.ac.jp

Y. Himiyama et al. (eds.), *Human Geoscience*, Advances in Geological Science,
https://doi.org/10.1007/978-981-32-9224-6_16

related to earthquake-induced landslides are not typically distinguished from other fatalities caused by earthquakes, such as building collapse. Even if landslides do not lead to fatalities, important infrastructure such as roads and railways are susceptible to damage. The 2015 Gorkha Earthquake in Nepal induced numerous landslides (Kargel et al. 2015; Tsou et al. 2017) and damaged roads and hydroelectric power stations, as well as residential houses, although the landslides were rather shallow and small. In recognition of the danger posed by earthquake-related landslides, the Japanese government changed its policies regarding earthquake prediction research after the 2011 Tohoku Earthquake to also include earthquake-induced hazards such as landslides. This reflects the fact that research solely on earthquakes cannot mitigate against the associated hazards that contribute to the scale of such disasters.

Landslides vary in speed from very slow (less than a few millimeters per year) to >100 km/h (Cruden and Varnes 1996), and the resultant damage depends on the speed. Very slow landslide movement may not be problematic, because we can adapt to such ground deformation (Fig. 16.1). However, if the landslide movement is rapid, there is no time to mitigate against disaster and the landslide may potentially lead to fatalities.

Landslides such as rockslides and rock avalanches commonly form a landslide dam (Costa and Schuster 1988), which then ponds water upstream and when breached causes downstream flooding and severe damage. The 1933 Diexi Earthquake in Sichuan (China) created a landslide dam that was breached after 45 days and flooded downstream, causing 8000 fatalities (Fig. 16.2; Huang 2008). This was one of the worst landslide-dam breaching events in recent times. The 2008 Wenchuan

Fig. 16.1 Cracks produced by a slow-moving landslide at Kundasang in Borneo, Malaysia

Fig. 16.2 Lake created by a landslide dam following the 1933 Diexi Earthquake, China

Earthquake-induced >60,000 landslides and created 800 landslide dams, none of which failed catastrophically (Wang et al. 2015). The 2016 Kaikoura Earthquake in New Zealand triggered many landslides and generated a large number of landslide dams, which may possibly breach.

Landslides generally occur on hillslopes, but landslides that occur in quick clay (Sect. 16.4.2) can form on nearly flat land with or without a triggering earthquake. Quick clays are widely distributed in North America and northern Europe, where glaciomarine clays have been uplifted and interstitial water has been replaced with fresh water, which reduces the clay strength (Mitchell 1993). Landslides produced by the 1964 Alaskan Earthquake caused widespread devastation over highly populated residential areas in Anchorage (Seed and Wilson 1967). The slip surface was along a quick clay layer with sand, suggesting that clay weakening and sand liquefaction were the main causes of the landslide.

Landslides not only have adverse effects on human society, but can also have positive aspects. Firstly, landslides reduce the slope gradient, thereby providing new agricultural land for use such as rice paddy (Fig. 16.3), as well as land suitable for residential areas. Although the landslide body may still be moving, during long dormant periods the area is suitable for habitation. Landslide deposits on narrow valley bottoms sometimes form a wide flat area, which may also be suitable for human use. Hewitt (2016) reported that catastrophic landslide deposits provided a good foundation for a village along the upper Indus River.

Landslide susceptibility is strongly dependent on rainfall patterns, which may change in the future due to global warming. Given that global warming is expected

Fig. 16.3 Paddy fields on slow-moving or dormant landslides in Niigata, Japan. A rock avalanche at upper right was induced by the 2004 mid-Niigata Prefecture Earthquake

to result in a higher frequency and magnitude of heavy precipitation (Fowler and Hennessy 1995), this may increase the likelihood of rainfall-induced landslides. Future landslide behavior has been discussed in relation to probable climate change (Comegna et al. 2013). The amount of rainfall needed to trigger a landslide is site-specific. For example, only a few tens of millimeters of rainfall can induce landslides in Italy (Peruccacci et al. 2012), but rainfall events totaling >400 mm or rainfall events with an hourly rainfall of >50 mm are generally necessary to produce landslides in Japan. If rainfall patterns change with global climate change, then the distribution of slopes susceptible to landslides will also change. Less frequent landslides due to lower amounts of rainfall in a given area may suggest that the area contains a large amount of unstable material on slopes, which could then be mobilized by future increase in rainfall amounts.

16.2 Landslide Classification

The most commonly used classification scheme for landslides is a series of papers by Varnes (Varnes 1978; Cruden and Varnes 1996) in which the classification is based on a combination of material and movement type. The materials considered are rock, debris, and earth. The types of movement are falls, topples, slides, lateral spreads, flows, and complex types. Landslides in a comprehensive sense were once referred to as slope movements (Varnes 1978) because debris flows are not "slides" but "flows". However, the term "landslides" was a simple one-word and so widely recognized that it came back into usage in Cruden and Varnes (1996). Hungr et al. (2014) published an update to the classification by Cruden and Varnes (1996) (Table 16.1), proposing simple terms such as rock avalanches rather than combining two terms such as rockslide–debris avalanches. As such, the most important mode of movement can be used for classification, even though the actual phenomenon is a combination of several movement modes. Hungr et al. (2014) also proposed to expand the material classification from rock, debris, and earth to rock, boulder, debris, and silt, etc., which would properly take into account the geotechnical properties of the materials. Another major advance on Cruden and Varnes (1996) was that the term "rock avalanche" or "debris avalanche" was included in the classification scheme, taking account

Table 16.1 Classification of landslides (Hungr et al. 2014)

Type of movement	Rock	Soil
Fall	***Rock/ice* fall**	***Boulder/debris/silt* fall**
Topple	***Rock block topple***	***Gravel/sand/silt topple***
	Rock flexural topple	
Slide	Rock rotational slide	*Clay/silt* rotational slide
	Rock planar slide	*Clay/silt* planar slide
	Rock wedge slide	***Gravel/sand/debris* slide**
	Rock compound slide	*Clay/silt* compound slide
	Rock irregular slide	
Spread	Rock slope spread	***Sand/silt* liquefaction spread**
		Sensitive clay spread
Flow	***Rock/ice* avalanche**	*Sand/silt/debris* dry flow
		***Sand/silt/debris* flowslide**
		Sensitive clay flowslide
		Debris flow
		Mudflow
		Debris flood
		Debris avalanche
		Earthflow
		Peatflow
Slope deformation	Mountain slope deformation	Soil slope deformation
	Rock slope deformation	Soil creep
		Solifluction

Movement types that usually reach extremely rapid velocities as defined by Cruden and Varnes (1996) are shown in bold style. The words in italics are placeholders (use only one)

of landslides involving extremely rapid flows of fragmented rock or debris. Such avalanches are quite common and correspond to *Houkai* in Japanese or *Ponta* in Chinese. Other landslide classification schemes in Europe and America have been proposed by Heim (1932), Sharpe (1960), Nemcok et al. (1972), and Hutchinson (1988).

Landslide classification is closely related to human society and is country-dependent. Compared with North American and European schemes, the Japanese classification of landslides places greater importance on velocity as a factor in determining the risk to those in the path of the landslide. Classification of landslides in Japan, where many landslide disasters have occurred, began with Wakimizu (1919) who classified slope movements by their behavior. Sudden, paroxysmal, and rapid movements were classified as landslides proper, whereas slow, continuous movements were classified as land creep. Wakimizu (1919) then subdivided landslides based on the materials, rocks, and earth involved. Presently, a classification that distinguishes landslides, slope failures (*Houkai*), debris flows, and creep is most commonly used in Japan. Using this classification system, "landslides" are typically defined as slow-moving slides of rock or debris with little displacement, comprising a master sliding surface where most of the sliding material remains in its source area. Dramis and Sorriso-Valvo (1994) defined deep-seated gravitational slope deformation (DGSD) as a process of large-scale mass movement with or without a master sliding surface and with little displacement. The process of DGSD includes most landslides (*Jisuberi*) in the Japanese classification system. In contrast, "slope failure" describes rapid failure events during which most of the material moves out of its source area. In this sense, slope failures may be compared to the "disintegrating landslides" of Keefer (1984) or "debris or rock avalanches" of Hungr et al. (2014). Oyagi (2004) further proposed that *Jisuberi* can be used as an all-encompassing term equivalent to "landslide" in English.

16.3 Landslide Mobility

Rock or debris avalanches are highly mobile with long runout distances, and so frequently result in major disasters. Their high mobility may result from several mechanisms, including the air cushion hypothesis (Shreve 1966), undrained loading (Sassa and Lee 1993), acoustic fluidization (Melosh 1979), and rock fragmentation energy (Davies and McSaveney 2009). Landslide mobility is generally expressed by the equivalent coefficient of friction (H/L) or apparent friction angles (arctan H/L), in which H and L are the elevation difference and horizontal distance, respectively, between the crown and toe of a landslide (Heim 1932). The equivalent coefficients of friction are known to decrease with increasing landslide volume (Fig. 16.4) (Scheidegger 1973; Hsu 1975), but debris avalanches of pyroclastic fall deposits induced by earthquakes are exceptionally mobile despite their small volumes (Fig. 16.4). Landslide velocities vary widely, from less than a few millimeters a year to >100 km/h

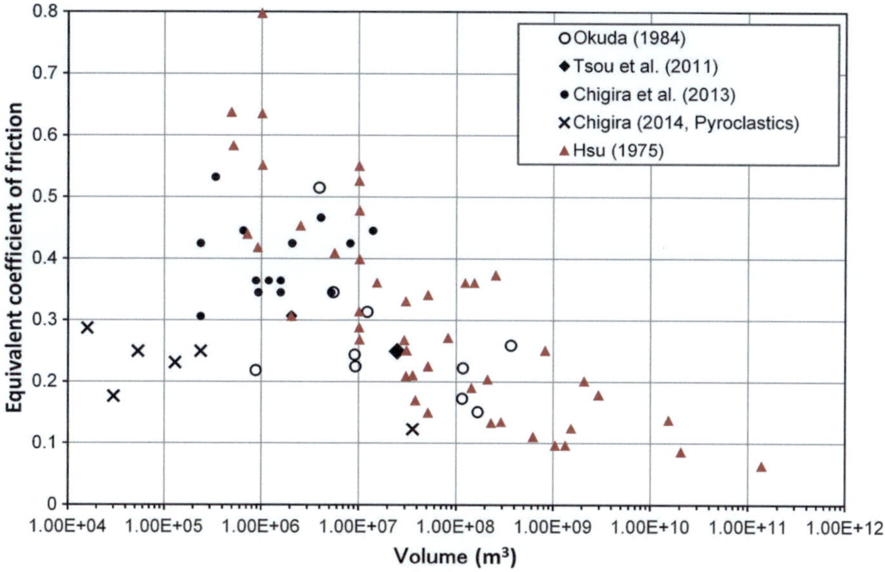

Fig. 16.4 Relationship between the equivalent coefficient of friction and landslide volume. Landslides of pyroclastic fall deposits plot far below the trend defined by other landslides

(Cruden and Varnes 1996). Debris flow velocities vary along its flow path and are dependent on gully gradients, but are mostly <8 m/s (29 km/h) (Takahashi 1991).

16.4 Earthquake-Induced Landslides

Earthquake-induced landslides were reviewed by Keefer (2002), and subsequently many other earthquakes and landslides have occurred yielding more data. Recent earthquakes, such as the 2016 Kumamoto Earthquake and 2011 Tohoku Earthquake in Japan, the 2009 Padang Earthquake in Indonesia, the 2008 Wenchuan Earthquake in China, the 2008 Iwate–Miyagi inland earthquake in Japan, the 2005 northern Pakistan Earthquake, and the 2004 mid-Niigata Prefecture Earthquake in Japan induced numerous landslides. These events revealed geomorphological and geological features of landslides and far more data than were previously available. Many of these landslides were shallow rock falls and debris avalanches, but some were very large rock avalanches with long runout distances.

The Ninety-nine peaks landslides induced by the 1999 Chi-Chi Earthquake in Taiwan were the most spectacular example of how mountain peaks are strongly shaken due to the amplification of seismic tremor (Wang et al. 2003b). These peaks were all green and vegetated prior to the earthquake, but this cover was removed by the earthquake along with the Quaternary surface layers of unconsolidated conglomerate (Fig. 16.5). Landslide debris continued to be supplied to the valley floor for many

Fig. 16.5 Landslides induced by the 1999 Chi-Chi Earthquake, Taiwan. Photograph taken by H. Suwa

years after the Chi-Chi Earthquake. The 2008 Wenchuan Earthquake also induced many shallow landslides, and the debris transported to the valley floors was mobilized by subsequent rainfall events, producing debris flows (Tang et al. 2011). The amplification of seismic waves on mountain slopes has been recently monitored and validated by Hata et al. (2015).

16.4.1 Rock Avalanches Preceded by Gravitational Slope Deformation

A rock avalanche is an extremely rapid, massive, flow-like motion of fragmented rock from a large rock slide or fall (Hungr et al. 2014). Typical rock avalanches triggered by earthquakes include the Chiu-fen-erh-shan and Tsaoling landslides associated with the 1999 Chi-Chi Earthquake (Chigira et al. 2003; Wang et al. 2003a) and the Daguanbao landslide triggered by the 2008 Wenchuan Earthquake (Fig. 16.6; Chigira et al. 2010). Analyses of these landslides suggest that the sites of large rock avalanches can be predicted from topographic features and geological structures. The Chiu-fen-erh-shan landslide was on a dip slope of stratified sandstone and shale, which had been buckled prior to the landslide event and formed a linear depression with a steep step. The Daguanbao landslide occurred on a dip slope of carbonate rocks, which had also been buckled before the landslide. Buckling is a type of DGSD (Chigira 1992). The Tsaoling landslide occurred on an undercut dip slope of shale and sandstone, which had been displaced slightly to leave a depression upslope prior to the landslide.

During the Chiu-fen-erh-shan and Tsaoling landslides, houses on the landslide bodies moved >1 km and landed on top of the depositional area, suggesting that the

Fig. 16.6 The Daguanbao landslide, which was the largest landslide induced by the 2008 Wenchuan Earthquake (AVNIR-2 image by JAXA). The red solid line is the trace of the surface fault rupture (Chigira et al. 2012a, b)

movement was not completely chaotic. In the case of the Chiu-fen-erh-shan landslide, three people in an affected house were only slightly injured.

16.4.2 Landslides Preceded by Rock Weathering

Debris avalanches with high mobility in pyroclastic fall deposits have been induced during many previous earthquakes (Chigira 2014; Chigira and Suzuki 2016). More than six large earthquakes in Japan have produced landslides of this type, and the 2001 El Salvador Earthquake and 2009 Padang Earthquake also triggered numerous examples of such landslides (Fig. 16.7; Table 16.2). The most recent such landslide was associated with the 2016 Kumamoto Earthquake (Fig. 16.8).

Stratigraphic successions have been studied for the landslides listed in Table 16.2, and the stratigraphic horizon of the sliding surface has been identified for each case. All of the layers accommodating a sliding surface were confirmed to contain a clay mineral (halloysite), apart from the landslides induced by the 2001 El Salvador Earthquake for which the mineralogy has not been reported (Chigira and Suzuki

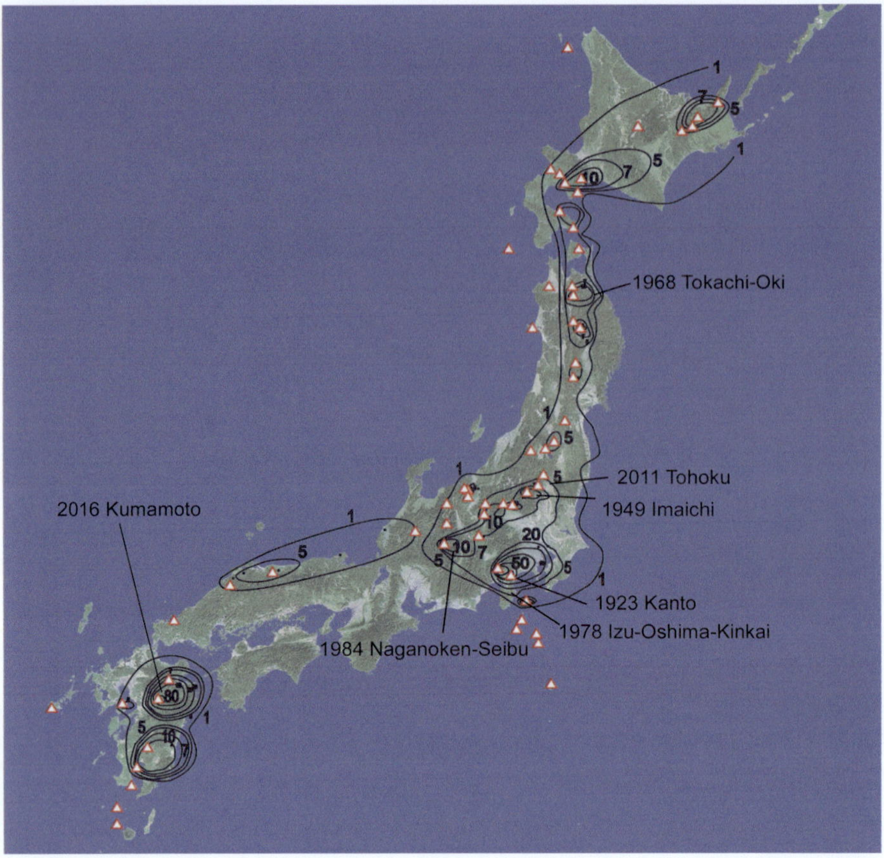

Fig. 16.7 Isopach map (m) of pyroclastic fall deposits younger than 80 ka and the locations of landslides induced by major earthquakes. The 2016 Kumamoto Earthquake has been added to the data from Chigira and Suzuki (2016). Triangles are active volcanoes

2016). Sliding surfaces formed within several types of layers. The most common sliding layer is of volcanic soil (i.e., paleosol) just beneath pumice or scoria beds. The second type of layer that accommodates sliding is a weathered pumice or scoria bed beneath an andesitic lava flow or volcanic soil. The third type is a mixture of pumice clasts and underlying volcanic soil (paleosol), which is generally similar to the first type. The sliding surfaces described above suggest that potential sliding surfaces can be identified by studying volcanic stratigraphy and weathering mechanisms. We are now investigating the landslides induced by the 2016 Kumamoto earthquake, but we found that tens of debris avalanches of tephra occurred with their sliding surfaces in black volcanic soil with halloysite (Sato et al. 2017).

Another rock type susceptible to earthquake shaking is carbonate rock, which is easily dissolved by groundwater. The 2008 Wenchuan Earthquake-induced numerous landslides in carbonate rocks (Huang 2011; Huang and Fan 2013). Many of

Table 16.2 A list of catastrophic landslides of pyroclastic fall deposits. Chigira and Suzuki (2016)

Earthquake		1923 Kanto		1949 Imaichi	1968 Tokachi-Oki	1978 Izu-Oshima-Kinkai	1984 Naganoken-Seibu	2011 Tohoku	2001 El Salvador	2009 Padang
Date		1 Sept.		26 Dec.	16 May	14 Jan.	14 Sept.	11 March	13 Jan.	30 Sept.
Magnitude		Mjma 7.9		Mjma 6.4	Mjma 7.9 (Mw8.2)	Mjma 7.0	Mjma 6.8	Mw 9.0	Mw 7.7	Mw 7.5
Seismic intensity at landslide sites (JMA)		6		5–6	5	5–6	6	6– to 6+	MM 6, 7 4–5-(JMA)	MM 8 (USGS) 5 + (JMA)
Antecedent rain (mm)	Rain gage			Utsunomiya	Hachinohe	Inatori	Ontakesan	Shirakawa	–	–
	10 days			22.5	181	12	183	12.5	No data (Nov.–Apr.: dry season)[h]	Unknown (occurred during a rainstorm)
	30 days			80.8	292	172	555	83.5		
	60 days			255	307	334	839	93.5		
Number of collapsing landslide		2 in Nebukawa	1 in Hadano (Shinseiko)	88[a]	152[b]	7[d] (controlled by the material distribution)	5[j]	<10[e]	>1000[g]	160[i]
Material of sliding surface		Weathered pumice[n] Halloysite	Weathered pumice[m] Halloysite	Weathered pumice[a], Lapilli[m] Halloysite[a)]	Paleosol (Sandy ash) Halloysite[c]	Paleosol Halloysite[d]	Weathered pumice and scoria Halloysite[k]	Paleosol Halloysite[e]	Paleosol[f] No report of clay minerals	Mixed layer of paleosol and pumice Halloysite[i]

(continued)

Table 16.2 (continued)

Earthquake	1923 Kanto		1949 Imaichi	1968 Tokachi-Oki	1978 Izu-Oshima-Kinkai	1984 Naganoken-Seibu	2011 Tohoku	2001 El Salvador	2009 Padang
Stratigraphic name of the bed of sliding surface	Komekami lava Group[u]	Tokyo Pumice (60–65 ka)[r]	Kanuma pumice (32 ka), Ogawa lapilli, Imaichi pumice[m]	Towada-Hachinohe tephra (15 ka)[c]	Just below the Hachi-noyama tephra (29 ka)[q]	Senbonmatsu scoria (84–76 ka)[p]	Sr10 (scoria), Takaku pumice (330 ka)[o]	Tobas Color Café deposits	The base of Qhpt (younger than 70–80 ka)[t]
Slid material	Andesite lava, lapilli from Hakone volcano[n]	Tephra from Fuji volcano[m]	Tephra from Nantai volcano	Towada-Hachinohe tephra[b, c]	East Izu monogenic volcanic tephra[d]	Scoria, lava, agglutinate, terrace deposits,	Tephra from Nasu volcano[e]	Pumice etc. of the Tierra Blanca and the Tobas Color Café deposits	Pumice (Qhpt)
Source of the slid materials	Hakone volcano	Hakone and Fuji volcanoes	Nantai volcano	Towada volcano	Higashi-Izu mono-genetic volcanoes	Ontake volcano	Nasu volcano	Unknown	Tandikat volcano[i]
Sliding surface depth (m)	70 m, 30 m	17 m	3–5 m[a]	<3 m[b], –2.5 m[c]	2–6 m[d]	5–200 m (Ontake)[j]	3–9 m[e]	ca.20 m (Las Colinas)[f]	3.5–5.5 m[i]

(continued)

Table 16.2 (continued)

Earthquake	1923 Kanto		1949 Imaichi	1968 Tokachi-Oki	1978 Izu-Oshima-Kinkai	1984 Naganoken-Seibu	2011 Tohoku	2001 El Salvador	2009 Padang
Slope-parallel bedding	Yes	Yes	Yes	Yes	Yes	Yes	Yes	Yes	Yes
Undercut	Yes	Yes	Yes	Unknown	Yes	Yes	Yes	Yes	Mostly yes
Fatality	447[n]	2	8	33	7	29	13	844[g]	600?
Reference	[a]Morimoto (1951), [b]Inoue et al. (1970), [c]Yoshida and Chigira (2012), [d]Chigira (1982), [e]Chigira et al. (2012, b), [f]Crosta et al. (2005), [g]Jibson et al. (2004), [h]Evans and Bent (2004), [i]Nakano et al. (2015), [j]Hirano et al. (1985), [k]Tanaka (1985), [l]Suzuki (1993), [m] Chigira (unpublished data), [n]Kamai (1990), [o]Suzuki (1992), [p]Kobayashi (1987), [q]Hayakawa and Koyama (1992), [r]Kasama and Yamashita (2008), [s]Takeuchi et al. (1998), [t]Tjia and Muhammad (2008), [u]Takahashi (2007)								

Fig. 16.8 Debris avalanche deposits, comprising pyroclastic fall material, induced by the 2016 Kumamoto Earthquake. Photograph by Kokusai Kogyo Company Ltd. and Pasco Corporation. The largest debris avalanche deposit had a sliding surface dipping ca. 10° along the base of a weathered pumice bed

these landslides occurred on dip slopes of well-stratified carbonate rocks with sliding surfaces formed along bedding planes. The sliding surfaces commonly have rough surfaces with dimple-like depressions and fractured protrusions. These depressions were formed by the dissolution of carbonate (Chigira et al. 2010), whereby groundwater flow along a bedding plane dissolves carbonate rocks and decreases the area of contact between rock masses above and below the surface. These contacts along the bedding plane are eventually broken by seismic shaking. In contrast, the enlargement of pore spaces drains groundwater, meaning that a build-up of pore water pressure is unlikely during rainfall events.

16.4.3 Pore Pressure Build-up

A distinctive type of landslide was produced by the 1966 Matsushiro Earthquake, which caused groundwater to gush out of seismogenic faults, with the rise in water pressure inducing rotational landslides (Morimoto et al. 1967). Faults may locally cause groundwater pressure to build up by the accumulation of tectonic strain (Sibson 1996), and the affected areas are in some cases are >100 km from the source fault (Toda et al. 1995).

16.4.4 Reactivation of Rockslides that Collide with an Opposite Slope and Are then Undercut

If a landslide body that once ran out across a river from one side of the valley and collided with the opposing slope is undercut by erosion, then the support of the upslope material is removed, meaning the landslide body becomes unstable. Such landslide bodies can then be reactivated by later earthquakes. This type of landslide occurred in many locations during the 2004 mid-Niigata Prefecture Earthquake in Japan (Chigira and Yagi 2006). The Hattian landslide triggered by the 2005 northern Pakistan Earthquake also had similar characteristics (Chigira et al. 2007; Dunning et al. 2007; Schneider 2008).

16.5 Rain-Induced Landslides

Both shallow- and deep-seated landslides are induced by rainfall. Rain-induced shallow landslides comprise near-surface weathering products or debris and usually form very rapidly moving debris avalanches, which then flow into downslope channels and transform into debris flows. Typical examples of these landslides were the debris avalanches induced by the 2014 Hiroshima rainstorm event that led to 74 fatalities (Matsushi et al. 2015; Fig. 16.9). Rain-induced deep-seated landslides are typically rock avalanches that have been preceded by gravitational slope deformation. An example of a rain-induced rock avalanche is those induced by the 2011 Typhoon Talas in Japan (Chigira et al. 2013).

Rainstorm events that have caused landslides in Japan are listed in Table 16.3, which shows that shallow- and deep-seated landslides have generally occurred in different rainfall events. One rainfall event can induce numerous shallow landslides without any larger deep-seated landslides, whereas the opposite may also be true. This difference can be attributed to the weathering characteristics of near-surface rocks and prior gravitational slope deformation.

16.5.1 Shallow Landslides

Shallow landslides were not previously considered to be affected by bedrock type, but Table 16.3 suggests that at least in Japan they do have a preferential bedrock type. Most rock types involved in shallow landslides are granite (Fig. 16.10), along with volcanic materials such as volcanic ash, volcanic soil, and pyroclastic materials. These landslides have a distinctive contrast between a loose surface layer and denser and less permeable rocks below it (Chigira 2001; Chigira et al. 2002; Miyabuchi et al. 2004). The surface layer becomes immediately saturated with intense rainfall and pore pressure increases to induce a landslide. Another example of a rain-induced

Fig. 16.9 Debris flows that originated upslope from shallow landslides in Hiroshima. These were induced by intense rain of >150 mm in 3 h on August 20th, 2014

Table 16.3 A list of rainstorms that induced deep-seated catastrophic landslides or many shallow landslides in Japan (Chigira 2014)

Year	Date	Trigger (T: typhoon)	Place (prefecture)	Geology	Deep-seated landslide	Many shallow landslides
1998	26 to 31 August	Rain (Front)	Fukushima	Vapor-phase crystallized ignimbrite	–	✓
1999	29 June	Rain (Baiu front)	Hiroshima	Granite	–	✓
2000	28-29 July	Rain (Front)	Rumoi (Hokkaido)	Soft sedimentary rocks	–	✓
	11-12 Sept.	Rain (Front + T14)	Tokai (Aichi)	Granite	–	✓
2003	20 July	Rain (Baiu Front)	Minamata (Kumamoto) Hishikari (Kagoshima)	Andesite lava and pyroclastics	✓	✓
	9-10 Aug.	Rain (T10)	Hidaka (Hokkaido)	Sandstone and conglomerate	–	✓
	Ditto	Ditto	Ditto	Melange	–	✓
2004	13 July	Rain (Baiu front)	Nagaoka (Niigata)	Weak mudstone	–	✓
	Ditto	Ditto	Fukui	Volcanic rocks	–	✓
	28-29 Sept.	Rain (T21)	Miyagawa (Mie)	Accretional complex (Hard sedimentary rocks)	✓	–
	1 Aug.	Rain (T10)	Kisawa (Tokushima)	Accretional complex (Greeenstone and hard sedimentary rocks)	✓	–
	29 Sept.	Rain (T21)	Ehime-Kagawa	Heavily weathered hard sandstone and mudstone	–	✓

(continued)

Table 16.3 (continued)

Year	Date	Trigger (T: typhoon)	Place (prefecture)	Geology	Deep-seated landslide	Many shallow landslides
	29 Sept.	Rain (T21)	Saijo (Ehime)	Schist	✓	✓
2005	6 Sept.	Rain (T14 (Nabi))	Mimikawa (Miyazaki)	Accretional complex (Hard sedimentary rocks)	✓	–
2006	19 July	Rain (Baiu front)	Okaya (Nagano)	Volcanic soil	–	✓
2009	21 July	Rain (Baiu front)	Hofu (Yamaguchi)	Granite	–	✓
2010	16 July	Rain (Baiu Front)	Shobara (Hiroshima)	Soil	–	✓
2011	4 September	Rain (T12 (Talas))	Kii Mountains (Nara, Wakayama)	Accretional complex (Hard sedimentary rocks)	✓	–
2012	12 July	Rain (Baiu Front)	Aso (Kumamoto)	Volcanic ash	–	✓
2013	16 Oct.	Rain (T26 (Wipha))	Izu-Oshima (Tokyo)	Volcanic ash	–	✓
2014	20 Aug.	Rain	Hiroshima	Granite, Hornfels	–	✓

shallow landslide is the failure of unwelded ignimbrite caused by the weight increase of a surficial weathered layer due to the capillary barrier effect at its base and suction depletion (Chigira and Yokoyama 2005). A further example of a rain-induced shallow landslide is those due to piping, which result from preferential water flow and underground erosion (Sidle et al. 1995). These landslides are not easy to predict, given that the sizes and locations of the pipes are difficult to estimate.

16.5.2 Rock Avalanches Preceded by Gravitational Slope Deformation

Rock avalanches have been induced by rainfall in many locations worldwide, including the 2009 Shiaolin event in Taiwan (Tsou et al. 2011b), the 2006 Ginsaugon event in the Philippines (Evans et al. 2007), and the 2011 Kii Mountains event in Japan (Chigira et al. 2013). Such avalanches occur suddenly, move extremely rapidly, have long runout distances, and commonly form landslide dams. These catastrophic fail-

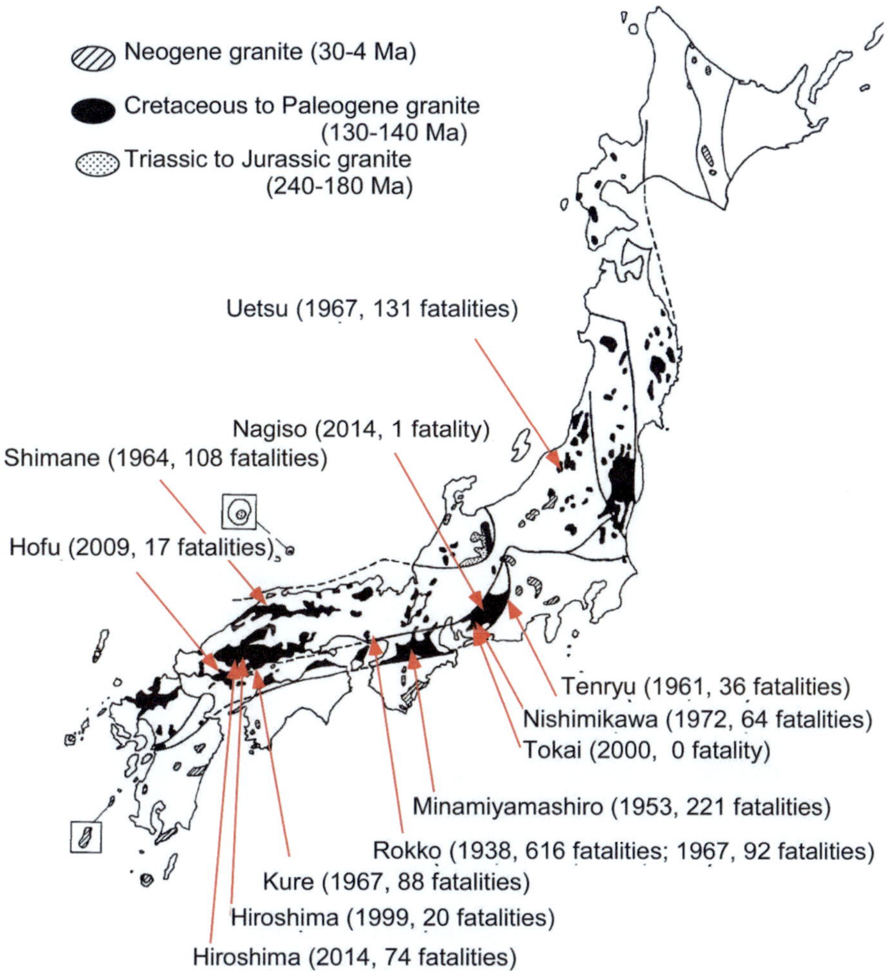

Fig. 16.10 Landslide disasters caused by heavy rain in granitic terrains of Japan (modified from Chigira 2001)

ures are preceded by DGSD (Crosta et al. 2006; Geertsema et al. 2006; Chigira 2009; Chigira et al. 2013). Chigira et al. (2013) analyzed high-resolution DEMs obtained by LiDAR before and after rock avalanches, and found that the landslides were preceded by gravitational slope deformation with small scarps upslope. This suggests that the slopes had already been displaced before the catastrophic failure. Arai and Chigira (2015) found that these small scarps are due to displacement along large thrust faults dipping downslope. Another type of gravitational deformation leading to catastrophic failure by rainfall is buckling on a dip slope (Tsou et al. 2015). Rain-induced deep-seated rock avalanches require large amounts of rainfall rather than intense rainfall (Tsou et al. 2011a; Chigira et al. 2013).

16.6 Landslide Mapping and Predicting Sites of Catastrophic Failure

In mountainous countries, infrastructure such as railways and roads commonly passes along narrow valleys and is therefore susceptible to landslides. Even flat areas are usually located downstream of a river, meaning that a landslide dam and subsequent breach are hazards to large cities on flat areas. Therefore, site prediction of catastrophic landslides and the detection of landslide dam formation are important for society.

Several approaches have been developed to predict potential sites of landslides, including the use of physical modeling, stochastic modeling, and the indexing of geological and geomorphological features (Guzzetti et al. 1999). Physical modeling, such as slope stability analysis under conditions of heavy rainfall or an earthquake, is possible if information related to hydrological or geotechnical factors and subsurface structures is available (Jibson et al. 1998; Montgomery et al. 2000). However, such data are rarely available in many areas. Stochastic modeling can be effective for investigating relatively shallow landslides, which typically occur repeatedly in areas with particular geological characteristics, such as certain types of weathering profiles. In contrast, the occurrence of deep-seated catastrophic landslides is highly dependent on local and specific geological and/or geomorphological conditions. These features make it possible to predict their potential sites, as described in Sects. 16.4.1 and 16.5.2. Deep-seated and slow-moving or catastrophic landslides can also be predicted using aerial photography, airborne LiDAR data, and geological information. Most slow-moving landslides are reactivated landslides with topographic features created by previous movements.

Landslide mapping programs have been conducted in countries such as Japan, Italy, and the USA (California). The California Geological Survey (CGS) started systematic landslide mapping in the 1970s and this has continued to the present day. The CGS produces four types of landslide maps: landslide-inventory, landslide-hazard, landslide-risk, and landslide-zone maps. Due to the 1998 landslides in Sarno that led to >150 fatalities, Italy started landslide mapping by the Geological Survey of Italy, and their results are provided on a website (http://193.206.192.136/cartanetiffi/default_nosso.asp).

The islands of Japan have been exposed to slope movement hazards for a long time, and in response to this risk, Japan has established four Acts to mitigate landslide disasters (Table 16.4). Three of these Acts have specifically targeted debris flows, slow-moving landslides, failure of slag heaps, and rapid slope failure of steep slopes, and have set out "hard" engineering countermeasures. The fourth Act, established in 2000, was the first to focus on "soft" countermeasures, such as evacuation and early warning systems, as well as land-use regulations. This "soft" approach followed two rain-induced landslide disasters in 1998 (Fukushima) and 1999 (Hiroshima) in which numerous shallow landslides and debris flows affected residential areas and welfare facilities. This Act recognized that a countermeasures approach based solely on hard engineering solutions is not sufficient to mitigate against landslide

Table 16.4 Landslide-related Acts in Japan (Chigira 2014)

Year	Act	Target
1897	Erosion Control Act	Debris flow
1958	Landslide Prevention Act	Slow-moving landslide and failure of slag heaps
1969	Act on Prevention of Disasters Caused by Steep Slope Failure	Rapid slope failure (shallow)
2000	Act on Sediment Disaster Countermeasures for Sediment Disaster Prone Areas	Promotion of soft measures such as evacuation warning systems as well as land usage regulation

hazards, because increasing numbers of people live in landslide-susceptible areas. Today, every prefectural government in Japan is undertaking investigations to identify landslide-susceptible areas so that appropriate evacuation procedures and warning systems can be developed. However, large rock avalanches with long runout distances and debris avalanches occurring on gentle slopes are not included in these investigations. Besides the above governmental activity, since 1982, the National Research Institute of Earth Science and Disaster Prevention in Japan has been publishing landslide maps based on interpretations of aerial photographs (Shimizu 1983). These maps provide basic data for land management and now cover almost all of Japan. Landslide mapping used to be performed by aerial photograph interpretations, which cannot be applied to heavily vegetated areas. Since the 2000s, airborne LiDAR has been applied to mountain slopes and successfully used to identify landslides and related phenomena (Chigira et al. 2004; Guzzetti et al. 2012; Jaboyedoff et al. 2012). Chigira et al. (2013) found that catastrophic rock avalanches induced by Typhoon Talas in 2011 were preceded by small scarps upslope, which were on the order of meters in size and could be detected by LiDAR, even beneath vegetated areas.

Feng (2011) and Yamada et al. (2012) analyzed seismic records when large landslides occurred and found that landslide dam formation can be detected in seismic records, and thus that an earthquake monitoring network can be used to detect rain-induced landslides. However, an earthquake-induced landslide cannot be detected, owing to the seismic shaking associated with the earthquake itself.

16.7 Prediction of Landslides

Prediction of landslides is not only scientifically important but is essential information for human society to assure the safety of populated areas. For rain-induced landslides, alarms can be issued on the basis of rainfall amount and intensity. Pioneering work to evaluate the amounts of rainfall in relation to landslide initiation was performed by Caine (1980) who constructed a diagram of the intensity–duration (ID) rainfall threshold for possible landslide occurrence. Cumulated rainfall–rainfall event duration (ED) thresholds were proposed by Guzzetti et al. (2007), and these are equivalent to the ID rainfall threshold. One of the most sophisticated alarm systems is based on the soil water index in Japan (Okada 2007), which is an index of the amount of water in conceptual underground storage tanks and is calculated hourly (in 1 km^2 grid cells) using Automated Meteorological Data Acquisition System (AMeDAS) radar data. When the calculated water amount exceeds levels of the previous 10 years, the probability of rain-induced landslides is considered to be high and an alarm is issued by the Japan Meteorological Agency. However, the tank model is simplistic and uses just one set of parameters for all areas.

Another method of landslide prediction is based on the monitoring of ground displacement (Saito 1965; Fukuzono 1985). Saito's (1965) model uses an empirical law where the logarithms of fracturing time are expressed as a linear function of logarithms of strain rate during secondary creep. He extended this model to the tertiary creep stage and successfully applied it to forecast landslide failure (Saito 1969). Fukuzono (1985) found that the timing of landslide failure could be predicted using the inverse of displacement velocity, based on landslide flume tests. Both methods have been applied to many landslide case studies, and they can be applied to specific landslide sites that are being monitored.

16.8 Conclusions

Landslides have been reviewed in this chapter from a geoscience and societal perspective. Landslides are important natural phenomena that degrade the Earth's surface and are also closely connected to human society. Therefore, understanding landslide mechanisms and their potential location and timing of formation are not only scientifically important but also significant for society. Given that rain-induced landslides are strongly dependent on rainfall events, climate change may strongly affect landslide occurrence in the future. Potential sites of some landslide types can be predicted but more study is required. Rain-induced shallow landslides can be physically modeled, but this requires an understanding of geotechnical properties and subsurface structures, and such data are not available for many areas. Deep-seated landslides occur in association with specific geological conditions and topographic features, which can be used to identify potential sites for their formation. Pyroclastic fall deposits commonly slide catastrophically during an earthquake event and have a long runout distance. Although such slides have no topographic features, the stratigraphic nature of these slides is becoming better understood. Landslide inventory or susceptibility mapping is now underway in a number of countries.

References

Arai N, Chigira M (2015) Rain-induced rock avalanches with a sliding surface along an out-of-sequence thrust. Paper presented at the 10th IAEG Asian regional conference, Kyoto

Caine N (1980) Rainfall intensity-duration control of shallow landslides and debris flows. Geogr Ann Ser A Phys Geogr 62A:23–27

Chigira M (1982) Dry debris flow of pyroclastic fall deposits triggered by the 1978 Izu-Oshima-Kinkai earthquake: the "collapsing" landslide at Nanamawari, Mitaka-Iriya, southern Izu Peninsula. J Nat Disaster Sci 4:1–32

Chigira M (1992) Long-term gravitational deformation of rocks by mass rock creep. Eng Geol 32:157–184

Chigira M (2001) Micro-sheeting of granite and its relationship with landsliding specifically after the heavy rainstorm in June 1999, Hiroshima Prefecture, Japan. Eng Geol 59:219–231

Chigira M (2009) September 2005 rain-induced catastrophic rockslides on slopes affected by deep-seated gravitational deformations, Kyushu, Southern Japan. Eng Geol 108(1–2):1–15

Chigira M (2014) Geological and geomorphological features of deep-seated catastrophic landslides in tectonically active regions of Asia and implications for hazard mapping. Episodes 37:284–294

Chigira M, Suzuki K (2016) Prediction of earthquake-induced landslides of pyroclastic fall deposits. In: Aversa S et al (eds) Landslides and engineered slopes. Experience, Theory and Practice, vol 1. Associone Geotecnica Italiana, Rome, pp 93–100

Chigira M, Yagi H (2006) Geological and geomorphological characteristics of landslides triggered by the 2004 Mid Niigata prefecture Earthquake in Japan. Eng Geol 82:202–221

Chigira M, Yokoyama O (2005) Weathering profile of non-welded ignimbrite and the water infiltration behavior within it in relation to the generation of shallow landslides. Eng Geol 78:187–207

Chigira M, Nakamoto M, Nakata E (2002) Weathering mechanisms and their effects on the landsliding of ignimbrite subject to vapor-phase crystallization in the Shirakawa pyroclastic flow, Northern Japan. Eng Geol 66:111–125

Chigira M, Wang W-N, Furuya T, Kamai T (2003) Geological causes and geomorphological precursors of theTsaoling landslide triggered by the 1999 Chi-Chi Earthquake, Taiwan. Eng Geol 68:259–273

Chigira M, Duan F, Yagi H, Furuya T (2004) Using an airborne laser scanner for the identification of shallow landslides and susceptibility assessment in an area of ignimbrite overlain by permeable pyroclastics. Landslides 1:203–209

Chigira M, Yagi H, Kausar AB (2007) Landslides induced by the 2005 northern Pakistan Earthquake and long-term gravitational slope deformation. Geophys Res Abstr (Eur Geosci Union) 9:05938

Chigira M, Wu X, Inokuchi T, Wang G (2010) Landslides induced by the 2008 Wenchuan earthquake, Sichuan, China. Geomorphology 118(3–4):225–238. https://doi.org/10.1016/j.geomorph.2010.01.003

Chigira M, Nakasuji A, Fujiwara S, Sakagami M (2012a) Soil slide-avalanches of pyroclastic fall deposits induced by the 2011 off the Pacific Coast of Tohoku Earthquake. J Jpn Soc Eng Geol 52:222–230

Chigira M, Wang G, Wu S (2012b) Landslides induced by the Wenchuan earthquake. In: Clague JJ, Stead D (eds) Landslides types, mechanisms and modeling. Cambridge University Press, Cambridge, pp 383–392

Chigira M, Tsou CY, Matsushi Y, Hiraishi N, Matsuzawa M (2013) Topographic precursors and geological structures of deep-seated catastrophic landslides caused by Typhoon Talas. Geomorphology 201:479–493. https://doi.org/10.1016/j.geomorph.2013.07.020

Comegna L, Picarelli L, Bucchignani E, Mercogliano P (2013) Potential effects of incoming climate changes on the behaviour of slow active landslides in clay. Landslides 10(4):373–391. https://doi.org/10.1007/s10346-012-0339-3

Costa JE, Schuster RL (1988) The formation and failure of natural dams. Geol Soc Am Bull 100:1054–1068

Crosta GB, Imposimato S, Roddeman D, Chiesa S, Moia F (2005) Small fast-moving flow-like landslides in volcanic deposits: the 2001 Las Colinas Landslide (El Salvador). Eng Geol 79:185–214
Crosta GB, Chen H, Frattini P (2006) Forecasting hazard scenarios and implications for the evaluation of countermeasure efficiency for large debris avalanches. Eng Geol 83:236–253
Cruden DM, Varnes DJ (1996) Landslide types and processes. In: Turner AK, Schuster RL (eds) Special Report 247: landslides, investigation and mitigation. TRB, National Research Council, Washington, D.C., pp 36–75
Davies TR, McSaveney MJ (2009) The role of rock fragmentation in the motion of large landslides. Eng Geol 109(1–2):67–79. https://doi.org/10.1016/j.enggeo.2008.11.004
Dramis F, Sorriso-Valvo M (1994) Deep-seated gravitational slope deformations, related landslides and tectonics. Eng Geol 38(3–4):231–243. https://doi.org/10.1016/0013-7952(94)90040-x
Dunning SA, Mitchell WA, Rosser NJ, Petley DN (2007) The Hattian Bala rock avalanche and associated landslides triggered by the Kashmir Earthquake of 8 October 2005. Eng Geol 93(3–4):130–144. https://doi.org/10.1016/j.enggeo.2007.07.003
Evans SG, Bent AL (2004) The Las Colinas landslide, Santa Tecla: a highly destructive flowslide triggered by the January 13, 2001, El Salvador earthquake. In: Rose WI, Bommer JJ, Lopez DL, Carr MJ, Major JJ (eds) Natural hazards in El Salvador. Geological Society of America Special Paper, Boulder, Colorado, pp 25–37
Evans SG, Guthrie RH, Roberts NJ, Bishop NF (2007) The disastrous 17 February 2006 rockslide-debris avalanche on Leyte Island, Philippines: a catastrophic landslide in tropical mountain terrain. Nat Disaster Earth Syst Sci 7:89–101
Feng Z (2011) The seismic signatures of the 2009 Shiaolin landslide in Taiwan. Nat Hazards Earth Syst Sci 11:1559–1569
Fowler AM, Hennessy KJ (1995) Potential impacts of global warming on the frequency and magnitude of heavy precipitation. Nat Hazards 11:283–303
Fukuzono T (1985) A method to predict the time of slope failure caused by rainfall using the inverse number of velocity of surface displacement. J Japan Landslide Soc 22(2):8–13
Geertsema M, Clague JJ, Schwab J, Evans SG (2006) An overview of recent large catastrophic landslides in northern British Columbia, Canada. Eng Geol 83:120–143
Guzzetti F, Carrara A, Reichenbach P, Cardinali M (1999) Landslide hazard evaluation: a review of current techniques and their application in a multi-scale study, Central Italy. Geomorphology 31:181–216
Guzzetti F, Peruccacci S, Rossi M, Stark CP (2007) Rainfall thresholds for the initiation of landslides in central and southern Europe. Meteorol Atmos Phys 98:239–267
Guzzetti F, Mondini AC, Cardinali M, Fiorucci F, Santangelo M, Chang KT (2012) Landslide inventory maps: new tools for an old problem. Earth Sci Rev 112(1–2):42–66. https://doi.org/10.1016/j.earscirev.2012.02.001
Hata Y, Wang G, Kamai T (2015) Preliminary study on contribution of predominant frequency components of strong motion for earthquake-induced landslide. In: Engineering geology for society and territory—volume 2: landslide processes, pp 685–689. https://doi.org/10.1007/978-3-319-09057-3_114
Hayakawa Y, Koyama M (1992) Eruptive history of the Higashi Izu monogenetic volcano field 1: 0-32 ka. Bull Volcanol Soc Japan 4:167–181
Heim A (1932) Bergsturz und Menschenleben. Fretz und Wasmuth, Zurich
Hewitt K (2016) Gifts and perils of landslides: catastrophic landslides and related landscape developments are an integral part of human settlement along upper Indus streams. Am Sci 98(5):410
Hirano M, Ishii T, Fujita T, Okuda S (1985) Geomorphological and geological characteristics of 1984 landslide hazard in Ohtaki village, Nagano Prefecture, Japan. Ann Disaster Preventin Res Inst Kyoto Univ 28:519–532
Hsu KJ (1975) Catastrophic debris streams (sturzstroms) generated by rockfalls. Geol Soc Am Bull 86:129–140
Huang RQ (2008) General review of catastrophic landslides in China. In: Huang RQ, Xu Q (eds) Catastrophic landslides in China. Science Publisher, Beijin, pp 1–36

Huang RQ (2011) Geo-engineering lessons learned from the 2008 Wenchuan earthquake in Sichuan and their significance to reconstruction. J Mt Sci 8(2):176–189. https://doi.org/10.1007/s11629-011-2103-x
Huang RQ, Fan XM (2013) The landslide story. Nat Geosci 6(5):325–326
Hungr O, Leroueil S, Picarelli L (2014) The Varnes classification of landslide types, an update. Landslides 11(2):167–194
Hutchinson JN (1988) General report: morphological and geotechnical parameters of landslides in relation to geology and hydrogeology. In: Bonnard C (ed) Proceedings of 5th International Symposium on Landslides. Balkema, Rotterdam, Netherlands, pp 3–35
Inoue Y, Honsho S, Matsushima M, Esashi Y (1970) Geological and soil mechanical studies on the slides occurred during the 1968 Tokachioki earthquake in southeastern area of Aomori Prefecture. Bulletin of the Central Research Institute of Electric Power Industry (69086), 27p
Jaboyedoff M, Oppikofer T, Abellan A, Derron MH, Loye A, Metzger R, Pedrazzini A (2012) Use of LIDAR in landslide investigations: a review. Nat Hazards 61(1):5–28. https://doi.org/10.1007/s11069-010-9634-2
Jibson RW, Harp EL, Michael JA (1998) A method for producing digital probabilistic seismic landslide hazard maps: an example from the Los Angeles, California, Area. US Geological Survey, pp 98–113
Jibson RW, Crone AJ, Harp EL, Baum RL, Major JJ, Pullinger CR, Escobar CD, Martinez M, Smith ME (2004) Landslides triggered by the 13 January and 13 February 2001 earthquakes in El Salvador. In: Rose WI, Bommer JJ, Lopez DL, Carr MJ, Major JJ (eds) Natural hazards in El Salvador. Geological Society of America Special Paper, Boulder, Colorado, pp 69–88
Kamai T (1990) Failure mechanism of deep-seated landslides caused by the 1923 Kanto earthquake, Japan. In: Proceedings of the sixth international conference and field workshop on landslides, pp 187–198
Kargel JS, Leonard GJ, many (2015) Geomorphic and geologic controls of geohazards induced by Nepal's 2015 Gorkha earthquake. Sciencexpress https://doi.org/10.1126/science.aac8353:1-18. https://doi.org/10.1126/science.aac8353
Kasama T, Yamashita H (2008) Tokyo Pumice from Hakone volcano, Central Japan. Res Rep Kanagawa Prefect Mus Nat Hist 13:91–110
Keefer D (1984) Landslides caused by earthquakes. Geol Soc Am Bull 95:406–421
Keefer D (2002) Investigating landslides caused by earthquakes—a historical review. Surv Geophys 23(6):473–510
Kobayashi T (1987) Geological aspect of the 1984 Denjo landslide on Ontake volcano. Trans Jpn Geomorphol Union 8(2):113–125
Matsushi Y, Watakabe T, Tsou CY, Hirata Y, Chigira M (2015) Triggering mechanisms and rainfall threshold of shallow landslides: cases in the Hiroshima Disaster on 20 August 2014, vol 58. Annual Report of the Disaster Prevention Research Institute, Kyoto University, pp 24–33
Melosh HJ (1979) Acoustic fluidization: a new geologic process? J Geophys Res 84(B13):7513–7520
Mitchell JK (1993) Fundamentals of soil behavior. Wiley, New York
Miyabuchi Y, Daimaru H, Komatsu Y (2004) Landslides and lahars triggered by the rainstorm of June 29, 2001, at Aso Volcano, Southwestern Japan. Trans Jpn Geomorphol Union 25:23–43
Montgomery DR, Schmidt KM, Greenberg HM, Dietrich WE (2000) Forest clearing and regional landsliding. Geology 28–4:311–314
Morimoto R (1951) Geology of Imaichi district with special reference to the earthquakes of Dec. 26th., 1949 (II). Bull Earthq Res Inst 29:349–358
Morimoto R, Nakamura K, Tsuneishi Y, Ossaka J, Tsunoda N (1967) Landslides in the epicentral area of the Matsushiro earthquake swarm—their relation to the earthquake fault. Bull Earthq Res Inst 45:241–263
Nakano M, Chigira M, ChounSian L, Sumaryono G (2015) Geomorphological and geological features of the collapsing landslides induced by the 2009 Padang earthquake. Paper presented at the 10th Asian Regional Conference of IAEG, Kyoto

Nemcok A, Pasek J, Ryb J (1972) Classification of landslides and other mass movements. Rock Mech 4:71–78
Okada K (2007) Current state of disaster generation risk forecast by SOIL WATER INDEX. Tsuchi-to-Kiso 55(9):4–6
Oyagi N (2004) Classification and definition of landslides. In: Landslides–geomorphological and geological identification and terminloogy. Jpn Landslide Soc 3–15
Peruccacci S, Brunetti MT, Luciani S, Vennari C, Guzzetti F (2012) Lithological and seasonal control on rainfall thresholds for the possible initiation of landslides in Central Italy. Geomorphology 139–140:79–90. https://doi.org/10.1016/j.geomorph.2011.10.005
Petley D (2012) Global patterns of loss of life from landslides. Geology 40:927–930
Saito M (1965) Forecasting the time of occurrence of a slope failure. In: 6th international conference on soil mechanics and foundation engineering, Montreal, pp 315–318
Saito M (1969) Forecasting the time of occurrence of a slope failure by tertiary creep. In: 7th international conference on soil mechanics and foundation engineering, Mexico City, pp 677–683
Sassa K, Lee HL (1993) Measurement of the apparent friction angle during motion by the high-speed ring shear apparatus. J Jpn Landslide Soc 30:1–10
Sato T, Chigira M, Matsushi Y (2017) Topographic and geological features of landslides induced by the 2016 Kumamoto earthquake in the western Part of the Aso caldera. DPRI Annuals 60B:431–452
Scheidegger AE (1973) On the prediction of the reach and velocity of catastrophic landslides. Rock Mech Felsmechanik Mécanique des Roches 5(4):231–236. https://doi.org/10.1007/BF01301796
Schneider JF (2008) Seismically reactivated Hattian slide in Kashmir, Northern Pakistan. J Seismolog 13:387–398
Seed HB, Wilson SD (1967) The Turnagain Heights landslide, Anchorage, Alaska. J Soil Mech Found Div Proc Am Soc Civ Eng 325–353
Sharpe CFS (1960) Landslides and related phenomena—a study of mass-movements of soil and rocks. Pageant Books Inc., New Jersey
Shimizu F (1983) Aerial-photograph interpretation for landslide mapping in the Shinjo Area, northeast Japan. J Jpn Landslide Soc 19:10–18
Shreve RL (1966) Sherman landslide, Alaska. Science 154:1639–1644
Sibson RH (1996) Structural permeability of fluid-driven fault-fracture meshes. J Struct Geol 18:1031–1042
Sidle RC, Kitahara H, Terajima T, Nakai Y (1995) Experimental studies on the effects of pipeflow on throughflow partitioning. J Hydrol 165(1–4):207–219
Suzuki T (1992) Tephrochronological study on Nasu Volcano. Bull Volcanological Soc Jpn 37(5):251–263
Suzuki T (1993) Stratigraphy of middle Pleistocene tephra layers around Nasuno Plain, in north Kanto, central Japan. J Geogr 102:73–90
Takahashi T (1991) Debris flow. Taylor & Francis, New York
Takahashi M (ed) (2007) Hakone Volcano. Geological leaflet of national parks. Geological Society of Japan, Tokyo
Takeuchi M, Nakano S, Harayama S, Otsuka T (1998) Geology of the Kiso-Fukushima district with Geological Sheet Map at 1:50,000. Geological Survey of Japan
Tanaka K (1985) Some earthquake induced landslides in Japan. Paper presented at the international conference and field workshop on landslides, Tokyo
Tang C, Zhu J, Qi X, Ding J (2011) Landslides induced by the Wenchuan earthquake and the subsequent strong rainfall event: a case study in the Beichuan area of China. Eng Geol 122(1–2):22–33. https://doi.org/10.1016/j.enggeo.2011.03.013
Tjia HD, Muhammad RFH (2008) Blasts from the past impacting on Peninsular Malaysia. Geol Soc Malays Bull 54:97–102
Toda S, Tanaka K, Chigira M, Miyakawa K, Hasegawa T (1995) Coseismic behavior of groundwater by the 1995 Hyogo-ken nanbu earthquake. Jishin 48:547–553

Tsou, C-Y, Chigira M, Higaki D, Sato G, Yagi H, Sato HP, Wakai A, Dangol V, Amatya SC, Yatagai A (2017) Topographic and geologic controls on landslides induced by the 2015 Gorkha earthquake and its aftershocks: an example from the Trishuli Valley, Central Nepal. Landslides. https://doi.org/10.1007/s10346-017-0913-9
Tsou C-Y, Feng Z-Y, Chigira M (2011a) Catastrophic landslide induced by Typhoon Morakot, Shiaolin, Taiwan. Geomorphology 127:166–178
Tsou C-Y, Feng ZY, Chigira M (2011b) Catastrophic landslide induced by Typhoon Morakot, Shiaolin, Taiwan. Geomorphology 127(3–4):166–178
Tsou C-Y, Chigira M, Matsushi Y, Chen S-C (2015) Deep-seated gravitational deformation of mountain slopes caused by river incision in the Central Range, Taiwan: spatial distribution and geological characteristics. Eng Geol 196:126–138. https://doi.org/10.1016/j.enggeo.2015.07.005
Varnes DJ (1978) Slope movement types and processes. In: Schuster RL, Krizek RJ (eds) Landslides analysis and control, vol 176. TRB. National Research Council, Washington, pp 11–33
Wakimizu T (1919) The causes and types of landslides. Jpn Soc Civ Eng Mag 5:19–49 (in Japanese)
Wang W-N, Furuya T, Chigira M (2003a) Geomorphological precursors of the Chiu-fen-erh-shan landslide triggered by the Chi-chi earthquake in Central Taiwan. Eng Geol 69:1–13
Wang W-N, Wu HL, Nakamura H, Wu SC, Ouyang S, Yu MF (2003b) Mass movements caused by recent tectonic activity: the 1999 Chi-Chi earthquake in central Taiwan. Island Arc 12:325–334
Wang G, Huang R, Kamai T (2015) The internal structure and stability of some large landslide dams induced by the 2008 Wenchuan (Mw7.9) Earthquake, China. In: Engineering geology for society and territory—volume 2: landslide processes, pp 1163–1169. https://doi.org/10.1007/978-3-319-09057-3_204
Yamada M, Matsushi Y, Chigira M, Mori J (2012) Seismic recordings of landslides caused by Typhoon Talas (2011), Japan. Geophys Res Lett 39(13):5. https://doi.org/10.1029/2012GL052174
Yoshida M, Chigira M (2012) The relation between weathering of pyroclastic fall deposits and the collapses caused by the 1968 Tokachi-Oki earthquake. J Jpn Soc Eng Geol 52:213–221

Chapter 17
Nuclear Disaster and Human Geoscience

Akihiko Kondoh and Akira Hama

Abstract The 2011 earthquake off the Pacific coast of Tohoku, Japan, led to core meltdowns of the nuclear power generators at the Fukushima Daiichi Nuclear Power Plant (FDNPP) of Tokyo Electric Power Company (TEPCO). A large amount of radioactive material was released into the environment, and consequently, *anökumene* got deposited in the ground of Fukushima. By observing ongoing nuclear disasters from the perspective of refugees, the relationship between science and society can be better understood. Focusing on the Yamakiya district, one of the former evacuation areas, we describe the situation before and after the accident. A detailed map of dose rate was created using a ground-based survey in the range of livelihood. To attain reconstruction of the Yamakiya district as a mountain village, an understanding of the relationship between life and nature is necessary. We describe this relationship from the standpoint of miner subsistence and then consider measures taken against radioactivity and necessary items to restore daily life. Finally, we propose an idea of restoration based on an understanding of both human and natural dimensions originating from the regional and temporal characteristics of the Yamakiya district.

Keywords Fukushima · Nuclear disaster · Yamakiya · Satoyama watershed

17.1 Introduction

Science and technology experienced tremendous development in the twentieth century. People have received much benefit from the modern civilization; however, the development of technology has also weakened people's willingness to respect civilization itself. Based on the hypothesis that the civilization may decline in such a situation, as proposed by Ortega, a Spanish philosopher, Kobayashi (1991) advocates the hypothesis of "barbarians in civilized society" to describe people who enjoy the

A. Kondoh (✉)
Center for Environmental Remote Sensing, Chiba University, Chiba, Japan
e-mail: kondoh@faculty.chiba-u.jp

A. Hama
Graduate Course of Sciences, Chiba University, Chiba, Japan
e-mail: a4k4i8r8a@yahoo.co.jp

Y. Himiyama et al. (eds.), *Human Geoscience*, Advances in Geological Science,
https://doi.org/10.1007/978-981-32-9224-6_17

merits of the civilization but have no interest in the process that leads to the creation of a civilized society. Kobayashi (1991) has shown that people in Japan may be "barbarians in civilized society," using a questionnaire survey. Can this hypothesis expand to nuclear technology?

An incident that will be recorded in the human history occurred in Fukushima, Japan, on March 11, 2011. Core meltdowns took place in the nuclear power generators of Fukushima Daiichi Nuclear Power Plant (FDNPP) of Tokyo Electric Power Company (TEPCO), and a large amount of radioactive material was released into the environment. As a result, *anökumene* appeared in the ground of Fukushima. Evasion of agricultural products has taken place in areas distant from Fukushima, and there still remains off-limit areas, though more than eight years have passed since the accident. The nuclear disaster in Fukushima revealed that the Japanese are the "barbarians in civilized society" and raised an issue about the future of Japan as a civilized society. The relationship between the risk and benefit of nuclear power generation should be understood from the spatial and temporal viewpoints of human geoscience.

After the incidents, scientists performed intensive and detailed scientific research on the behavior of radioactive materials in the environment. Although many reports and scientific papers have been published, a standard of action for people in contaminated areas is based not only on scientific facts but also on various situations of individuals in the human dimension. Human geoscience is the field of study that examines the relationship between human activities and attributes in areas of human life. To make use of scientific results for the restoration of the environment and also for the restoration of the normal life of victims, human geoscience will serve in Fukushima toward the reconstruction of the lost homeland. This article discusses not only the natural environment but also the progression in an evacuation area from the human dimension standpoint. Recently, two books were published describing the progress of nuclear disaster, the human response to radioactive contamination, and the process of recovery from the nuclear disaster (Yamakawa and Yamamoto 2016, 2017). This article describes the results of the author's research and support activities in the Yamakiya district of Kawamata, one of the evacuation areas in Fukushima, over the past seven years.

Figure 17.1 shows the location of the Fukushima Prefecture in Japan and the topography of Fukushima and the surrounding area. The FDNPP is located on the terrace facing the Pacific Ocean to the east. The area along the coast is called *Hamadoori*. The Abukuma Mountains, its height of more than 1000 m, run from north to south, separated from the ocean by a plain, with a width of roughly 10 km. The gentle slope of the Abukuma Mountains, created through the long history of the weathering of granite, has promoted the development of dairy farms and the production of branded beef cattle. A deciduous, broadleaved forest has provided resources such as material for compost and mountain products and has brought qualitative richness to the mountain village. Figure 17.2 shows the typical landscape of Abukuma Mountains. To the west of the Abukuma Mountains lies a rift valley called *Nakadoori*, where large cities such as Fukushima and Koriyama are located, forming the trunk line of the Tohoku district in Japan. Radioactive plume emitted from FNDPP on

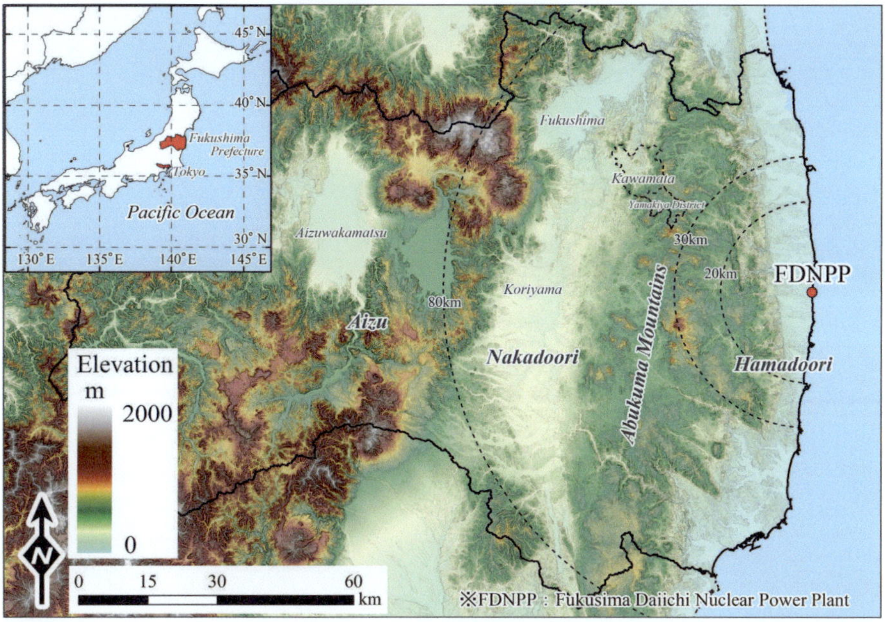

Fig. 17.1 Location of Fukushima Prefecture and Fukushima Daiichi Nuclear Power Plant (FDNPP), with concentric circles of 20, 30, and 80 km from FDNPP. The map shows topography of Fukushima Prefecture and surrounding area

March 15, 2011, covered villages in the Abukuma Mountains, separating human life from nature. Human geoscience is concerned with the relationship between humans and nature. Thus, science should be used to describe the forced separation between humans and nature and the reconstruction of human life.

17.2 The Accident and Progress to the Present

17.2.1 Accident of FDNPP and Deposition of Radioactive Materials

The 2011 earthquake off the Pacific coast of Tohoku occurred on March 11, 2011. The resulting Tsunami caused severe damage to FDNPP, and the nuclear reactor lost control after the power supply was turned off. The buildings that housed the reactors exploded one after the other on March 12, and a great amount of radioactive material was released into the environment. A major radioactive plume was emitted in the afternoon on March 15, flowing to the northwest, and radioactive material was deposited into the ground in the Abukuma Mountains with the spring snowfall. This

Fig. 17.2 Northern part of Yamakiya district in Kawamata. Landscape includes gentle slope with cleared land and pasture ground, deciduous, broadleaved forest as *Satoyama*, paddy field in lowland, and so on (photo taken on June 1, 2014)

deposition necessitated the setting of an evacuation area more than 40 km away from the FDNPP.

Just after the accident, an airborne dose rate survey was carried out by the US Department of Energy (DOE) and the Ministry of Education, Culture, Sports, Science, and Technology (MEXT) of Japan in order to visualize the dose rate distribution in the area within 80 km from FDNPP. The first monitoring was carried out on March 17, and the results were made public through the DOE homepage. The results were also conveyed to the local residents of Fukushima by researchers and supporters. Airborne dose rate monitoring was repeatedly conducted by MEXT and the dose rate map was first introduced to the public on May 8, 2011. Figure 17.3 shows the dose rate distribution map in the area within 80 km from FDNPP as announced by the government(Nuclear Regulation Authority 2011). Japanese citizens learned the dose rate distribution two months after the accident. Since then, the Japanese government has continued monitoring using airborne and driving surveys and has made the results available to the public.

Parallel to the dose rate survey conducted by the government, a survey carried out by individual scientists also revealed the dose rate distribution around the Fukushima area. In many cases, researchers went to different areas, not only in those neighboring FDNPP but also in the mountain villages of Abukuma, and measured the dose rate

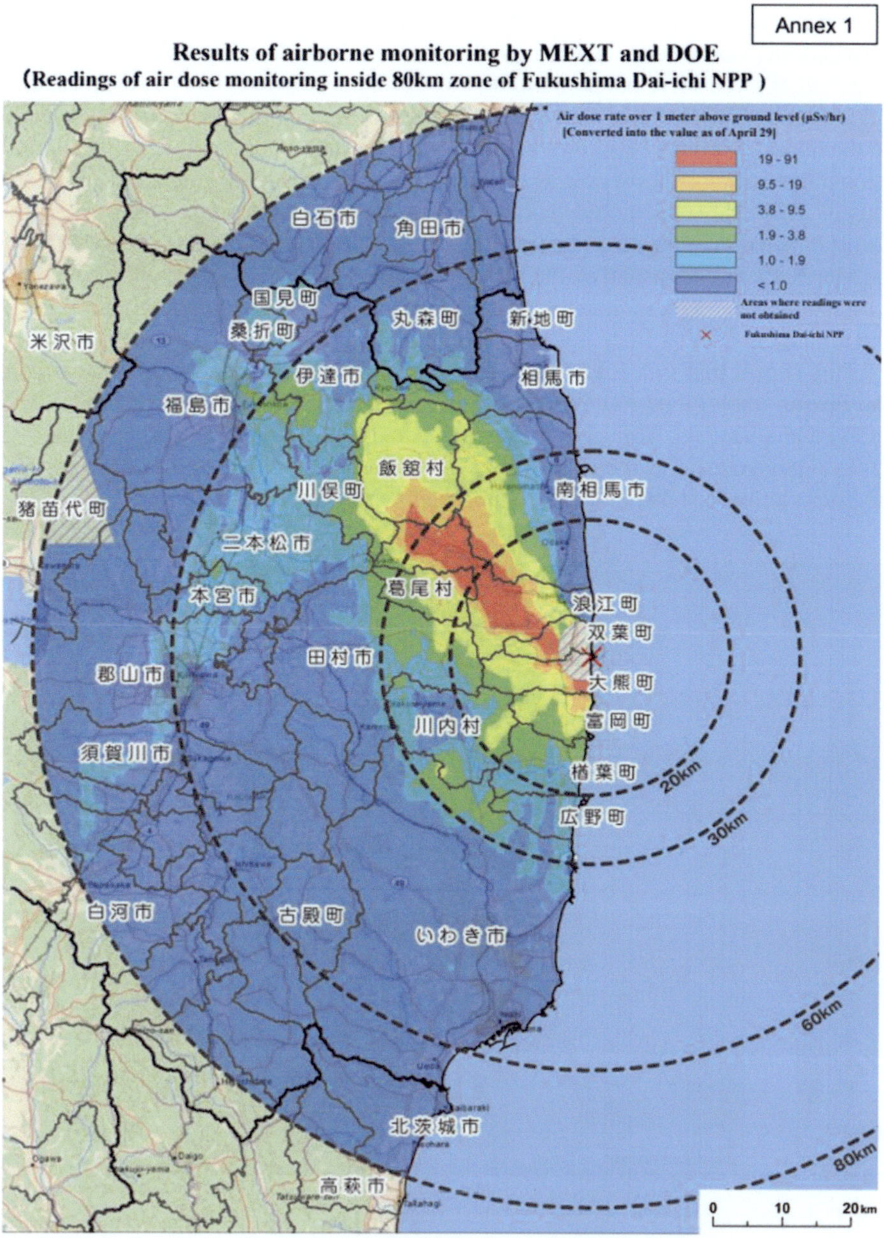

Fig. 17.3 The result of airborne survey by Department of Energy (DOE) and Ministry of Education, Culture, Sports, Science and Technology (MEXT) announced on May 8, 2011. (http://radioactivity.nsr.go.jp/en/contents/4000/3180/24/1304797_0506.pdf)

with the cooperation of local residents. Their purpose was to know the dose rate distribution in areas connected to daily life and to inform the refugees about the actual situation of radioactive contamination. For example, Kondoh et al. (2012) measured the dose rate distribution of the northern Abukuma Mountains and reported the results to the region. Figure 17.4 shows the dose rate distribution of the northern Abukuma Mountains and *Nakadoori* area as measured by a driving survey held during June and July of 2011. The track included not only the main trunk but also the path through the forest on the mountain slope. In the figure, radioactive material emitted from FDNPP on March 15 was deposited on the valley bottom of the eastern slope of the Abukuma Mountains. However, radioactive material was also deposited on the higher parts of the mountain slope at Iitate and the Yamakiya district.

This shows that two routes existed for getting information on radioactive contamination: the official route through the government and the private route through individual scientists. This also means the existence of a hierarchy of stakeholders. One scientist considers the government as a stakeholder, and the other considers the disaster victims a stakeholder to collaborate.

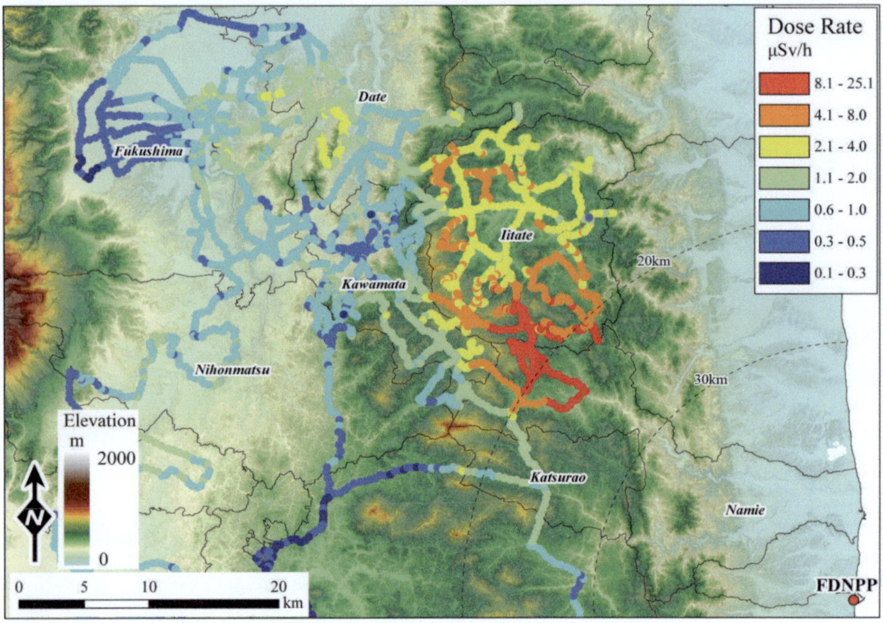

Fig. 17.4 Dose rate distribution map measured by driving survey held in June and July of 2011

17.2.2 Evacuation Order and the Sequence of Evacuation

After the accidents, the statement, "Not immediately affecting health," was broadcasted repeatedly by the government. At that time, preparations for farming and for the new school term had begun in the Yamakiya district like in the other evacuation districts in the Abukuma Mountains. The government designated a planned evacuation zone on April 12, which was officially announced on April 22. This did not mean, however, that shelters were prepared for the evacuees. Residents had to look for the destination, and several months were required for preparation. As a result, these individuals were forced to live in areas of high exposure to radioactivity.

For example, Mr. Kanno, a resident of the Yamakiya district, started preparation for planting for the spring season just after the accident. The designation of the planned evacuation zone forced him to look for shelter. He visited relatives and friends living outside an evacuation area to find shelter. Once the shelter was sought, the refuge first started accommodating children, and the adults moved in later, so the family was separated. Cattle could not evacuate immediately. Residents who waited for the construction of temporary housing had to stay in high exposure of radioactivity for several months (Kanno 2012). Although circumstances were different across regions and families, people made it through the difficulties with mutual help even in the absence of any wide-area evacuation plans for nuclear disaster.

17.2.3 Progress After Evacuation

Figure 17.5 shows the land use map depicting the division of the evacuation area right after the evacuation notice was issued. The area within 20 km from FDNPP was designated as a "restricted zone." The "planned evacuation zone" and "emergency evacuation preparation zone" were designated depending on the dose rate. The planned evacuation zone, extending northwestward from FDNPP, was at risk of an additional dose of more than 20 mSv/year. The division was revised gradually and most of the planned evacuation zone was canceled on April 2017, except for the "difficult-to-return zone." Evacuation order cancelation was made with the conflicts or painful decisions between different stakeholders in different circumstances.

For example, a consensus on the reconsideration of evacuation areas was achieved in the Yamakiya district two years after most of the evacuation was completed, and the area was divided into "evacuation order release preparation areas" and "residential restricted area" on August 8, 2013. This was the last re-designation among peripheral evacuation municipalities because of an imbroglio due to differences in the degree of radioactive contamination. Views on their origin are complicated and varied. This means that human geoscience should consider a sense of values of the refugees in order to achieve transdisciplinarity among local residents who are victims of the disaster.

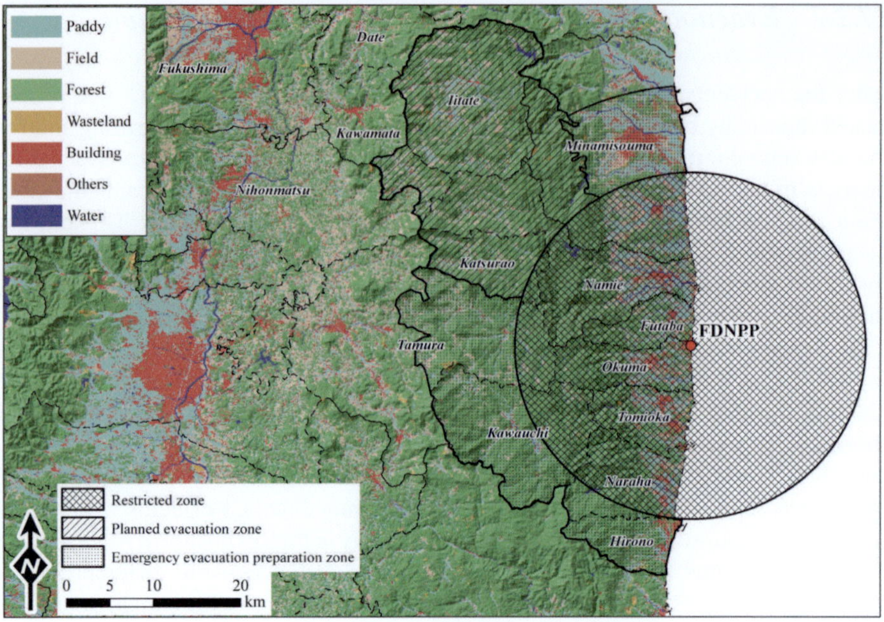

Fig. 17.5 The extent of initial evacuation areas showing the land use map

17.3 Decontamination of Radioactive Materials and Its Verification

Decontamination of radioactive materials was carried out by the Japanese government in the special decontamination zone as described in the *Act on Special Measures Concerning Nuclear Emergency Preparedness* enacted on August 30, 2011. The goals were as follows: (1) the area with an additional dose of over 20 mSv/year should be reduced promptly and incrementally; (2) an additional dose of less than 1 mSv/year should be aimed for in the long term for areas with an additional dose of lower than 20 mSv/year; and (3) decontamination should be prioritized for living environments with children, such as schools and parks.

Decontamination was made to the extent of 20 m from the farmland, roads, residential sites, and periphery of the forest. Mountain forest was excluded because of its vast extent. Decontamination directly controlled by the Japanese government was carried out for seven out of eleven evacuated municipalities by May 2016. Materials removed by decontamination were placed in flexible container bags and temporarily placed at specific locations to await transportation to an interim storage facility, which was still under construction in the *Hamadoori* region near FDNPP. Figure 17.6 shows one of the temporal storage sites assigned on a former paddy field. A delay in the completion of the interim storage facility caused the stagnation of decontam-

Fig. 17.6 Temporal storage site of decontaminated materials awaiting transportation to interim storage facility (photo taken on November 11, 2015, at Yamakiya)

inated materials, creating a mental obstacle to the refugees for reconstruction in the region.

Committees for the verification of decontamination were formed in each municipality toward the completion of the process of decontamination. The committees summarized the results of decontamination depending on the circumstances of each municipality, presented the basis of the evacuation order cancelation, and offered opinions on measures to be taken after the return. The committee members included not only officials but also resident representatives and scientists. It verified decontamination, considered measures against radioactivity, and proposed a plan for the future from various viewpoints.

The verification committee on decontamination and other related things in the Yamakiya district was established in Kawamata in April 2015, when evacuation orders were canceled in other municipalities. Completion of the decontamination operation was scheduled for the end of 2015, and the purposes of the committee were as follows:

1. To analyze and verify the results of decontamination
2. To collect research results on radiation
3. To propose a plan for environmental restoration.

The final report was issued in March 2016, and it summarized matters concerning decontamination, radiation protection measures, and other things.

The standard to judge the effect of decontamination and the health impact was set to the annual additional dose of 20 mSv, which is the target value of decontamination set by the Japanese government. Official data acquired by governmental organizations, such as the dose rate around decontaminated houses, the dose rate at monitoring posts, and the latest data from the driving dose rate survey were inspected, confirming the annual accumulated dose rate to be under 20 mSv/year, which means the short-range goal of decontamination was attained. However, an area with more than 1 mSv/year of dose rate still remained, which is the long-term goal of decontamination. The necessity of continuous monitoring was described in the final report.

There are many ways of thinking about 20 mSv/year, which is the threshold of annual additional dose rate. The committee should take a "scientific" position; however, only an additional dose of 100 mSv/year has evidence of carcinogenicity. The value of 20 mSv/year is only the recommended value of annual additional dose by the International Commission on Radiological Protection (ICRP). The idea behind this value is the Linear Non-Threshold (LNT) hypothesis. The role of the committee was to check whether the annual additional dose was less than 20 mSv. However, the value of 20 mSv/year is not an indication of safe return but the policy threshold. It is impossible to scientifically determine the safety level for return. Nevertheless, we have to set a standard, and it will be at the level of general consensus. Since an annual additional dose of over 1 mSv is caused by exposure to radioactivity from the artificial accident, the consensus level could be determined by the balance between an annual additional dose of 1 mSv and 20 mSv. The degree of compensation by TEPCO and the government may change the consensus level for additional dose. Yet, the local government hesitates to confront the government, which determines the budget, and at the same time, refugees who have the will to return to their homeland are unwilling to take the time for negotiation. If all the stakeholders have a common goal, a consensus level may be determined; however, the hierarchy among stakeholders and their internal conflicts prevent them from reaching a specific agreement level. Multiple agreement levels are desirable essentially because many different circumstances exist for the different regions, families, and perspectives. All the different decisions should be respected and should incorporate the agreement levels of refugees with the promotion of "a double-track reconstruction" (Science Council of Japan 2014).

The final part of the report for the Yamakiya district includes opinions on measures to restore normal daily life in Yamakiya. There are five possible measures:

1. Creation of a map that is useful for undertaking realistic radiation protection measures under the existing exposure situation
2. Radioactivity measures of the forest, which people use on a daily basis in the mountain villages
3. Holistic counselor system
4. Acceleration of environmental recovery
5. Measures on market revival.

These opinions only indicate the fundamental ways of restoring daily life, they do not guarantee the realization of this goal. It is important to strengthen the partnership between specialists and people in the region and that is the essential role of human geoscience.

17.4 Distribution of Dose Rate in *Satoyama* Watershed

The need for measures to be undertaken in the forest is described in the final report on decontamination in the Yamakiya district as one of the proposals. Although forest decontamination is not included in the official measures of the government, usage of the forest's ecosystem is inevitable for residents of the mountain villages. *Satoyama* is the ecosystem formed by the interaction between human activity and natural systems in the mountain forest. A watershed is a spatial unit of hydrologic and material cycles in the mountains. Therefore, the establishment of measures to be taken in the *Satoyama* watershed should be considered an important subject in human geoscience.

A fundamental activity called miner subsistence forms a part of life in the mountain villages and includes edible wild plant picking, mushroom picking, and so on (e.g., Kaneko 2015). An important problem in Fukushima is that no legal framework exists on compensation for miner subsistence. Compensation in Japan prioritizes an economic point of view, and the estimate for compensation is usually paid using depreciation and the amortization principle. Current compensation seems to be made based on the idea of life in urban areas, and accordingly, such a compensation is difficult to be acceptable for the agricultural world because the whole landscape consists of homeland for residents of the mountain village. Reconstruction of homeland, including the field of miner subsistence, is necessary for the revival of Fukushima, especially in the *Satoyama* region of the Abukuma Mountains. This is the subject of human geoscience that considers the sound relationship between human and nature. Thus, we should first know the distribution of radioactive materials and characteristics of movement in the mountain forest.

The distribution of dose rate for the mountain slope can be made clear by a ground-based walking survey (e.g., Kondoh et al. 2012; Kondoh 2012). Figure 17.7 shows the map of dose rate distribution by walking in the mountains with a gamma-ray spectrometer. In the northern part of the Yamakiya district, dose rate is high in the upper part of the slope. Hotspots or hot zones of relatively high dose rate correspond to the small topography, such as the slope facing FDNPP, and the evergreen trees. At the fallout of the radioactive materials in March of 2011, deciduous trees lost their leaves and the evergreen trees received more radioactive materials in the canopy. These effects indicate the need for continuous large spatial scale surveys of dose rate. The footprint of the airborne survey is several hundred meters long, and it is difficult to express a detailed distribution of dose rate in the mountain slopes. The small-scale map created by the airborne survey delivered images of the radioactive contamination all over Fukushima and brings a calm eyes from a distance.

Radioactive materials (most of them are radioactive cesium) stayed within the upper 5 cm of soil layer, especially in the organic layer. The litter and bio-mat layers, part of the A-horizon consisting of fine root, contained large amounts of radioactive material and had very low mobility. Images from a time-lapse camera installed on the mountain slopes showed that litters in the autumn defoliation were covered by snow thereafter, and wet litters displayed less mobility, thus starting to decompose.

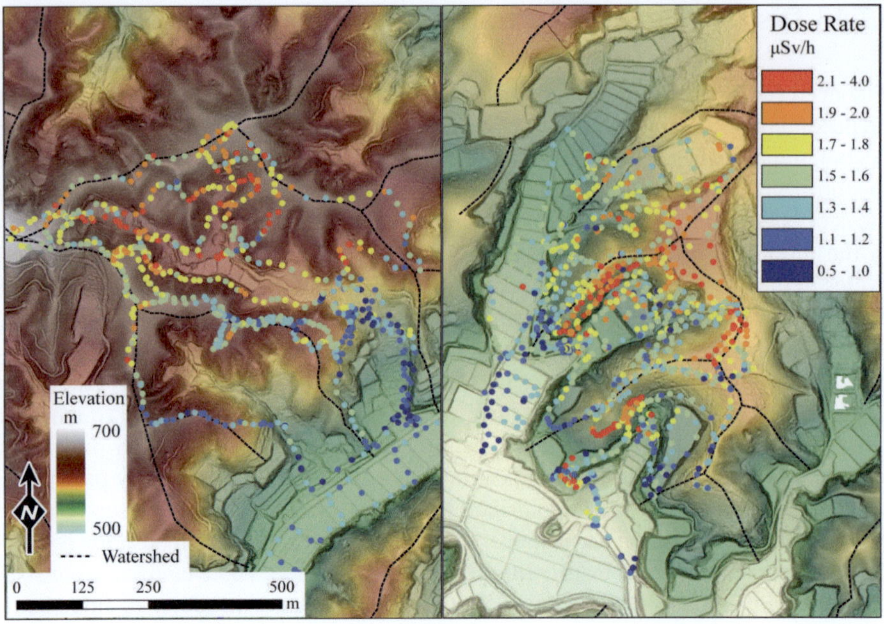

Fig. 17.7 The distribution of dose rate in mountain watershed of Yamakiya district measured by ground-based walking survey in 2013

Thus, mobility of the surface organic layer is the key to creating measures in the mountain slopes.

The general shape of a mountain watershed is like the leaves of a tree, which have a narrow outlet and wide interior. The path through which the material flows out of the watershed is a stream. Therefore, the obstruction of materials from flowing into the stream prevents the movement of radioactive materials out of the watershed. Although decontamination of the whole watershed is difficult, insolation or containment of the radioactive materials at hotspots or hot zones after measuring the dose rate distribution will be a possible way of determining measures to be taken and securing the minds of residents of the mountain villages. As a method, small technology that has been cultivated in Japan may be useful, such as lateral stripes or fascine used in erosion control. Small technologies are the ones which are applicable by individuals as well as communities. *Satoyama*, with its strong relationship between humans and nature, should be highly prioritized, and the watershed, with its rare relationship to human activity, should be isolated. Erosion, flood controls, and greening may be the means of maintaining the relationship between local residents and *Satoyama*. Making policy recommendations based on scientific knowledge is a necessary task for human geoscience to become a design science.

17.5 Rebirth of the Pride of the Region

Tohoku district in Japan, where Fukushima is located, has a cold climate in winter. Even in summer, north-easterly cold wind, called *Yamase*, often blows from Okhotsk high pressure. *Yamase* has sometimes brought cold weather damage, causing the reduction of grain production in the area. Paddy field maintenance and progress in agricultural technology in the period of high economic growth in Japan around 1970s allowed for the creation of the Yamakiya district granary in Kawamata. People fostered pride in the Yamakiya district through the hard time, but the accident of FDNPP led a sudden loss of pride of the homeland. Yet, the restoration of the pride of the region should be a driving force for recovering the homeland. Human geoscience as a transdisciplinary science focuses on the pride that lies in the existence of a sound relationship between humans and nature.

The Yamakiya district in Kawamata was originally the Yamakiya village formed in 1889. In 1955, one town and seven villages merged to form Kawamata. The Yamakiya village of the Meiji era was prone to cold weather damage by *Yamase*. Crop failure occurred roughly once every three years. Promulgation of a basic agricultural low in Japan promoted the development of agricultural infrastructure, which led to the disappearance of leaky paddy fields. Ownership of agricultural machines raised the production of rice from around 2600 bales to 10,000 bales until the late 1980s. The Yamakiya district was responsible for the shipment of 90% of the rice in Kawamata. A poor mountain village was transformed into a rich agricultural area (Kawamata Town Agricultural Cooperatives 1989).

Satoyama supports agriculture in the mountain village through the supply of water, litter as material of compost, mountain herbs, mushrooms, and many rich resources of the forest. Sato et al. (2015) took ortho-aerial photos for different years after 1947 and described changes in the land use to explain the relationship between residents and *Satoyama*. Figure 17.8 shows the distribution of narrow work roads in the mountain area and reclaimed land with modifications in topography, as revealed by a high-resolution Digital Elevation Model (DEM) created by an aerial Light Detection and Ranging (LiDAR) survey by Geographical Survey Institute, Japan. The map revealed the people's close relationship with the mountain forested area. Such a relationship is not experienced among residents of urban areas. Respecting the relationship between humans and nature in rural areas and contributing to the reconstruction of the mountain village in the Abukuma Mountains are important subjects of human geoscience. The accident of FDNPP reminded scientists of the subject of the relationship between the urban and rural societies.

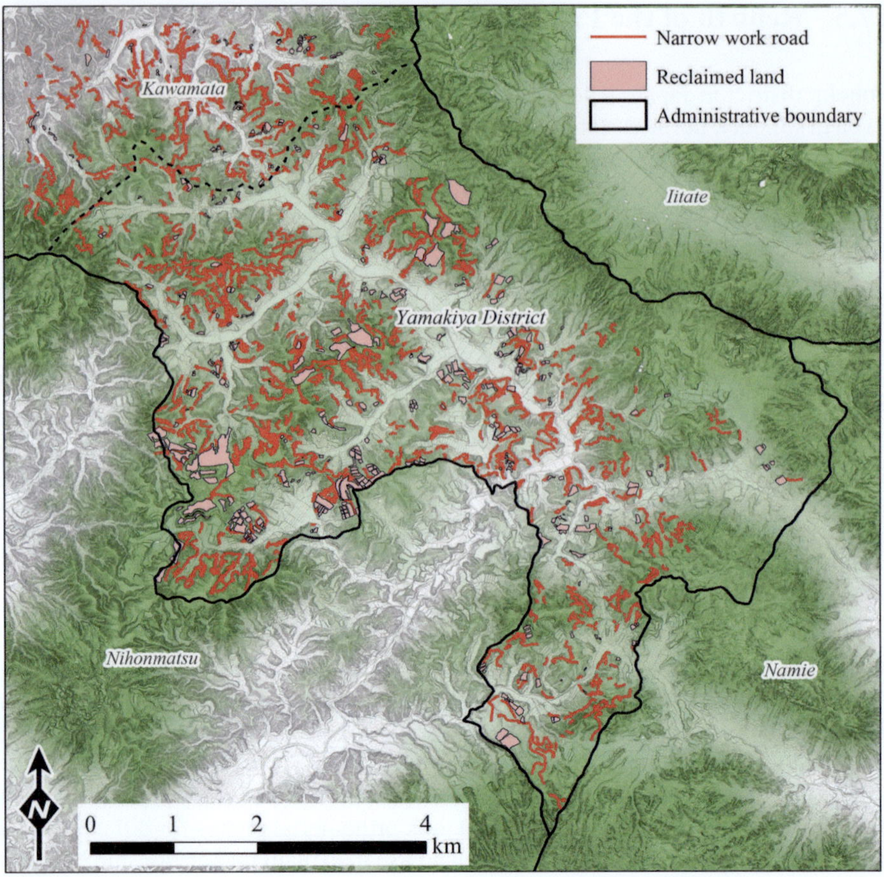

Fig. 17.8 Distribution of narrow work roads in mountain area and reclaimed land with modification of topography using 1 m resolution Digital Elevation Model (DEM) created by aerial Light Detection and Ranging (LiDAR) survey

17.6 Restoration of Livelihood

Former evacuation areas, except for "difficult-to-return zone," were released until April 2017. The most essential subject for sufferers is the rebuilding of daily life and livelihood. In the final report on the Yamakiya district, the following subjects were listed:

1. Appropriate methods of radiation protection
2. Maintenance of basic infrastructure, which accelerates restoration concerning daily life
3. Establishment of a comprehensive and continuous independent support system
4. Recovery of livelihood.

These subjects are the ones to be solved by human geoscience as a science of the relationship between humans and nature; however, the nuclear disaster in Fukushima is still in progress.

Continuous dose rate monitoring and construction of the Yamakiya Geographic Information System (GIS) for land management (Kondoh 2012) are necessary for radiation protection. Under cooperation between local residents and scientists, a basic computer system was installed in the local NPO. One problem is that the operation of software and hardware systems should be handled by the residents themselves. The other problem is the construction of radiation monitoring system. Evacuation order cancelation creates a tendency for returnees to manage their land; however, this will take some time because recovery of livelihood is their first priority. Human geoscience scientists have to prepare for transdisciplinary works.

Measures to recover the lost homeland have just begun to function. There are many difficulties in reclaiming human well-being in former evacuation areas; however, if all the stakeholders, including human geoscience scientists, have a common purpose to be achieved, it promises to open a new way to make a renewed region. It is the realization of transdisciplinarity that *Future Earth*, a research initiative on global environmental change and sustainability, is aiming at.

17.7 Problems to Be Solved for Future Fukushima

More than seven years have passed since the FDNPP accident of March 11, 2011. Recovery plans have gradually been completed; however, questions remain as to whether the plans have the flexibility to reflect various way of thinking for the different people and regions. The time axis for the plan is another important perspective because the effect of the radioactive materials deposited in the ground will stay for a long time. This is a present subject for human geoscience when *Future Earth* and *SDGs, Sustainable Development Goals*, are in the execution stage.

To tackle these serious and complicated problems, a transdisciplinary approach is inevitable. We have to realize collaboration among stakeholders, including scientists, even though the stakeholders have a complex configuration. Stakeholders have a hierarchy, and sometimes the idea of restoration is different from one stakeholder (e.g., the government) to another (e.g., an evacuee). Conflict among the stakeholders may prevent implementation of the measures. Under the nuclear disaster, there are various opinions on the allowed limit of exposure to radiation. The reconstruction of Fukushima is a formation of independent involvement by stakeholders and must pave way for consensus building. The Science Council of Japan released an opinion on "multi-track reconstruction" (Science Council of Japan 2014). The varied status of each individual, such as "return," "immigration," and "continued evacuation," should be respected, and the government should guarantee restorations of life in Fukushima. As for the Abukuma Mountains area, we have to understand the community of the mountain village and come up with measures based on the characteristics of the region.

Society in Japan and in the world at large tends to be controlled by the urban community. Yet, the rural world forms a major part of the global community. Building a good relationship between urban and rural areas is also a main subject of *Future Earth* and *SDGs*. Although the Fukushima nuclear disaster may be a local problem in the eastern part of Asia, we can consider global issues through this incident, such as the urban–rural relationship and a reconsideration of modern civilization. Although the target of human geoscience is practically local problems, we can produce new knowledge by sharing the outcomes globally. Then, the discussion will sublimate to the upper issues to be solved. These will be achievements toward realization of *Future Earth* and *SDGs*.

As of the spring of 2018, about 30% of former residents have returned to the Yamakiya district. Many issues still remain to be solved to reconstruct a new Yamakiya district. This is one of the situations wherein human geoscience should work to become science in society and the one for society.

References

Kaneko H (2015) Radioactive contamination of mountains and its damage to local communities: from the view point of miner subsistence. J Environ Sociol 21:106–121 (in Japanese with English Abstract)

Kanno G (2012) Thinking from Planned Evacuation Zone—move forward in order to go home. Mod Agric 360–362 (in Japanese)

Kawamata Town (2016) Report of the committee on decontamination in Yamakiya District, Kawamata Town, http://www.town.kawamata.lg.jp/uploaded/attachment/7709.pdf (referred in November 1, 2017) (in Japanese)

Kawamata Town Agricultural Cooperatives (1989) Progress in the cooperation in Yamakiya District, 130 pp (in Japanese)

Kobayashi S (1991) Examination of the hypothesis "Barbarians in civilized society"—a study over the correlation between science and technology, and culture and society. J Sci Policy Res Manage 6(4):247–260 (in Japanese)

Kondoh A (2012) Construction of Geographic Information System aimed at radioactivity measures in the *SATOYAMA* watershed. J Jpn Soc Revegetation Technol 28(2):274–277 (in Japanese)

Kondoh A, Kobayashi T, Tang C, Suzuki H (2012) Radioactivity survey in Yamakiya District, Kawamata Town, toward radioactivity measures in a watershed scale. J Rural Planning 30(4):528–529 (in Japanese)

Nuclear Regulation Authority (2011) Results of airborne monitoring by the Ministry of Education, Culture, Sports, Science and Technology and the U.S. Department of Energy (May 6, 2011). Accessed 29 Oct 2018. http://radioactivity.nsr.go.jp/en/contents/4000/3180/24/1304797_0506.pdf

Sato S, Hama A, Kondoh A (2015) Long-term changes in *SATOYAMA* in Yamakiya District, Kawamata Town, Fukushima Prefecture, and forced decoupling by accident of nuclear power plant. In: Proceedings of 2015 scientific assembly of Japan Society of Rural Planning, pp 28–29 (in Japanese)

Science Council of Japan (2014) Opinion on the reconstruction of life and home for long-term evacuees by the accident of Fukushima Daiichi Nuclear Power Plant, 30 pp. http://www.scj.go.jp/ja/info/kohyo/pdf/kohyo-22-t140930-1.pdf (referred in November 1, 2017)

Yamakawa M, Yamamoto D (eds) (2016) Unravelling the Fukushima disaster. Routledge, London

Yamakawa M, Yamamoto D (eds) (2017) Rebuilding Fukushima. Routledge, London

Part V
Global Environmental Problems

Chapter 18
Population Dynamics and the Environment

Yoshiki Wakabayashi

Abstract World population has been increasing geometrically, accelerating the speed of it being doubled, which is called "population explosion". It has become clear that the human population has an important effect on the natural environment. On the other hand, the spatial distribution of population is much affected by environmental conditions. In this chapter, we examine the dynamics of human population as well as population distribution and structure, with a special focus on spatial variations and implications. The major topics are as follows: population distribution and its change, components of natural change (fertility and mortality), human migration, population structure, and the relationship between population and the natural environment.

Keywords Population dynamics · Spatial distribution · Demographic Transition Model · Human migration · Environmental change

18.1 Introduction

One of the global issues in the modern age is the population explosion brought about mainly by better hygiene and sanitation and improved health care. Specifically, rapid population growth has been triggered by marked reductions in mortality, especially in less developed regions, where average life expectancy at birth increased during the second half of the twentieth century. Until recently, the world population has been increasing geometrically, accelerating the speed of it being doubled. At the dawn of agriculture, about 8000 BC, the population of the world was approximately five million. Over the 8000-year period up to the first century, it grew to 200 million, with a growth rate of under 0.05% per year.

A tremendous change occurred after the Industrial Revolution (Fig. 18.1). Whereas it had taken all of human history up to beginning of the 19th century for world population to reach one billion, two billion was achieved in 1927, three billion 30 years later in 1960, four billion in 1974, five billion in 1987, and six billion

Y. Wakabayashi (✉)
Department of Geography, Tokyo Metropolitan University, Tokyo, Japan
e-mail: wakaba@tmu.ac.jp

Y. Himiyama et al. (eds.), *Human Geoscience*, Advances in Geological Science,
https://doi.org/10.1007/978-981-32-9224-6_18

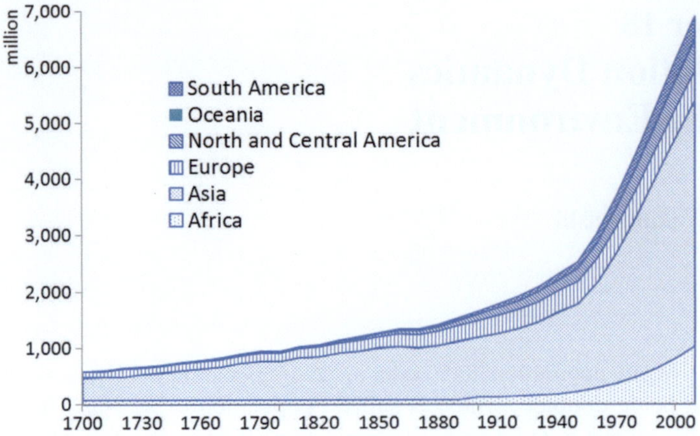

Fig. 18.1 World population growth by world regions after the eighteenth century. *Source* History Database of the Global Environment (HYDE)

only 10 years later, in 1999 (United Nations 2001). However, the population growth rate declined after the beginning of the twentyfirst century, especially in developed countries.

Figure 18.1 provides a breakdown of world population growth, displaying differences in population increase between world regions. From the eighteenth century to the first half of the twentieth century, Europe and the New World experienced marked population growth while the population of Asia and Africa grew slowly. However, the population of Asia and Africa rapidly increased after World War II. As a result, Asia has been the most populous region in the world for a long time and has had a significant impact on the world population trend.

It has become increasingly clear that the human population has an important effect on the environment. On the other hand, the spatial distribution of population is much affected by environmental conditions. In this chapter, we examine the dynamics of human population as well as population distribution and structure. The major topics are as follows: population distribution and its change, components of natural change, human migration, population structure, and the relationship between population and the environment.

18.2 Population Distribution and Its Change

World population is now more than seven billion. If all the inhabitants of the Earth were evenly distributed across the land area, the population density would be about 50 persons per square kilometer. However, people are unevenly distributed, creating huge regional disparities (Fig. 18.2). For example, almost 90% of the world's population lives north of the equator and two-thirds of that total live in the mid-latitudes

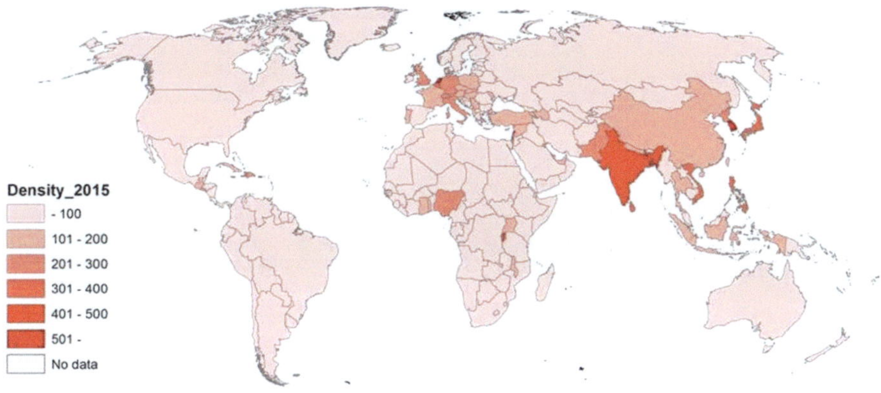

Fig. 18.2 Population density by country, 2015 (unit: person per square kilometer). *Source* United Nations

between 20 and 60 degrees north. A majority of the world's inhabitants occupy only a small part of its land surface. More than half the people inhabit about 5% of the land, two-thirds about 10%, and almost nine-tenths less than 20%.

As people congregate in lowland areas, population density decreases sharply with an increase in elevation. Though low-lying areas are preferred as settlement locations, not all such areas are equally favored. According to United Nations estimates, nearly 50% of the world's population lives within 200 km of the coastline, most of them on alluvial lowlands and river valleys. Figure 18.2 shows areas containing great clusters of population in East Asia, South Asia, and Europe.

The rate of population change also varies by country. Figure 18.3 is based on the estimates obtained from a report of United Nations (2015). This shows a contrast between tropical areas and the middle- and upper-latitude areas. In several places, countries with very slow growth border areas with extremely high growth.

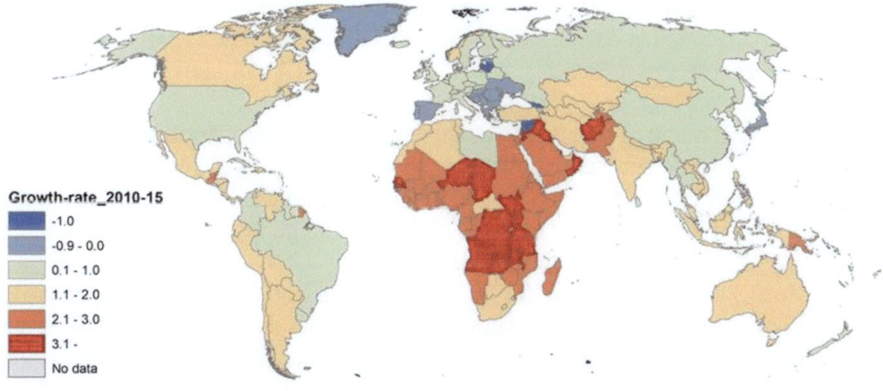

Fig. 18.3 Population growth rate by country, 2010–2015 (unit: %). *Source* United Nations

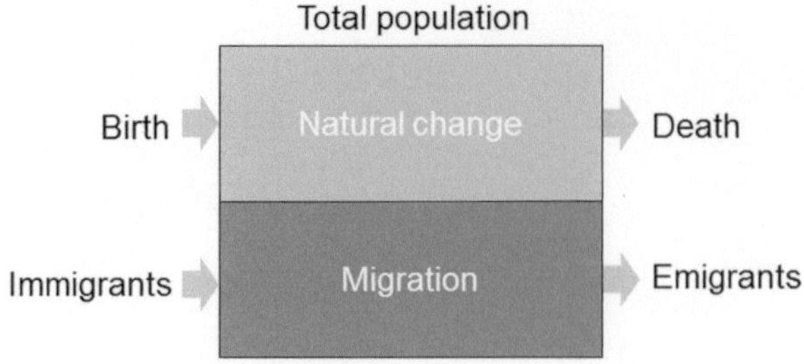

Fig. 18.4 Elements of population change. After Haggett (2001)

Population dynamics is composed of two elements: natural change which is the difference between total births and deaths and migration which is the difference between the numbers of immigrants and emigrants (Fig. 18.4). The subsequent two sections deal with geographical aspects of these elements.

18.3 Elements of Natural Change: Fertility and Mortality

Birth rate can be measured by several methods. The old and simple method of measuring fertility is to calculate crude birth rate (CBR), which is the annual number of live births per thousand people. However, CBR relates birth to total population without considering the age or sex composition of that population. Since the birth rate of a country is strongly influenced by the age and sex structure of its population, the total fertility rate (TFR) is a more refined measure than CBR. TFR is measured as the average number of children born per woman during her reproductive lifetime, considered to be from 15 to 49 years of age (Fellmann et al. 2008).

As shown in Fig. 18.5, TFR varies markedly from place to place in the world. In the countries of southern and eastern Europe and Japan, the TFR is less than 1.5. Countries with a TFR of 2.0 or lower are expected to experience population decline. By contrast, sub-Saharan Africa has the highest TFR of any sizable part of the world. However, during the past decades, TFRs have reduced in all sub-Saharan African nations.

Another component of population change is the death rate. The number of deaths per year per thousand people by country is shown in Fig. 18.6. Some of the world's highest death rates occur in sub-Saharan Africa, the poorest world region and most afflicted by life-threatening diseases and civil strife. In general, death rates of more than 20 per thousand people are rare today.

The world's highest death rate for 2013 occurred in South Africa and is primarily the result of a high prevalence of infectious diseases, such as HIV/AIDS. High death rates are also found in eastern European nations, due to a collapsing public care system in the post-Soviet era, environmental contamination and the increased

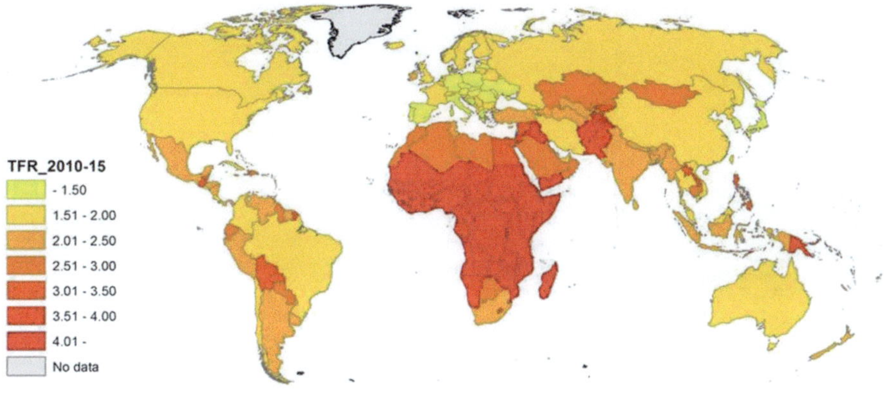

Fig. 18.5 Total fertility rate by country, 2010–2015. *Source* United Nations

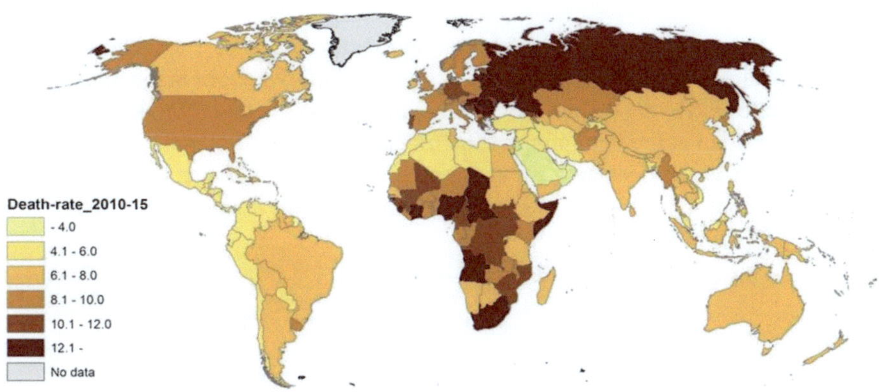

Fig. 18.6 Death rate by country, 2010–2015. *Source* United Nations

incidence of cancer, poor health choices including smoking and alcohol consumption and the high rate of disease.

In contrast, the American tropics generally have rather low death rates, as does the desert belt across North Africa, the Middle East, and central Asia. In these regions, the predominantly young population lowers the death rate. Because of its older population structure, the average death rate of the European Union is 10.5 per mil. In contrast, Australia, Canada, and the United States, which continue to attract young immigrants, have lower death rates than Europe.

As the maps of fertility and mortality suggest, the relationship between fertility and mortality can change with economic development. The Demographic Transition Model (Notestein 1945) traces the changing levels of human fertility and mortality presumably associated with industrialization and urbanization. The model assumes that high birth and death rates will gradually be replaced by low rates as shown in Fig. 18.7. In Stage 2, the death rate declines rapidly, causing a population explosion

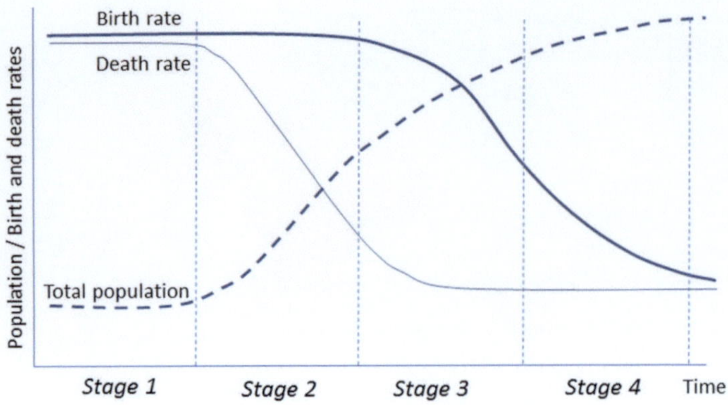

Fig. 18.7 Schematic representation of the Demographic Transition Model. After Knox and Marston (2010), Short (2015)

as the gap between the number of births and deaths widens. Then, the birth rate begins to decline sharply in Stage 3. The transition ends when, in Stage 4, both birth and death rates have reached low levels, by which time the total population is many times greater than at the beginning of the transformation.

This model is applicable to the demographic change in Japan after the end of nineteenth century (Fig. 18.8). At the beginning of the modern era, death rates

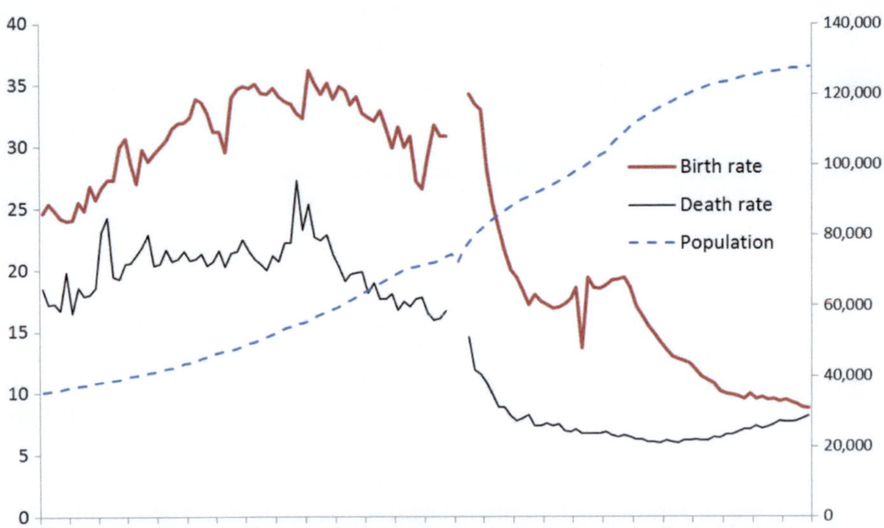

Fig. 18.8 Demographic change in Japan after 1875. (Left axis: fertility and mortality per 1000 persons, right axis: total population in thousands of persons). *Source* Vital statistics and population census of Japan (Fertility and mortality data for 1944–46 are not available)

and birth rates are high and roughly in balance in Japan (Stage 1). The death rate quickly declined after the 1920s owing to improvements in food supply and sanitation, increasing life expectancy and reducing disease (Stage 2). The birth rate also declined due to various fertility factors such as the spread of family planning and contraception after World War II (Stage 3). Since the end of the period of high economic growth in Japan, the birth rate has dropped to below replacement level, leading to a shrinking population (Stage 4).

18.4 Human Migration

Since many countries have entered the last stage of the Demographic Transition Model, the effects of natural change have become minimal. Instead, human migration is often the major cause of population change in countries with low birth and death rates. Human migration is divided into two types: internal and international migrations. Usually, internal migration refers to the movement between rural and urban regions, leading to urbanization (see next chapter). In this section, we focus on the international migration.

According to the United Nations Population Division, the number of international migrants worldwide has continued to grow rapidly over the past 15 years reaching 244 million in 2015. Between 2000 and 2015, the number of migrants grew in 167 countries worldwide. As a result of this growth rate, the share of migrants in the total population reached 3.3% in 2015. However, there were considerable differences between major areas. In Europe, Northern America, and Oceania, international migrants accounted for at least 10% of the population. By contrast, in Africa, Asia, and Latin America and the Caribbean, less than 2% of the population consisted of international migrants.

As shown in Table 18.1, high-income countries host more than two-thirds of all international migrants, whereas most migrants originate from middle-income countries. This reflects the dominance of the migration from the Global South to the Global North, which is often driven by migrants' desire to access better employment opportunities and improved standards of living. In addition, the number of refugees worldwide in recent years has reached the highest level since World War II. In 2014, the total number of refugees in the world was estimated at 19.5 million, representing about 8% of all international migrants (UNHCR (United Nations High Commissioner for Refugees) (2015)).

18.5 Population Structure and Characteristics

Population dynamics are also much affected by the population age structure of a region. A useful graphic device for comparing age characteristics is the population pyramid, which represents the patterns of population composition, namely, age

Table 18.1 Major destination or origin countries in international migration, 2015

Rank	Destination	N^a	Origin	N^a
1	USA	47	India	16
2	Germany	12	Mexico	12
3	Russia	12	Russia	11
4	Saudi Arabia	10	China	10
5	UK	9	Bangladesh	7
6	UAE	8	Pakistan	6
7	Canada	8	Ukraine	6
8	France	8	Philippines	5
9	Australia	7	Syria	5
10	Spain	6	UK	5

[a]Number of migrant (in million)
Source United Nations (2016) International Migration Reports 2015

structure. Figure 18.9 shows population pyramids of selected countries. Tanzania and India, which are positioned in the first or second stage in the Demographic Transition Model, display the classic stepped pyramid of a rapidly expanding population,

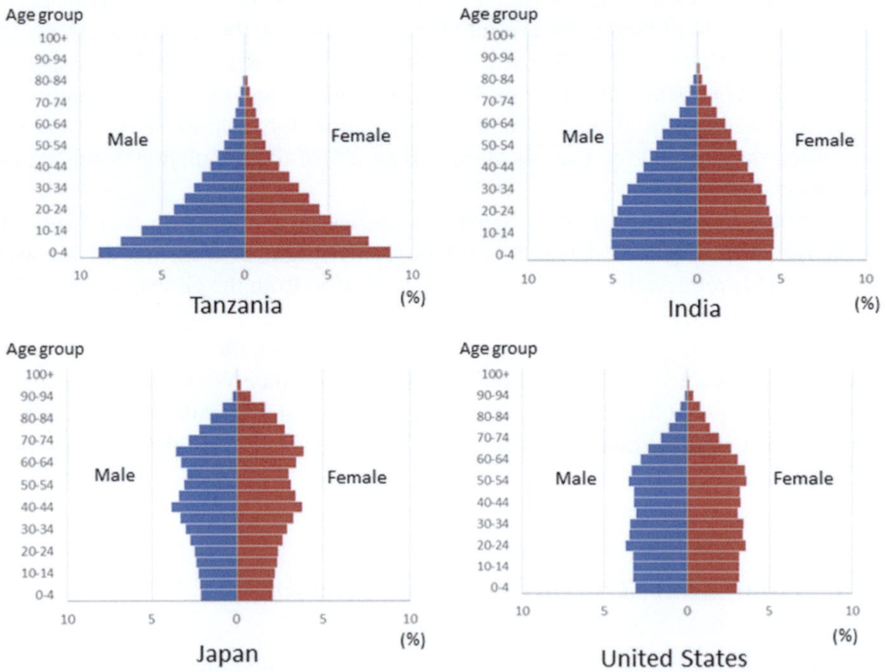

Fig. 18.9 Population pyramid: age distribution for selected countries, 2015. *Source* United Nations

whereas the US pyramid looks more balanced. In contrast, Japan's population pyramid reflects the lowered numbers of young people as result of its declining birth rate, which characterizes countries at the fourth stage in the Demographic Transition Model.

Careful study of such pyramids not only reveals the past progress of birth control but also allows us to predict future population trends. Youth-weighted pyramids suggest the rapid growth typical of population explosion. Population pyramids with more of a cylindrical shape represent counties approaching population stability or those in demographic decline.

Although the population is divided almost evenly between males and females, geographical differences are observed in the sex ratio, the ratio between men and women in a population (Fig. 18.10). Slightly more boys than girls are born, but infant boys have slightly higher mortality rates than infant girls. Beyond such a natural pattern, gender often influences demographic traits in specific ways. Often gender roles are closely tied to how many children are produced by couples. In general, many cultures demonstrate a marked preference for males as shown in the population pyramid of India, in which the bottom of the pyramid shows slightly asymmetric shape (Fig. 18.9). A cultural premium is placed on producing male children in many countries. This has resulted in pressures on couples to have boys in some countries, particularly in India, China and Islamic countries (see Fig. 18.10). On the other hand, the high sex ratio indicating male predominance in Arabian Peninsula can also be explained by immigration of foreign male workers who came from Asian countries.

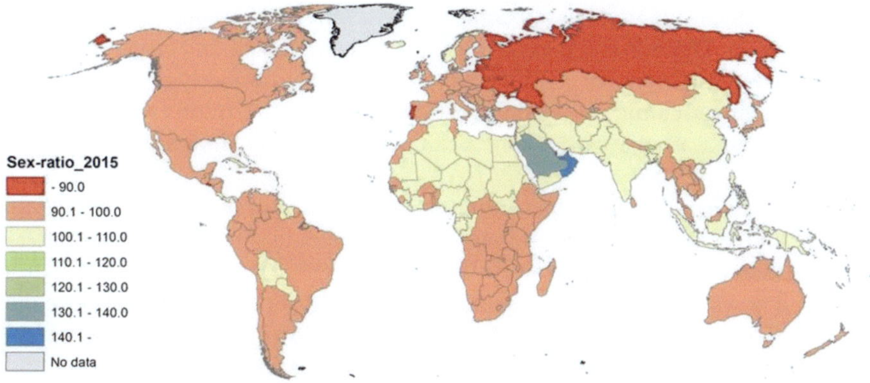

Fig. 18.10 Sex ratio by country, 2015. *Source* United Nations

18.6 Population and the Environment

It has become apparent that population, poverty, environmental degradation, and resource shortages were increasing at a rate that could not long be continued (Bartlett 1994). Most prominent among the publications that identified these problems in quantitative terms to provide extrapolations into the future as well as recommendations for corrective actions was the book entitled "Limits to Growth" (Meadows et al. 1972). After the publication of the book, the word "sustainable" has become popular and are used to describe a wide variety of activities which are generally ecologically laudable.

Characteristics of local population are often influenced by the availability of resources. In the middle latitudes, population densities tend to be greatest where the terrain is level, the climate is mild and humid, the soil is fertile, mineral resources are abundant, and the sea is accessible. Conversely, population tends to thin out with excessive elevation, aridity, coldness, ruggedness of terrain, and distance from the coast. Climatic factors also influence where people settle. Most of the sparsely populated zones in the world have difficult climates from the viewpoint of human habitation.

However, humans are remarkably adaptive, and our cultures include strategies that allow us to live in many different geographic environments. In addition, people modify their habits through their adaptive strategies. Particularly in areas where population density is high, radical alterations often occur. This can also happen in fragile environments even at relatively low population densities because the Earth's carrying capacity varies greatly with place and culture. For example, the need for food for the increasing number of people in environmentally fragile arid lands, such as sub-Saharan Africa, has led to overuse of the land grazing and crop production. When land overuse occurs in combination with a draught, formerly productive lands may decline in agricultural productivity (Raven et al. 2015). In this way, the extent of inhabited areas of the world, called ecumene, is largely limited by the natural environment.

On the other hand, many of our adaptive strategies are not sustainable. Population pressures and local ecological crises are closely related. In short, overpopulation relative to resource availability can precipitate environmental destruction, which in turn results in a downward cycle of worsening poverty, with an eventual catastrophe that is both ecological and demographic. However, the worldwide ecological crisis is not solely a function of overpopulation. A relatively small percentage of the Earth's population controls much of the industrial technology and consumes a disproportionate percentage of the world's resources each year.

In this way, population dynamics and environment are interrelated. The relationship between population and environment can be summarized as Fig. 18.11. Among four theoretical and methodological approaches in population–environment, viz., linear, multiplicative, mediated, and system-theoretical, the conceptual framework of Fig. 18.11 is based on the mediated approach characterized by the mediating factors

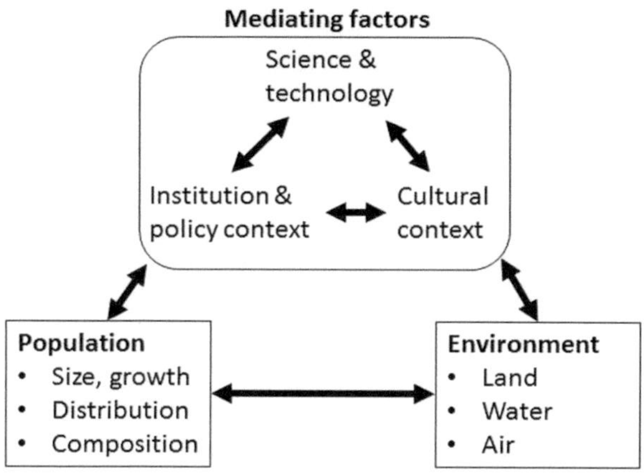

Fig. 18.11 A conceptual framework for the relationship between population and environment. After Hummel et al. (2013) and Hunter (2000)

such as policy context, science and culture link population factors with environmental outcomes (Hummel et al. 2013).

According to Hunter (2000), the size or growth of the population affects the limits on such global resources as arable land, drinking water, forests, and fisheries. Likewise, population size also influences pollution levels. The distribution of people around the globe has three implications for environmental change. First, as less-developed regions cope with an increasing share of global population, pressures will intensify on already dwindling resources within many of these areas. Second, the redistribution of population through migration shifts the relative pressures exerted on local environments, perhaps easing the strain in some areas and increasing it in others. Finally, the trend toward urbanization poses particularly complex environmental problems. Population composition can also have an effect on the environment because different population subgroups behave differently. Within countries and across households, the relationship between income and environmental pressure is different: environmental pressures can be greatest at the lowest and highest income levels.

Several factors mediate the relationship between human population dynamics and the environment: current technology, institutions, policy, and culture alter the ways in which demographic and environmental factors interact. For example, the technological changes that have most affected environmental conditions relate to energy use. Institutions and policy responses can be significant mechanisms through which humans react to environmental change and, in so doing, affect subsequent environmental change. These mechanisms can operate for good or ill. Cultural factors also influence how populations affect the environment: for example, cultural variations in attitudes toward wildlife and conservation.

The environmental implications of demographic dynamics are obviously complicated and can sometimes be controversial; many causes underlie contemporary environmental degradation, and only some are demographic in nature. Yet population does matter, and increased attention to the environmental implications of demographic dynamics can improve policy capacity to respond to contemporary environmental change.

References

Bartlett AA (1994) Reflections on sustainability, population growth, and the environment. Popul Environ 16(1):5–35

Fellmann JD, Getis A, Getis J (2008) Human geography: landscapes of human activities, 10th edn. McGraw-Hill Higher Education, Boston

Haggett P (2001) Geography: a global synthesis. Pearson Education Limited, Harlow

Hummel D, Adamo S, de Sherbinin A, Murphy L, Aggarwal R, Zulu L, Liu J, Knight K (2013) Inter- and transdisciplinary approaches to population environment research for sustainability aims: a review: a review and appraisal. Popul Environ 34:481–509

Hunter LM (2000) The environmental implications of population dynamics. RAND, Santa Monica

Knox LP, Marston SA (2010) Human geography: places and regions in global context, 5th edn. Pearson Education, New York

Meadows DH, Meadows DL, Randers J, Behrens WW (1972) The limits to growth: a report for the Club of Rome's Project on the predicament of Mankind. Universe Books, NY

Notestein FW (1945) Population: the long view. In: Schultz TW (ed) Food for the world. University of Chicago Press, Chicago, pp 36–57

Raven HR, Hassenzahl DH, Hager MC, Gift NY, Berg LR (2015) Environment, 9th edn. Wiley, USA

Short JR (2015) Human geography: a short introduction. Oxford University Press, New York

United Nations (2001) Population, environment and development, the concise report. United Nations, New York

United Nations (2015) World population prospects, the 2015 revision. https://esa.un.org/unpd/wpp/

United Nations (2016) International migration report 2015. United Nations, New York

UNHCR (United Nations High Commissioner for Refugees) (2015) UNHCR statistical yearbook 2014, 14th edn. UNHCR, Geneva, Switzerland

Chapter 19
Urbanization: Concept, Mechanism, and Global Implications

Yuji Murayama and Ronald C. Estoque

Abstract Urbanization has two types of impacts: positive and negative. On the one hand, urbanization brings positive impacts to society as it helps improve the social and economic aspects of people's lives. On the other hand, urbanization brings negative social and ecological impacts, both locally and globally. This chapter discusses the concept of urbanization in the context of human geoscience, including its history, mechanism, and trend in the modern world. It presents an overview of the negative ecological and social impacts associated with urbanization, such as urban heat island, air pollution, flooding, health, urban poverty, crimes and violence, and traffic congestion. This chapter also discusses the inclusion of urbanization to the global sustainable development agenda, as well as the importance of geospatial technologies, such as geographic information systems (GIS) and remote sensing, for urban monitoring towards sustainable urban development.

Keywords GIS · Impacts of urbanization · Remote sensing · Sustainable urban development · Urbanization

19.1 Concept of Urbanization

Urbanization generally refers to the process whereby "urban elements" replace "non-urban elements." However, the idea of urban aspects differs according to researchers' perspectives, making it difficult to establish an academic definition.

In urban sociology, urbanization refers to the process wherein urban living patterns supersede rural living patterns. Urban sociology emphasizes the social changes that occur during this process, including changes in lifestyles and attitudes toward life.

Y. Murayama (✉)
Faculty of Life and Environmental Sciences, University of Tsukuba, Tsukuba City, Ibaraki 305-8572, Japan
e-mail: mura@geoenv.tsukuba.ac.jp

R. C. Estoque
National Institute for Environmental Studies, 16-2 Onogawa, Tsukuba City, Ibaraki 305-8506, Japan
e-mail: estoque.ronaldcanero@nies.go.jp

Y. Himiyama et al. (eds.), *Human Geoscience*, Advances in Geological Science,
https://doi.org/10.1007/978-981-32-9224-6_19

In urban economics, which underscores economic activities, urbanization refers to the process through which an area's primary economic base shifts from agriculture to manufacturing or a tertiary industry. In demography, urbanization refers to the process of population concentration whereby populations move from a rural area to an urban one, leading to a relative rise in the number of city dwellers. Urban geography considers urbanization a local phenomenon. In this field, urbanization refers to the expansion of the urban area; that is, the amount of land that is functionally connected to the urban core.

How does *human geoscience* define urbanization? Human geoscience is a discipline that explores the various phenomena that occur on Earth's surface, focusing on the relationship between the natural environment and human activities. Accordingly, when researching urbanization, it is especially important to integrate the understandings of the various disciplines outlined above and to ascertain urbanization phenomena comprehensively from the angle of geoscience. In other words, human geoscience should seek to dynamically clarify the mechanisms and driving forces of urbanization with a focus on nature, scenery, and land use, as well as social, economic, and cultural factors. Another important task of human geoscience is to elucidate how urbanization affects the environment at the global and local levels.

19.2 History of Urbanization

As is well known, the Industrial Revolution occurred in Britain in the late eighteenth century. Before that, most societies across the world were focused on the primary industry, with many people engaged in agriculture and mainly dependent on land resources for their livelihoods. Since a subsistence economy prevailed at the time, people would establish a livelihood by securing nearby farmland. The main way of moving from one place to another was on foot, which placed a significant restriction on how far people could live from their land resources. Most settlements were small in scale.

The Industrial Revolution changed the population distribution and led to the rise of factory-based manufacturing. A large workforce was required to operate factories. This prompted the mass movement of people from rural areas to sites where the factories were located. Housing for factory employees sprang up around their workplaces, causing a densely populated, round urban area to form. During the early stages of urbanization, people would travel to work on foot; thus, factories and residential areas would be no more than two to three kilometers apart.

The Industrial Revolution brought a daily rhythm to people's lives. People would work for a set number of hours and rest on the weekend. Before the Industrial Revolution, work patterns depended on the weather. For example, people would not work when it rained or snowed. Furthermore, their working hours also depended on the available hours of sunlight. People did more work on long summer days compared to shorter winter days. The Industrial Revolution changed these circumstances by establishing a new social norm.

By the end of the nineteenth century, suburbanization had spread in Western cities. The driving force behind it was the emergence of tramways. Commuting distances became nearly four times as long compared to when people walked. Accordingly, housing complexes sprang up along the tramlines extending outward from city centers, leading to the formation of star-shaped urban areas. The development of tramways encouraged spatial separation between one's residence and workplace.

The turn of the century saw the proliferation of public transport such as subways and buses. The development of transport networks weaving across the length and breadth of urban areas gradually brought an end to the star-shaped urban pattern, which depended on tramways. By the late twentieth century, suburbanization had progressed a step further following the proliferation of private cars, leading in turn to expansive urban areas. While factories had been located in city centers, they now started relocating to the suburbs and nearby interchanges, which offered sites with optimal conditions and excellent transport access. The vacuums in centers were filled by commercial facilities and offices, leading to a segregation of functions in urban settings. Over time, the "donutization" phenomenon became increasingly conspicuous; in city centers, the gap between the nighttime and daytime populations widened. Meanwhile, on the frontline of urbanization, an increasing urban sprawl resulted in ever more land changing from agricultural to urban.

19.3 Mechanisms of Urbanization and Urban Trend in the Modern World

19.3.1 Urbanization in Developed Countries

In the late 1970s, migration from city centers to suburbs became a prevailing trend in Western countries, leading to the rise of counter-urbanization, which describes a decline in *urban population*. In this chapter, urban population refers to the number of people living in urban areas as defined by the United Nations (UN 2015a). The results of the 1970 United States Census presaged this phenomenon. The suburbs became attractive to families and the middle class because they offered low-cost open spaces. On the other hand, city centers became increasingly marked by decaying buildings and influxes of low-income groups resulting in some of them becoming slums. Overcrowding in cities led to many urban problems including spiraling land prices, gridlock, air and water pollution, and crime epidemics.

From the 1980s onward, counter-urbanization became a universal phenomenon among the major cities in the West. The manufacturing industries and offices, which had been concentrated in city centers, started moving to the cheaper suburbs or out of urban areas altogether. When employment stopped being concentrated in city centers, this prompted a spatial dispersal of populations and commercial functions. At the same time, growing numbers of city dwellers proactively moved outside urban locales, preferring the greenery and abundant living space afforded by rural areas or smaller towns, versus urbanized cores, which contained degraded living environments. Many of those who remained in city centers were immigrants and low-income groups. This situation exacerbated the deterioration of urban areas (Murayama et al. 2017).

Against this backdrop, municipal authorities pursued public sector-led redevelopment projects, but the costs of these schemes spiraled out of control, and they became unmanageable under the cities' limited budgets. Accordingly, the focus changed to the gentrification of inner-city areas, wherein architectural ventures and townscape improvements would be carried out at minimal cost with the support of local resident organizations and private capital. Such gentrification proved a success. Since inner cities were adjacent to urban centers and offered good transport access, they experienced a revival as residential and commercial spaces. The inner-city areas of many major North American metropolises have been revived as tourist zones that feature an array of ethnic restaurants.

Today, the spatial dispersal of Western urban populations has progressed to the extent that the frontlines of urbanization have reached as far as city peripheries and rural areas. This has led to the emergence of urban frontier zones called edge cities, and has rendered the border between city and countryside ambiguous. People's lifestyles and values have changed such that instead of living in densely populated cities, they prefer living around the city limits where there are large green spaces and good living environments. Telecommunications technology is a major factor driving the suburban dispersal of populations. The proliferation of the Internet and email has introduced flextime and the establishment of satellite offices. Increasing numbers of people work from home (telework) instead of commuting every day. Technological innovations have helped entrench the consumer/resident-centered society and expanded a residence-based work-style, particularly among young people. Today, major Western cities are being defined more loosely to include rural areas within their scope.

The twentieth century was a time of transport, but the twenty-first century is said to be one of information communication. In the twentieth century, commuting long distances to work was a prevailing trend, which encouraged the spatial separation between one's workplace and residence. Today, the development of information technology has helped break down this distance. In the future, we can expect that the concentrated distribution of the population will weaken, and spatial harmonization will increase. Such a trend might be considered a return to the circumstances that existed before the Industrial Revolution.

19.3.2 Urbanization in Developing Countries

Since the late twentieth century, developing countries have experienced a continued population explosion. Starting at the end of the twentieth century, the population growth rate in cities has been significantly higher than in rural areas. In some cases, a phenomenon called "urban explosion" has been apparent. Populations have especially concentrated in capitals and major cities (Murayama et al. 2016).

Such rapid population expansion results from a synergistic effect between social and natural increases. Large numbers of young people have left agricultural zones, which depend on agriculture and have a chronic shortage of employment opportunities, and moved to cities, which offer abundant job prospects. When these young people secure a stable livelihood in the city, they marry and have children. Economic globalization is a pull factor of major urban cores. In developing countries, the capital is often a hub of growth due to vibrant economic exchanges with developed countries. Since the national system of cities is often fragile in terms of its spatial organizational power, it is harder for cities (other than capitals) to receive the benefits of economic growth. Owing to underdeveloped transport and information networks, it is not possible to ensure a smooth distribution of goods and transmission of information to provincial areas. Moreover, the lack of a proper administrative framework for integrated control of the country's land makes it hard to efficiently disseminate innovation and growth-inducement strategies from the capital to other cities. Accordingly, economic disparities between capitals and other cities are widening.

Urban primacy leads to the concentration of capital and human resources in key cities, as well as the specialization of knowledge. It also causes transport networks to expand, and is a driving force behind the propagation of innovation. Inasmuch as the concentration of investment in the capital brings about economic effects in terms of scale and accumulation, it is economically effective for the country as a whole. On the other hand, the negative impacts are considerable. Four negative consequences are worth mentioning: (1) Urban primacy impedes the fair use of resources between the different parts of the country; (2) While it encourages trade with other countries, it reduces the distribution of goods within the nation, preventing balanced growth; (3) It produces regional inequalities in living standards; and (4) It marginalizes rural zones and widens regional gaps.

Economic globalization and advances in information and communications technology (ICT) are strengthening the mutual dependency between cities in developed countries and those in developing nations. In particular, the main metropolises of developing countries are hosting increasing numbers of branch and business offices of multinational companies, leading to a growing number of foreign business people. There are also rising numbers of tourists from developed nations. The inflow of foreign capital is strengthening the financial dominance of developed countries. The greater the dependency between cities of developing and developed states, the more cities in developing countries become directly influenced by the policies and economic situations of developed ones. In this way, developing nations are increasingly being incorporated into a global urban system (Murayama et al. 2017).

Today the growth, stagnation, or decline of major cities is governed not only by the power of cities themselves but by the strength of their relationships with other urban centers. With the incorporation of major cities around the world into a global system, important municipalities have started functioning as nodes of this network. New York, London, and Tokyo are key examples of this phenomenon; all are referred to as world cities. On the other hand, the economic dependency of megacities in developing countries to these world cities is increasing over time (Murayama and Estoque 2017).

19.3.3 Trends in Global Urbanization

In 1950, urbanization only affected 30% of the world's population. The subsequent advance of urbanization meant that by 2010, city populations would outnumber rural ones. According to the United Nations (UN) statistics, 2008 was the year when urban and rural populations reached the same level (UN 2015a). The pace of urbanization is accelerating. The urban population increased from 746 million in 1950 to 3.9 billion in 2014. It is estimated that by 2050, 66% of the world's population will live in cities. In 2014, the rural population was 3.4 billion, but this is expected to fall to 3.2 billion by 2050 (UN 2015a).

Looking at urbanization trends by region, in Europe, North America, and Oceania, the urban population was already outnumbering the rural one in 1950. Today, the urban population has exceeded 70% in those regions. In North America, 82% of the people live in cities. On the other hand, in Africa and Asia, the urban population has still not reached 50% (UN 2015a). It is expected that urbanization will advance rapidly in these regions in the future. The rise in urban populations is particularly apparent in large urban centers and their peripheries. According to UN statistics, there were only ten megacities (with a population of 10 million or more) in 1990, but in 2014, that figure had risen to 28 and is expected to increase to 41 by 2030 (UN 2015a). A majority will be in Asia and Africa.

19.4 Impacts of Urbanization

Generally, urbanization has two types of impacts: positive and negative. On the one hand, urbanization brings positive impacts to society as it helps improve the social and economic aspects of people's lives (Murayama and Estoque 2017). Cities are symbols of creativity and imagination, the power of humanity, cradles of innovation and knowledge creation, hearts of socio-cultural transformations, and engines of economic growth (Wu 2010). Globally, cities are connected through political, economic and technical systems, but more importantly through the Earth's biophysical life-support systems (Jansson 2013; Gómez-Baggethun et al. 2013). They are also part of the global sustainability agenda (Folke et al. 2011; Wu 2010; UN 2015b).

On the other hand, urbanization also brings negative impacts to the natural environment, both locally and globally (Bloom et al. 2008; Grimm et al. 2008; Wu 2010; Seto et al. 2012; Dahiya 2012; Estoque and Murayama 2014, 2015a, 2016). More so if urban "development" is poorly planned, or worst, unplanned at all, resulting in various negative socioeconomic and environmental impacts, such as poor urban environments and poor quality of life. Urbanization itself is arguably "the most drastic form of land transformation that results in irreversible landscape changes" (Estoque and Murayama 2014, p. 943). Here, we provide an overview of some of the issues and problems associated with urbanization, including urban heat island, air pollution, flooding, urban poverty, crimes and violence, and traffic congestion in the global context.

19.4.1 Urban Heat Island

Urban heat island, a phenomenon first described in 1818, refers to the phenomenon of higher atmospheric and surface temperatures occurring in urban areas than in the surrounding rural areas (Howard 1818; Oke and Hannel 1970; Voogt 2002; EPA 2008). Cities produce urban heat island regardless of size and location, though the effect often decreases as city size decreases (Oke 1973, 1982; Aniello et al. 1995; EPA 2008; Estoque and Murayama 2017).

There are two types of urban heat island: surface urban heat island and atmospheric urban heat island (EPA 2008). *Surface* urban heat islands are measured based on land surface temperature. While surface urban heat island is observed day and night, it tends to be strongest during the day due to solar radiation. *Atmospheric* urban heat islands, on the other hand, are measured based on air temperature, and are often classified into two: canopy layer urban heat islands and boundary layer urban heat islands. Canopy layer urban heat islands exist in the layer of air where people live, from the ground to below the tops of trees and roofs, while boundary layer urban heat islands start from the rooftop and treetop level and extend up to the point where urban landscapes no longer influence the atmosphere (EPA 2008).

The formation of urban heat islands is influenced by various factors, including the reduced vegetation in urban areas, properties of urban materials, urban geometry, anthropogenic heat, weather, and location. Some of the major negative impacts of urban heat islands include increased energy consumption, elevated emissions of air pollutants and greenhouse gasses, compromised human health and comfort, and impaired water quality (Voogt 2004; EPA 2008). Today, while scientists, engineers, and practitioners continue to make efforts in understanding the urban heat island phenomenon, they have also begun working on measures to mitigate its formation and impacts. A commonly discussed strategy is urban greening, in which urban green spaces are introduced, maintained or improved. The potentials of green roofs, cool roofs, and cool pavements for mitigating urban heat island effects are also being studied today (Hoverter 2012; Hoag 2015; Qin 2015; Sharma et al. 2016). For more information about the urban heat island phenomenon, the reader is referred to the academic and scientific literature (e.g., Voogt 2004; EPA 2008; Estoque et al. 2017).

19.4.2 Urban Air Pollution and Urban Health

In recent decades, pollution (air, water, and land) has intensified especially in the developing regions due to rapid urbanization, including industrialization and motorization. There are many unwanted "by-products" of urbanization in general that can cause serious and fatal illnesses, destroy ecosystems, and negatively affect living conditions, including vehicle exhaust, untreated wastewater, nitrogen fertilizer runoff, dirty fuel burning, industrial emissions, and toxic wastes such as e-waste (discarded electrical and electronic products) (The World Bank 2016a). In 2012, approximately nine million people died from air, water and land pollution, and about seven million people died from air pollution alone according to the Global Alliance on Health and Pollution and World Health Organization (WHO) (The World Bank 2016a). Here we highlight air pollution.

Particulate matters (e.g., PM_{10} and $PM_{2.5}$) are among the most important indicators for measuring the level of air quality. Exposure to these air pollutants has broad health effects, especially to the respiratory and cardiovascular systems. Recently, the WHO published the Ambient Air Pollution Database, which contains data on PM_{10} and $PM_{2.5}$ for the 2008–2015 period for about 3000 cities across 103 countries worldwide (WHO 2016a). The population size of the cities in the database ranges from a few hundred to more than nine million inhabitants, with about 25% of the cities having up to 20,000 residents (WHO 2016a).

The database shows that more than 80% of people living in urban areas that monitor air pollution are exposed to air quality levels that exceed WHO air quality guidelines (AQGs) (i.e., 20 $\mu g/m^3$ for PM_{10} and 10 $\mu g/m^3$ for $PM_{2.5}$) (WHO 2006, 2016a, b). Based on annual mean, the average PM_{10} and $PM_{2.5}$ values of all the cities included in the database are 37 and 21 $\mu g/m^3$, respectively. Of the world megacities (DEMOGRAPHIA 2016) that are included in WHO's Ambient Air Pollution Database, Delhi topped the list, followed by Beijing, Tianjin, London and Lima for PM_{10} (Fig. 19.1). For $PM_{2.5}$, Beijing topped the list, followed by Lima, Delhi, Paris, and Dhaka (Fig. 19.1). Overall, 98% of cities in low and middle-income countries with more than 100,000 inhabitants do not meet the WHO AQGs. In the high-income countries, this percentage decreases at 56% (WHO 2016b). This shows that while everybody is affected, those in the developing countries suffer the most. According to The World Bank (2016a), about 95% of children and adults who suffered from pollution-related illnesses live in low and middle-income countries.

While urbanization, including industrialization and motorization, is an important process towards socioeconomic progress, the problem of pollution must also be given proper attention because of its negative impacts on human health and economy. Upgrading and/or proper implementation of "standards" on environmental compliance for industries and the like, coupled with regular monitoring of air quality levels, especially in cities, is one way to help address the problem.

Today, some urban health assessment methodologies have been developed, including the development of the urban health index (UHI) (Rothenburg et al 2014; WHO 2014, 2016c). The UHI is a single metric and a composite measure of population

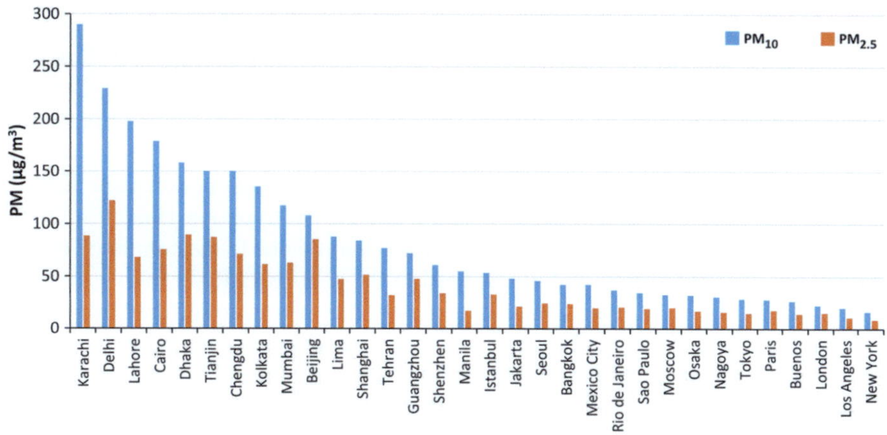

Fig. 19.1 Annual mean PM_{10} and $PM_{2.5}$ values (μg/m^3) of various megacities in the world. *Source* Authors' own figure based on WHO's 2016 Ambient Air Pollution Database (2008–2015) (WHO 2016a). Data for Jakarta and Sao Paulo were taken from the 2014 database (2008–2013) (WHO 2014)

health. It provides a standardized method for mapping and measuring the disparities in health determinants and outcomes in urban areas. The UHI can be used to examine the current status of urban areas, assess change and the effect of program interventions, and plan for urban improvements, and thus may be helpful to public health workers, evaluators, statisticians, program managers, academic researchers, and decision-makers (Rothenburg et al 2014; WHO 2014, 2016c).

19.4.3 Urban Flooding

Like urban heat island and air pollution, urban flooding is not exclusive to developing countries as it is also experienced in developed nations. Generally, urban areas can be inundated due to natural causes such as river flooding, coastal flooding, groundwater flooding, pluvial flooding, and human-induced causes such as artificial system failures.

River or fluvial flooding occurs when the volume of surface water runoff exceeds the capacity of the river channel, causing water runoff to overflow and spill out into adjacent, low-lying areas. Coastal flooding usually occurs from an expected sea-level rise due to storms or tsunamis. Groundwater flooding occurs when the level of water table rises and reaches the surface level. Finally, pluvial or overland flooding is caused by rainfall or snowmelt that flows over land due to the lack of permeability of the ground surface. Some of the human-induced causes of urban flooding include the saturation of drainage and sewerage capacity, lack of permeability due to increased concretization, and faulty drainage system and lack of management. More details of these causes of urban flooding can be found in Jha et al. (2012) and www.floodsite.net.

While flooding, in general, can cause a great damage to the society, it is more devastating in urban areas as cities contain more people and infrastructure. Cities are also the centers for various socioeconomic functions and urban services. Cities in coastal regions are likely to experience coastal flooding due to storms or tsunamis. Some of the recent major coastal floods that devastated many urban areas include those caused by the 2004 Indian Ocean earthquake and tsunami, hurricane Katrina in 2005, the 2011 Tohoku earthquake and tsunami, and typhoon Haiyan in 2013. The 2015 urban flooding in Joso City, Japan was mainly due to the combined effects of typhoon rains and structural failure, causing the Kinugawa River to overflow and burst its banks. In recent years, many other major cities have also suffered from major urban flooding due to the combined effects of storms, typhoons, heavy and persistent rains, fluvial and pluvial flooding, and artificial system failures (e.g., Dhaka in 2004, Hanoi in 2008, Bangkok in 2011, Beijing in 2012, Metro Manila in 2012, Jakarta in 2013, and Colombo in 2016).

In general, urban areas have limited pervious surfaces that can absorb rainwater. On top of this, many urban areas are also faced with rapid population growth, resulting in high demands for land. Unquestionably, there are many issues that urban development planners and decision makers must look into in every city. However, we contend that a proper drainage system and a well-crafted flood risk management plan are a must. This is not only to address the rainwater problem, but also to anticipate the combined effects of fluvial, pluvial and coastal flooding. Needless to say, most of major cities in the world are located along river banks or in coastal areas.

19.4.4 Urban Poverty

Although poverty, in general, is a global concern, it is common knowledge that it is mostly experienced in the poor and less-developed countries. Poverty issues and problems are found both in the rural and urban areas. Here we focus on the issues and problems associated with *urban poverty*.

Urban poverty is a phenomenon in which the urban poor live with the lack of material benefits considered to be basic necessities in a society. The day by day challenges that the urban poor face may include limited access to employment opportunities and income, inadequate and insecure housing and services, violent and unhealthy environments, little or no social protection mechanisms, and limited access to adequate health and education opportunities (The World Bank 2016b).

Generally, urban population grows due to natural increase (by birth) and rural-urban migration. People migrate and/or continue to live in urban areas because of the socioeconomic prospects they have on cities (e.g., employment opportunities and better social services). For some, rural-urban migration is an escape from constraining social and cultural traditions in rural areas (The World Bank 2016b). Consequently, however, the rapid growth of urban population also increases rapidly the level of demand for various urban services. And if the government is not able to cope up with the situation and fails to address the urban dwellers' demands for basic services, this rapid urban growth can have an unfavorable outcome, such as urban poverty, rather than urban prosperity.

One indicator of urban poverty is the proliferation of slums, including squatter or informal settlements. Slums are characterized by inadequate water supply, poor sanitation, overcrowded and dilapidated housing, hazardous locations, insecurity of tenure, fear, and social and economic exclusion and discrimination (UN-Habitat 2013). Slums are the opposite of prosperous cities, i.e., cities that provide productivity, infrastructure development, quality of life, equity and social inclusion, and environmental sustainability (UN-Habitat 2013).

According to the global assessment of slums, 828 million urban dwellers worldwide live in slums (UN-Habitat 2013). Worldwide, China (180.56 million), India (104.68 million), Nigeria (47.61 million), Brazil (44.95 million), and Pakistan (29.96 million) are the top five countries in terms of urban slum population (Fig. 19.2a).

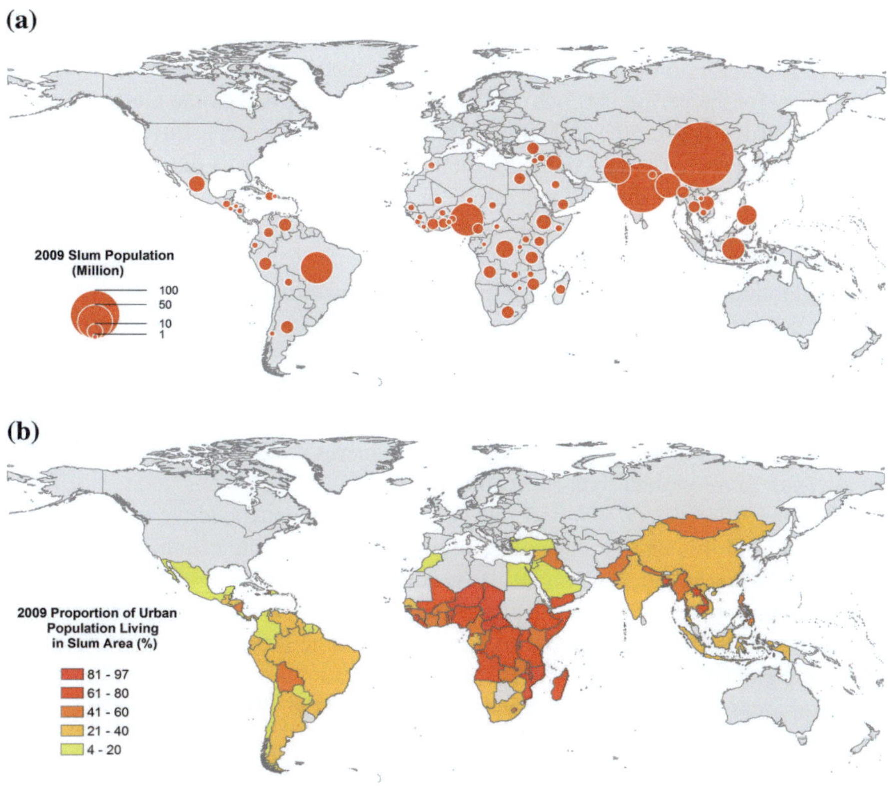

Fig. 19.2 Spatial distribution of urban slum population across the countries in the developing regions (Africa, Asia, and Latin America and the Caribbean): **a** 2009 urban slum population (i.e., for countries with urban slum population of ≥1 million); **b** proportion of urban population living in slum area (%). *Source* Authors' own maps based on the 2009 slum data (UN-Habitat 2013) and the country boundary layer (www.gadm.org)

However, in terms of proportion of urban population living in slum area, Sierra Leone (97.0%), Central African Republic (95.9%), Chad (89.3%), Guinea-Bisau (83.1%), and Niger (81.7%) are among the top five countries (Fig. 19.2b). For some information on the temporal pattern of slum during the past two decades, the reader is referred to UN-Habitat (2010, 2013).

19.4.5 Urban Crimes and Violence

Urban crimes and violence have become even more pressing with the substantial increase in urban population during the recent decades. Crimes and violence, in general, include assault, kidnapping, theft, robbery, burglary, sexual violence, human trafficking, illegal drugs, terrorism, and homicide (The World Bank 2011; UNODC 2014; https://data.unodc.org/). However, the monitoring of crimes and violence worldwide at the urban or city level is a challenging task. Here we highlight only the average rate of intentional homicide in the most populous city, by country/territory for the period 2005–2012 (Fig. 19.3), as compiled by the United Nations Office on Drugs and Crime (UNODC) (https://data.unodc.org/). Intentional homicide is defined as unlawful death inflicted upon a person with the intent to cause death or serious injury (https://data.unodc.org/).

Of the 125 cities (one for each country) with data on the rate of intentional homicide (2005–2012), Guatemala City (Guatemala), Caracas (Venezuela), Tegucigalpa (Honduras), Belize City (Belize), and Kingston (Jamaica) emerged as the top five cities (Fig. 19.3). Vaduz (Liechtenstein), Bandar Seri Begawan (Brunei), Tokyo (Japan), Hong Kong (China), and Lisbon (Portugal) are among the cities with the lowest rates of intentional homicide. Overall, out of the 125 cities, 84 (67%) had higher rates than their respective corresponding national rates (Fig. 19.3). This shows that based on intentional homicide data, urban areas were, by majority (but not always), more violent than rural areas.

The other 40 cities (33%) had rates lower than their respective national rates (Fig. 19.3), and this can be due to the high rates of the same crime in other cities and/or rural areas in the same country. One city (i.e., Zagreb, Croatia) had the same rate of intentional homicide at the city and country levels. The data for the country level do not show how much of the crime was contributed by rural areas. However, it has been argued in other studies that urban areas or cities are not always more violent than rural areas (e.g., The World Bank 2011; Myers et al. 2013).

19.4.6 Traffic Congestion

Traffic congestion is another major issue associated with rapid urbanization and motorization, posing a great challenge to urban planners and decision-makers. Research has shown that the likelihood of economic growth is higher when traffic

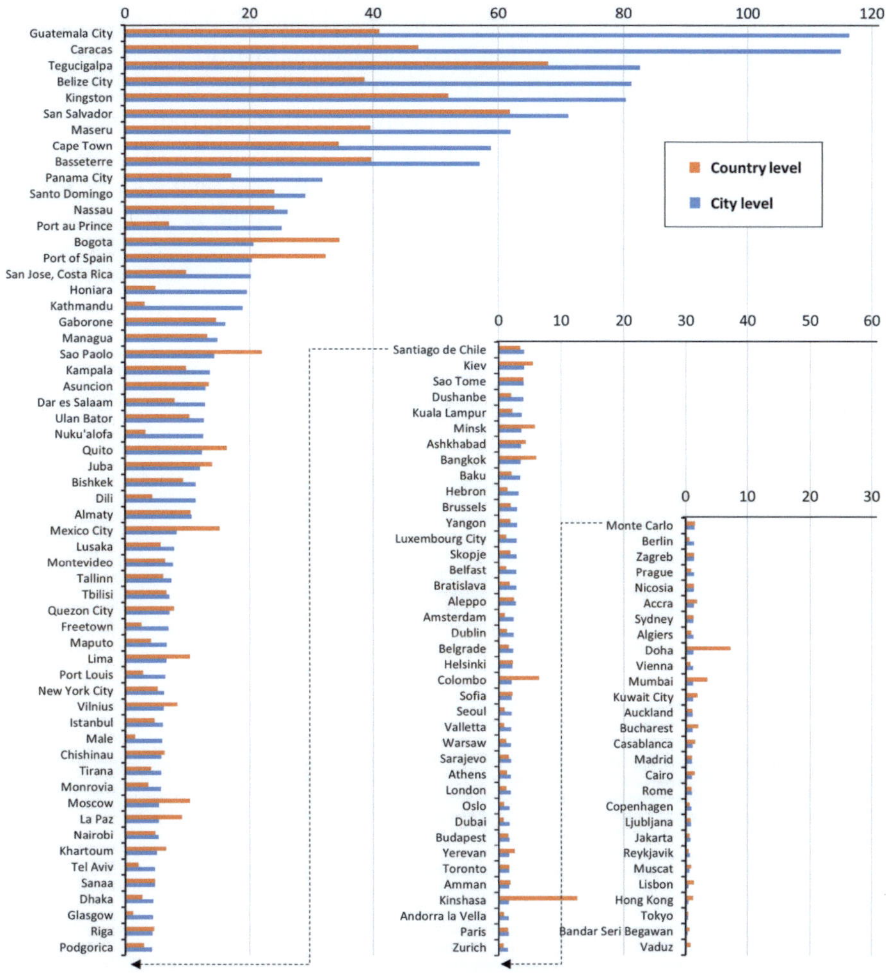

Fig. 19.3 Intentional homicide rates (per 100,000 population): City level versus country level, averages 2005–2012. *Source* Authors' own figure based on the data of UNODC (https://data.unodc.org/). *Note* City selection was based on the availability of data

flow is smoother (Weisbrod et al. 2003; Cox 2016). Traffic congestion also contributes to environment-related problems, such as higher energy consumption and greater air pollution. Fortunately, with the advances in technologies, vehicles today, which are already the cleanest in history, are getting even cleaner (Cox 2016).

Here we look at the state of traffic congestion worldwide using the TomTom Traffic Index data for 2015 (TomTom International 2016). The Tomtom Traffic Index measures congestion level, defined as an increase in overall travel time when compared to a free flow situation (e.g., a congestion level of 36% corresponds to 36%

extra travel time for any trip, anywhere in the city, at any time compared to what it would be in a free flow situation) (TomTom International 2016).

The 2015 TomTom Traffic database includes 295 cities across 38 countries worldwide. Figure 19.4 shows the top 50 cities in terms of the TomTom Traffic Index (extra travel time due traffic congestion). Among the top five cities are Mexico City (Mexico), Bangkok (Thailand), Lodz (Poland), Istanbul (Turkey), and Rio de Janeiro (Brazil). The data shows an overall average extra travel time of 25% due traffic congestion (Fig. 19.4).

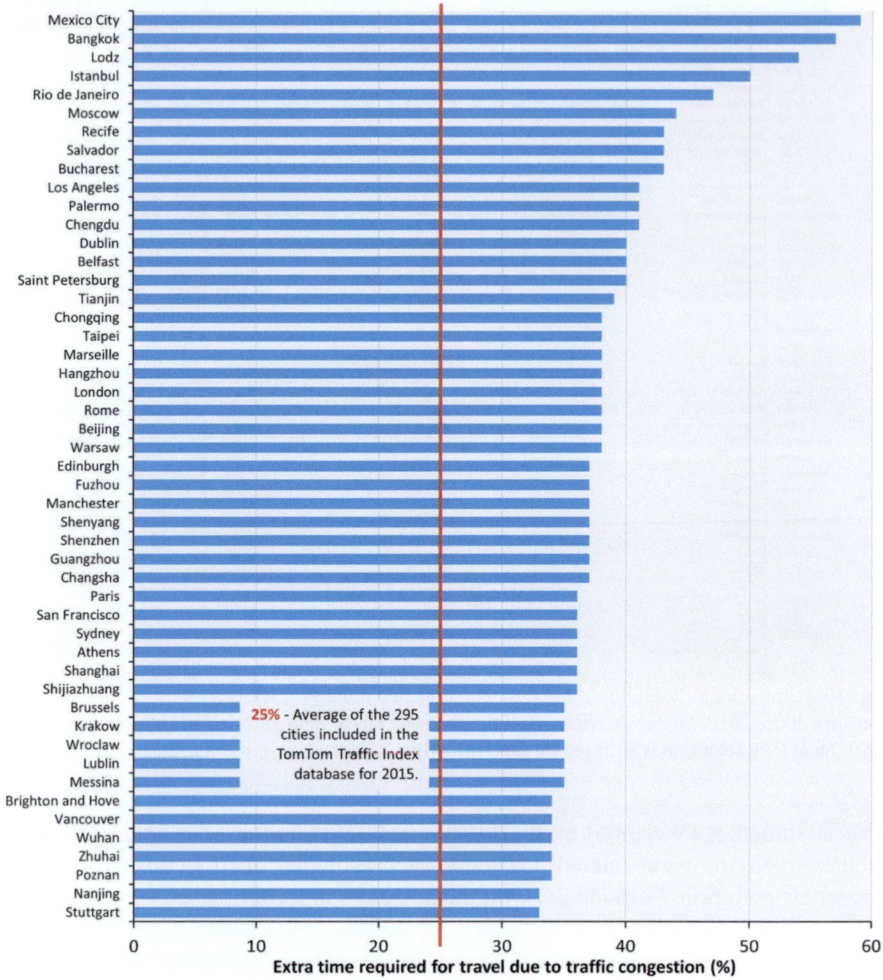

Fig. 19.4 Top 50 cities with the highest TomTom Traffic Index expressed in terms of extra time required for travel due to traffic congestion. *Source* Authors' own figure based on the 2015 TomTom Traffic Index data (TomTom International 2016)

In his analysis using TomTom's previous peak period traffic indexes, Cox (2014) found a significant positive association between population density and traffic congestion, which is also consistent with his previous findings, i.e., equating lower densities with faster travel times and an increasing automobile use in response to higher densities. In the 2015 TomTom Traffic database we used (Fig. 19.4), it can be noted that other major urban areas are not yet included. These include, among others, Jakarta, Manila, Hanoi, Tokyo, Zurich, and Munich, including Indian cities.

19.5 Toward Sustainable Urban Development

Sustainable urban development is a type of development that takes into account the three pillars of sustainability: economic prosperity, social justice, and environmental quality (Elkington 1997; Estoque and Murayama 2014). In principle, sustainable urban development has been part of the *United Nations' Millennium Development Goals (MDGs)* crafted in 2000 (UN 2015c). And due to its importance, it has been made more explicit in the more recent *United Nations' Sustainable Development Goals (SDGs)* (UN 2015b). Of the 17 SDGs declared in 2015, "to make cities and human settlements inclusive, safe, resilient and sustainable" is one of them, which is also known as Goal 11 (sustainable cities and communities). Overall, the 17 SDGs are a part of the new sustainable development agenda adopted by the UN-member nations. Each goal has specific targets, envisioned to be achieved over the next 15 years (2015–2030).

Goal 11 (sustainable cities and communities) provides the overall framework for achieving sustainable urban development worldwide. It has 10 specific targets, as follows (UN 2015b, p. 26):

i. By 2030, ensure access for all to adequate, safe and affordable housing and basic services and upgrade slums;
ii. By 2030, provide access to safe, affordable, accessible and sustainable transport systems for all, improving road safety, notably by expanding public transport, with special attention to the needs of those in vulnerable situations, women, children, persons with disabilities and older persons;
iii. By 2030, enhance inclusive and sustainable urbanization and capacity for participatory, integrated and sustainable human settlement planning and management in all countries;
iv. Strengthen efforts to protect and safeguard the world's cultural and natural heritage;
v. By 2030, significantly reduce the number of deaths and the number of people affected and substantially decrease the direct economic losses relative to global gross domestic product caused by disasters, including water-related disasters, with a focus on protecting the poor and people in vulnerable situations;

vi. By 2030, reduce the adverse per capita environmental impact of cities, including by paying special attention to air quality and municipal and other waste management;
vii. By 2030, provide universal access to safe, inclusive and accessible, green and public spaces, in particular for women and children, older persons and persons with disabilities;
viii. Support positive economic, social and environmental links between urban, peri-urban and rural areas by strengthening national and regional development planning;
ix. By 2020, substantially increase the number of cities and human settlements adopting and implementing integrated policies and plans towards inclusion, resource efficiency, mitigation and adaptation to climate change, resilience to disasters, and develop and implement, in line with the Sendai Framework for Disaster Risk Reduction 2015–2030, holistic disaster risk management at all levels;
x. Support least developed countries, including through financial and technical assistance, in building sustainable and resilient buildings utilizing local materials.

In its report, *State of the World's Cities 2012/2013*, the UN-Habitat (2013) also envisions that the city of the twenty-first century is one that: (1) reduces disaster risks and vulnerabilities for all, including the poor, and builds resilience to any adverse forces of nature; (2) stimulates local job creation, promotes social diversity, maintains a sustainable environment and recognizes the importance of public spaces; (3) creates harmony between the five dimensions of prosperity and enhances the prospects for a better future; and (4) comes with a change of pace, profile, and urban functions and provides the social, political and economic conditions of prosperity (p. xii). It pushes for the concept of *urban prosperity* based on five dimensions, namely productivity, infrastructures, quality of life, equity, and environmental sustainability, in which *a prosperous city* is one that (UN-Habitat 2013, p. xii):

i. contributes to economic growth through productivity, generating the income and employment that afford adequate living standards for the whole population;
ii. deploys the infrastructure, physical assets and amenities—adequate water, sanitation, power supply, road network, information communications technology, etc.—required to sustain both the population and the economy;
iii. provides the social services—education, health, recreation, safety security, etc.—required for improved living standards, enabling the population to maximize individual potential and lead fulfilling lives;
iv. has minimal incidence of poverty and inequalities. No city can claim to be prosperous when large segments of the population live in abject poverty and deprivation. This involves reducing the incidence of slums and new forms of poverty;
v. ensures that the creation and (re)distribution of the benefits of prosperity do not destroy or degrade the environment; instead, the city's natural assets are preserved for the sake of sustainable urbanization.

19.6 Concluding Remarks: The Importance of Geospatial Tools in Urban Monitoring

An important integral part of the above-mentioned targets for the UN SDG—Goal 11 and UN-Habitat's concept of urban prosperity is the monitoring of the spatial and temporal patterns of urbanization (e.g., population growth and built-up expansions) and its impact to the environment. Over the years, geospatial technologies such as remote sensing and geographic information systems (GIS) have become important tools for urban studies.

Remote sensing is a technique for acquiring information about an object or phenomenon without making physical contact with them. One common example of remote sensing is the use of aerial/satellite sensors to detect and classify objects on Earth (see Gibson 2000; Lavender and Lavender 2016). On the other hand, a GIS is a system that integrates software with hardware, data and a user for capturing, storing, updating and analysing spatial data to support analysis and decision-making (see Foresman 1998; Goodchild 2010).

It is not uncommon today that both remote sensing and GIS are used together to monitor the spatial and temporal patterns of urbanization and its impact to the environment. For example, remote sensing data and GIS techniques have been, and still are, used to study the urban heat island phenomenon (e.g., Voogt and Oke 2003; Weng et al. 2004; Estoque and Murayama 2017; Estoque et al. 2017; Ranagalage et al. 2017; Zhang et al. 2017). Other efforts on the use of remote sensing and GIS have focused on, among others, the classification of urban features (e.g., Blaschke 2010; Myint et al. 2011; Weng 2012; Estoque and Murayama 2015b; Estoque et al. 2015a; Pok et al. 2017; Simwanda and Murayama 2017), monitoring of urban expansions and their impacts on landscape patterns (e.g., Thapa and Murayama 2009; Bhatta 2010; Estoque and Murayama 2013, 2016; Bagan and Yamagata 2014; Angel et al. 2015; www.lincolninst.edu/subcenters/atlas-urban-expansion/; http://giswin.geo.tsukuba.ac.jp/mega-cities/), urban growth analysis and modeling (e.g., Murayama and Thapa 2011; Thapa and Murayama 2012; Estoque and Murayama 2012, 2016; Hou et al. 2016; Subasinghe et al. 2016; Murayama et al. 2017), urban flooding simulations (e.g., Chen et al. 2009; Voulgaris and Murayama 2014), urban volume measurement (built-up volume and/or green volume) (e.g., Koomen et al. 2009; Huang et al. 2013; Estoque et al. 2015b; Handayani et al. in press), urban walkability studies (e.g., Murayama 2012; Lwin and Murayama 2011), spatial analysis of people flow in urban areas (e.g., http://land.geo.tsukuba.ac.jp/persontrips.asia/), population estimation (e.g., Lwin and Murayama 2009; http://web.ornl.gov/sci/landscan/), and web-based urban monitoring (e.g., Gong et al. 2017; The World Urban Database and Access Portal Tools at www.wudapt.org/).

Another important growing field of urban studies today is the monitoring of *urban ecosystem services*. In general, ecosystem services refer to the benefits that ecosystems generate for and provide to people. Such benefits can be tangible (goods) or intangible (services), large or small and direct or indirect (Daily 1997; Costanza et al. 1997; MEA 2005; Estoque and Murayama 2013). Some examples of urban ecosystem services include food production, water flow regulation and runoff mitigation, urban temperature regulation, noise reduction, air purification, waste treatment, pollination and seed dispersal, and cultural services such as recreation, aesthetic benefits, place values and social cohesion, and cognitive development (Bolund and Hunhammar 1999; Elmqvist et al. 2013; Gómez-Baggethun and Baton 2013; Haase et al. 2014). Most, if not all, of these services depend on urban vegetation and green spaces. Some of which (e.g., urban temperature regulation, air purification) are also directly related to some of the above-mentioned issues associated with urbanization (e.g., urban heat island, air pollution). Thus, the monitoring of the spatiotemporal pattern of urban green spaces is important. While there have been efforts to quantify the impacts of urban expansions on urban ecosystem services at the city level (e.g., Estoque and Murayama 2016) or country level (e.g., Eigenbrod et al. 2011), such type of inquiry at the global level remains to be explored. In this context, the newly developed database on Global Urban Footprint (GUF) (Esch et al. 2017), alongside the advances in geospatial analysis, can make a valuable contribution.

Overall, the contributions of geospatial technologies such as remote sensing and GIS, not only to the field of urban studies, but also to the understanding of the coupled human-environment system in general towards sustainability, have been great and priceless. In the near future, with the continuous advances in technologies, the use and applications of geospatial tools and techniques are expected to expand across the broad spectrum of human geoscience.

References

Angel S, Blei AM, Civco DL, Sanchez NG, Lamson-Hall P, Madrid M, Parent J, Thom K (2015) Monitoring the quantity and quality of global urban expansion. Marron Institute of Urban Management—New York University, NY

Aniello C, Morgan K, Busbey A, Newland L (1995) Mapping micro-urban heat islands using Landsat TM and a GIS. Comput Geosci 21:965–969

Bagan H, Yamagata Y (2014) Land-cover change analysis in 50 global cities by using a combination of Landsat data and analysis of grid cells. Environ Res Lett 9:64015

Bhatta B (2010) Analysis of urban growth and sprawl from remote sensing data. Springer-Verlag, Heidelberg

Blaschke T (2010) Object based image analysis for remote sensing. ISPRS J Photogrammetry Remote Sens 65:2–16

Bloom DE, Canning D, Fink G (2008) Urbanization and the wealth of nations. Science 319:772–775

Bolund P, Hunhammar S (1999) Ecosystem services in urban areas. Ecol Econ 29:293–301

Chen J, Hill AA, Urbano LD (2009) A GIS-based model for urban flood inundation. J Hydrol 373:184–192

Costanza R, d'Arge R, de Groot R, Farber S, Grasso M, Hannon B et al (1997) The value of the world's ecosystem services and natural capital. Nature 387:253–260

Cox W (2014) Traffic congestion in the world: 10 worst and best cities. New geography. www.newgeography.com/content/004504-traffic-congestion-world-10-worst-and-best-cities (26 July 2016)

Cox W (2016) The TomTom Traffic Index: 2015. Wendell Cox Consultancy, Belleville, IL, USA

Dahiya B (2012) Cities in Asia, 2012: demographics, economics, poverty, environment and governance. Cities 29:S44–S61
Daily GC (1997) Nature's services: societal dependence on natural ecosystems. Island Press, Washington, DC
DEMOGRAPHIA (2016) Demographia: world urban areas, 12th annual edn. Wendell Cox Consultancy, Belleville, IL, USA
Eigenbrod F, Bell VA, Davies HN, Heinemeyer A, Armsworth PR, Gaston KJ (2011) The impact of projected increases in urbanization on ecosystem services. Proc R Soc B. https://doi.org/10.1098/rspb.2010.2754
Elkington J (1997) Cannibals with forks: the triple bottom line of the 21st century business. Capstone, Oxford
Elmqvist T, Fragkias M, Goodness J, Güneralp B, Marcotullio PJ, McDonald RI et al (eds) (2013) Urbanization, biodiversity and ecosystem services: challenges and opportunities. A global Assessment. Springer, Dordrecht
EPA (US Environmental Protection Agency) (2008) Reducing urban heat islands: compendium of strategies—urban heat island basics. US Environmental Protection Agency, Washington, DC
Esch T, Heldens W, Hirner A, Keil M, Marconcini M, Roth A et al (2017) Breaking new ground in mapping human settlements from space—the global urban footprint. ISPRS J Photogrammetry Remote Sens 134:30–42
Estoque RC, Murayama Y (2012) Examining the potential impact of land use/cover changes on the ecosystem services of Baguio city, the Philippines: a scenario-based analysis. Appl Geogr 35:316–326
Estoque RC, Murayama Y (2013) Landscape pattern and ecosystem service value changes: implications for environmental sustainability planning for the rapidly urbanizing summer capital of the Philippines. Landscape Urban Plan 116:60–72
Estoque RC, Murayama Y (2014) Measuring sustainability based upon various perspectives: a case study of a hill station in Southeast Asia. Ambio 43:943–956
Estoque RC, Murayama Y (2015a) Intensity and spatial pattern of urban land changes in the megacities of Southeast Asia. Land Use Policy 48:213–222
Estoque RC, Murayama Y (2015b) Classification and change detection of built-up lands from Landsat-7 ETM + and Landsat-8 OLI/TIRS imageries: a comparative assessment of various spectral indices. Ecol Ind 56:205–217
Estoque RC, Murayama Y (2016) Quantifying landscape pattern and ecosystem service value changes in four rapidly urbanizing hill stations of Southeast Asia. Landscape Ecol 31:1481–1507
Estoque RC, Murayama Y (2017) Monitoring surface urban heat island formation in a tropical mountain city using Landsat data (1987–2015). ISPRS J Photogrammetry Remote Sens 133:18–29
Estoque RC, Murayama Y, Akiyama CM (2015a) Pixel-based and object-based classifications using high- and medium-spatial-resolution imageries in the urban and suburban landscapes. Geocarto Int 30:1113–1129
Estoque RC, Murayama Y, Tadono T, Thapa RB (2015b) Measuring urban volume: geospatial technique and application. Tsukuba Geoenviron Sci 11:13–20
Estoque RC, Murayama Y, Myint SW (2017) Effects of landscape composition and pattern on land surface temperature: an urban heat island study in the megacities of Southeast Asia. Sci Total Environ 577:349–359
Folke C, Jansson A, Rockström J, Olsson P, Carpenter SR, Chapin FS III et al (2011) Reconnecting to the biosphere. Ambio 40:719–738
Foresman TW (ed) (1998) The history of geographic information systems: perspectives from the pioneers. Prentice Hall PTR, Upper Saddle River, NJ
Gibson P (2000) Introductory remote sensing: principles and concepts. Routledge, New York, NY
Gómez-Baggethun E, Baton DN (2013) Classifying and valuing ecosystem services for urban planning. Ecol Econ 86:235–245

Gómez-Baggethun E, Gren A, Barton DN, Langemeyer J, McPhearson T, O'Farrell P et al (2013) Urban ecosystem services. In: Elmqvist T, Fragkias M, Goodness J, Güneralp B, Marcotullio PJ, Robert I, McDonald RI (eds) Urbanization, biodiversity and ecosystem services: challenges and opportunities: a global assessment. Springer, Dordrecht, pp 175–251
Gong H, Simwanda M, Murayama Y (2017) An internet-based GIS platform providing data for visualization and spatial analysis of urbanization in major Asian and African cities. ISPRS Int J Geo-Inf 6:257
Goodchild MF (2010) Twenty years of progress: GIScience in 2010. J Spat Inf Sci 1:3–20
Grimm NB, Faeth SH, Golubiewski NE, Redman CL, Wu JG, Bai XM, Briggs JM (2008) Global change and the ecology of cities. Science 319:756–760
Haase D, Larondelle N, Andersson E, Artmann M, Borgström S, Breuste J et al (2014) A quantitative review of urban ecosystem services assessment: concepts, models and implementation. Ambio 43:413–433
Handayani HH, Estoque RC, Murayama Y (in press) Estimation of built-up and green volume using geospatial techniques: a case study of Surabaya, Indonesia. Sustain Cities Soc. https://doi.org/10.1016/j.scs.2017.10.017
Hoag H (2015) How cities can beat the heat. Nature 524:402–404
Hou H, Estoque RC, Murayama Y (2016) Spatiotemporal analysis of urban growth in three African capital cities: A grid-cell-based analysis using remote sensing data. J Afr Earth Stud 123:381–391
Hoverter SP (2012) Adapting to urban heat: a tool kit for local governments. Georgetown Climate Center, Washington, DC
Howard L (1818) The climate of London, vol 1. W. Phillips, London
Huang Y, Yu B, Zhou J, Hu C, Tan W, Hu Z, Wu J (2013) Toward automatic estimation of urban green volume using airborne LiDAR data and high resolution remote sensing images. Front Earth Sci 7:43–54
Jansson A (2013) Reaching for a sustainable: resilient urban future using the lens of ecosystem services. Ecol Econ 86:285–291
Jha AK, Bloch R, Lamond J (2012) Cities and flooding: a guide to integrated flood risk management for the 21st century. The World Bank, Washington, DC
Koomen E, Rietveld P, Bacao F (2009) The third dimension in urban geography: the urban-volume approach. Environ Plan B Plan Des 36:1008–1025
Lavender S, Lavender A (2016) Practical handbook of remote sensing. CRC Press, Boca Raton, FL
Lwin KK, Murayama Y (2009) A GIS approach to estimation of building population for micro-spatial analysis. Trans GIS 13:401–414
Lwin KK, Murayama Y (2011) Modelling of urban green space walkability: eco-friendly walk score calculator. Comput Environ Urban Syst 35:408–420
MEA (Millennium Ecosystem Assessment) (2005) Ecosystems and human well-being: current state and trends: findings of the condition and trends working group. Island Press, Washington, DC
Myers SR et al (2013) Safety in numbers: are major cities the safest places in the United States? Ann Emerg Med 62:408–418
Myint SW, Gober P, Brazel A, Grossman-Clarke S, Weng Q (2011) Per-pixel vs. object based classification of urban land cover extraction using high spatial resolution imagery. Remote Sens Environ 115:1145–1161
Murayama Y, Thapa RB (eds) (2011) Spatial analysis and modeling in geographical transformation process: GIS-based applications. GeoJournal Library Series, 100, Springer, Dordrecht
Murayama Y (ed) (2012) Progress in geospatial analysis. Springer, Tokyo
Murayama Y, Estoque RC (2017) Future of metropolitan areas in developing Asia and Africa. In: Murayama Y, Kamusoko C, Yamashita A, Estoque RC (eds) Urban development in Asia and Africa—geospatial analysis of metropolises. Springer Nature, Singapore, pp 415–420
Murayama Y, Estoque RC, Hou H, Gong H, Simwanda M, Subasinghe S and Zhang X (2016) Visualization of land-use/land-cover changes in major Asian and African cities. Annual Report on the Multi Use Social and Economic Data Bank, University of Tsukuba, Japan

Murayama Y, Kamusoko C, Yamashita A, Estoque RC (eds) (2017) Urban development in Asia and Africa—geospatial analysis of metropolises. Springer Nature, Singapore

Oke TR (1973) City size and the urban heat island. Atmos Environ 7:769–779

Oke TR (1982) The energetic basis of the urban heat island. Q J R Meteorol Soc 108:1–24

Oke TR, Hannel FG (1970) The form of the urban heat island in Hamilton, Canada. WMO Tech. Note No. 108, WMO No. 254 TP. 141:113–126

Pok S, Matsushita B, Fukushima T (2017) An easily implemented method to estimate impervious surface area on a large scale from MODIS time-series and improved DMSP-OLS nighttime light data. ISPRS J Photogrammetry Remote Sens 133:104–115

Qin Y (2015) A review on the development of cool pavements to mitigate urban heat island effect. Renew Sustain Energy Rev 52:445–459

Ranagalage M, Estoque RC, Murayama Y (2017) An urban heat island study of the Colombo Metropolitan Area, Sri Lanka, based on Landsat data (1997–2017). ISPRS Int J Geo-Inf 6:189

Rothenburg R, Weaver SR, Dai D, Strauber C, Prasad A, Kano M (2014) A flexible urban health index for small area disparities. J Urban Health: Bull NY Acad Med 91:823–835

Seto KC, Reenberg A, Boone CG, Fragkias M, Haase D et al (2012) Urban land teleconnections and sustainability. Proc Natl Acad Sci 109:7687–7692

Sharma A, Conry P, Fernando HJS, Hamlet AF, Hellmann JJ, Chen F (2016) Green and cool roofs to mitigate urban heat island effects in the Chicago metropolitan area: evaluation with a regional climate model. Environ Res Lett 11:064004

Subasinghe S, Estoque RC, Murayama Y (2016) Spatiotemporal analysis of urban growth using GIS and Remote Sensing: a case study of the Colombo Metropolitan Area, Sri Lanka. ISPRS Int J Geo-Inf 5:197

Simwanda M, Murayama Y (2017) Integrating geospatial techniques for urban land use classification in the developing Sub-Saharan African City of Lusaka, Zambia. ISPRS Int J Geo-Inf 6:102

Thapa RB, Murayama Y (2009) Examining spatiotemporal urbanization patterns in Kathmandu Valley, Nepal: remote sensing and spatial metrics approaches. Remote Sens 1:534–556

Thapa RB, Murayama Y (2012) Scenario based urban growth allocation in Kathmandu Valley, Nepal. Landscape Urban Plan 105:140–148

The World Bank (2011) Violence in the city: understanding and supporting community responses to urban violence. The World Bank, Washington, DC

The World Bank (2016a) Reducing pollution. The World Bank, Washington, DC

The World Bank (2016b) Urban poverty: an overview. The World Bank, Washington, DC

TomTom International (2016) TomTom traffic index: measuring congestion worldwide. www.tomtom.com/en_gb/trafficindex/ (26 July 2016)

UN (United Nations) (2015a) World urbanization prospects: the 2014 revision. United Nations, New York

UN (United Nations) (2015b) Transforming our world: the 2030 agenda for sustainable development. United Nations, New York

UN (United Nations) (2015c) The millennium development goals report 2015. United Nations, New York

UN-Habitat (2010) State of the world's cities 2010/2011: bridging the urban divide. Earthscan, London

UN-Habitat (2013) State of the world's cities 2012/2013: prosperity of cities. Routledge, New Yok

UNODC (United Nations Office on Drugs and Crime) (2014) UNODC Annual report 2014. UNODC, Vienna, Austria

Voogt J, Oke TR (2003) Thermal remote sensing of urban climates. Remote Sens Environ 86:370–384

Voogt J (2002) Urban heat island. In: Munn T (ed) Encyclopedia of global environmental change, vol 3. Wiley, Chichester, pp 660–666

Voogt JA (2004) Urban heat islands: hotter cities. www.actionbioscience.org/environment/voogt.html (17 July 2016)

Voulgaris G, Murayama Y (2014) Tsunami vulnerability assessment in the Southern Boso Peninsula, Japan. Int J Disaster Risk Reduction 10:190–200
Weisbrod G, Vary D, Treyz G (2003) Measuring economic costs of urban traffic congestion to business. Transp Res Record: J Transp Res Board 1839:98–106
Weng Q (2012) Remote sensing of impervious surfaces in the urban areas: requirements, methods, and trends. Remote Sens Environ 117:34–49
Weng Q, Lu D, Schubring J (2004) Estimation of land surface temperature-vegetation abundance relationship for urban heat island studies. Remote Sens Environ 89:467–483
WHO (World Health Organization) (2006) WHO air quality guidelines for particulate matter, ozone, nitrogen dioxide and sulfur dioxide: global update 2005. World Health Organization, Geneva, Switzerland
WHO (World Health Organization) (2014) The urban health index: a handbook for its calculation and use. Health Organization, Geneva, Switzerland
WHO (World Health Organization) (2016a) WHO global urban ambient air pollution database (update 2016). World Health Organization, Geneva, Switzerland
WHO (World Health Organization) (2016b) Air pollution levels rising in many of the world's poorest cities. World Health Organization, Geneva, Switzerland
WHO (World Health Organization) (2016c) Global report on urban health: equitable, healthier cities for sustainable development. World Health Organization, Geneva, Switzerland
Wu J (2010) Urban sustainability: an inevitable goal of landscape research. Landscape Ecol 25:1–4
Zhang X, Estoque RC, Murayama Y (2017) An urban heat island study in Nanchang City, China based on land surface temperature and social-ecological variables. Sustain Cities Soc 32:557–568

Chapter 20
Climate Change We Human Are Responsible for

Akio Kitoh

Abstract Global surface air temperature has warmed about one degree during the industrial era. Continuous emission of greenhouse gases will cause further warming and long-lasting changes in all components of the climate system. In particular, changes in extreme weather and climate events will have inevitable impacts on human system.

Keywords Climate change · Greenhouse gases · Global warming

20.1 What Is Climate Change?

The Glossary in the Intergovernmental Panel on Climate Change (IPCC) Working Group I (WGI) Fifth Assessment Report (AR5) states that "Climate change refers to a change in the state of the climate that can be identified (e.g., by using statistical tests) by changes in the mean and/or the variability of its properties, and that persists for an extended period, typically decades or longer. Climate change may be due to natural internal processes or external forcings such as modulations of the solar cycles, volcanic eruptions and persistent anthropogenic changes in the composition of the atmosphere or in land use. Note that the [United Nations] Framework Convention on Climate Change (UNFCCC), in its Article 1, defines climate change as: 'a change of climate which is attributed directly or indirectly to human activity that alters the composition of the global atmosphere and which is in addition to natural climate variability observed over comparable time periods'. The UNFCCC thus makes a distinction between climate change attributable to human activities altering the atmospheric composition, and climate variability attributable to natural causes" (IPCC 2013).

Climate change, particularly changes in extreme weather and climate events, will affect human and society on their vulnerable parts. Thus, addressing climate change is one of global goals in the 2030 Agenda for Sustainable Development Goals (SDGs) (UN 2015). Integrated research on climate change is an immediate issue in various

A. Kitoh (✉)
Japan Meteorological Business Support Center, Tsukuba, Japan
e-mail: akiokitoh@gmail.com

Y. Himiyama et al. (eds.), *Human Geoscience*, Advances in Geological Science,
https://doi.org/10.1007/978-981-32-9224-6_20

communities such as World Climate Research Program (WCRP) (WCRP 2015), The Global Programme of Research on Climate Change Vulnerability, Impacts and Adaptation (PROVIA) (UNEP 2013) and Future Earth (Future Earth 2014).

20.2 Observational Changes in the Climate System

There is multiple evidence of global climate change in the past century. IPCC (2013) assembles multiple indicators of a changing global climate (Fig. 20.1). Global average near-surface temperature for 2016 was the warmest on record, and was approximately 1 °C above the 1850–1900 average. It is noted that there exists substantial interannual, decadal and multi-decadal variability in the temperature record, including the mid-twentieth century and the early-twenty-first century warming hiatus

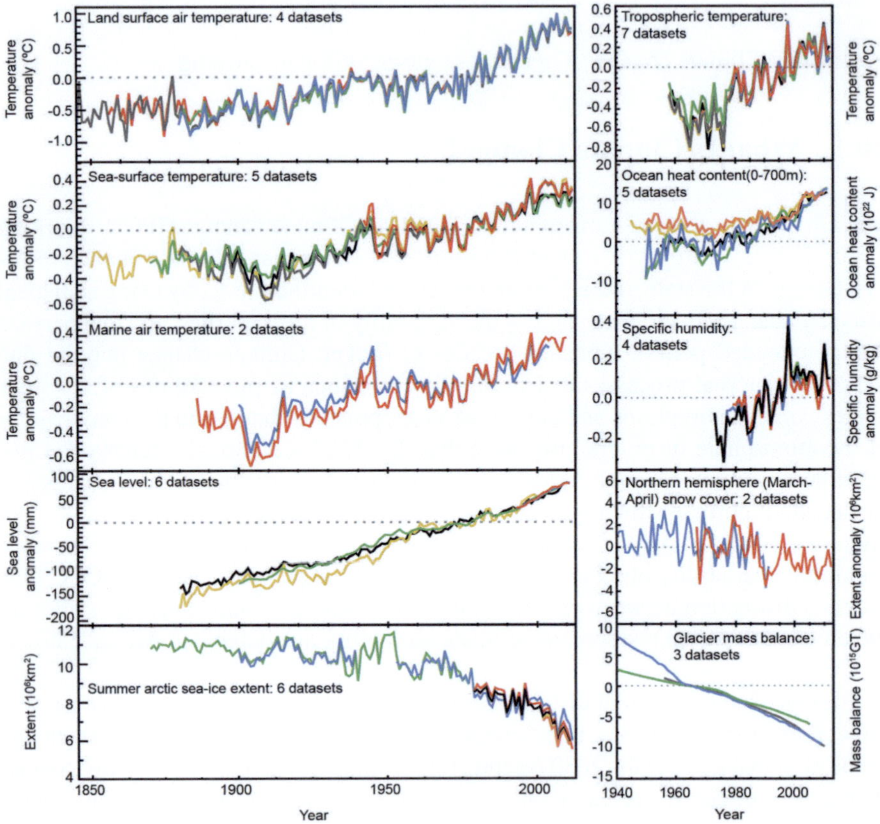

Fig. 20.1 Multiple complementary indicators of a changing global climate. Taken from IPCC WGI AR5 Figure TS.1, IPCC (2013)

(pause). Three-dimensional atmospheric temperature data shows the tropospheric warming and the stratospheric cooling since the mid-twentieth century. The upper ocean (from surface to 700 m) temperature has warmed from 1971. More recent data revealed the ocean warming below 700 m. The Arctic sea ice extent decreased over the period 1979–2012. The rate of the annual decrease during this period is between 3.5 and 4.1% per decade. The extent of Arctic perennial and multi-year ice decreased with a greater rate. Glacier length, area, volume, and mass are decreasing globally. The springtime snow cover extent has decreased in the Northern Hemisphere. Global mean sea level has risen by 0.19 m over the period 1901–2010. The mean rate of sea-level rise is accelerating during this period. The pH of ocean surface water has decreased by 0.1, corresponding a 26% increase in hydrogen ion concentration (Fig. 20.2b).

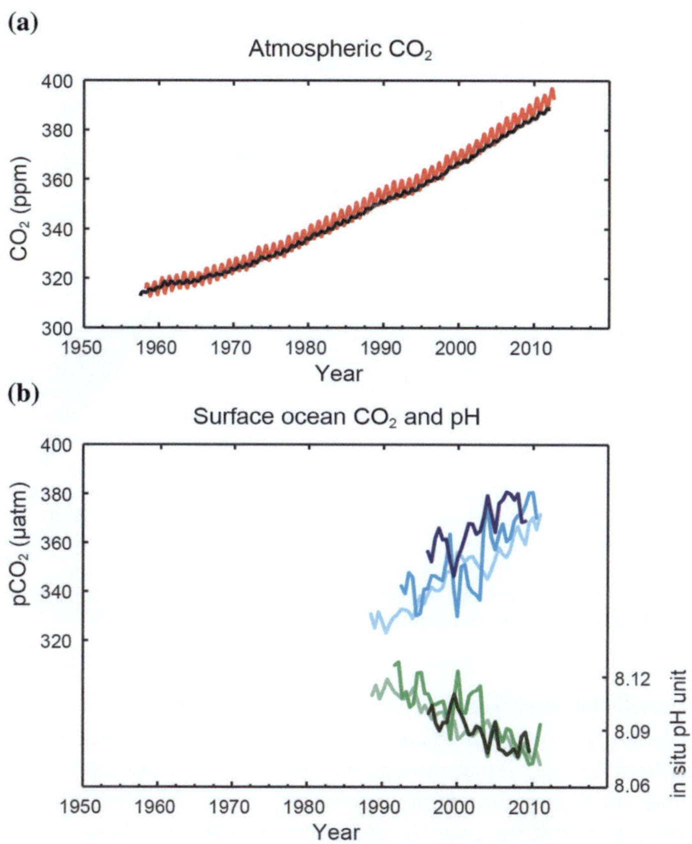

Fig. 20.2 **a** Atmospheric concentration of CO_2 from Mauna Loa and South Pole. **b** Partial pressure of dissolved CO_2 at the ocean surface and in situ pH. Taken from IPCC WGI AR5 Figure SPM.4, IPCC (2013)

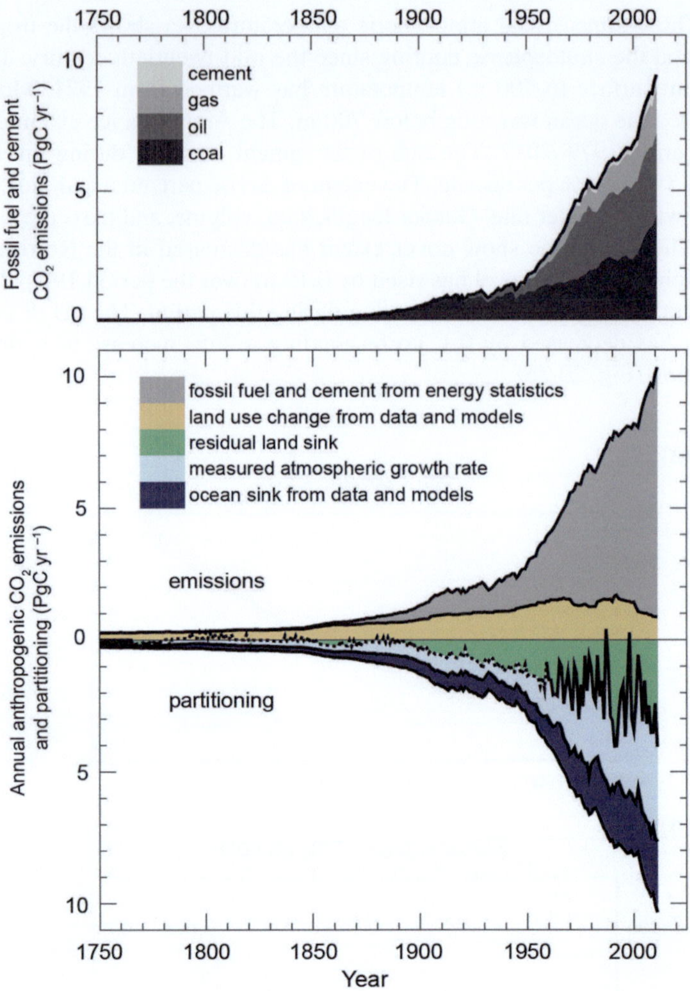

Fig. 20.3 Annual anthropogenic CO_2 emissions and their partitioning among the atmosphere, land, and ocean from 1750 to 2011. Taken from IPCC WGI AR5 Figure TS.4, IPCC (2013)

During the glacial-interglacial cycle in the past 800,000 years, atmospheric concentration of CO_2 was between 185 and 280 ppm. At the mid-Holocene (7000 years ago), atmospheric CO_2 level measured in Antarctica is about 260 ppm. It increased to about 280 ppm at the beginning of the Industrial Era in 1750. The reason of this CO_2 increase during the Holocene could be compensating effects of land use, oceanic and terrestrial processes (IPCC 2013). Atmospheric concentration of CO_2 increased 43% from 278 to 400 ppm during the period 1750–2015 (Fig. 20.2a). The main driver of this CO_2 increase is anthropogenic emissions.

Figure 20.3 shows temporal records of CO_2 emissions and partitioning. Since the industrial revolution, anthropogenic emission of CO_2 drastically increased and its speed is accelerating. The CO_2 emissions from fossil fuel and cement are 375 PgC between 1750 and 2011. The land-use change (mainly forest clearance) released 180 PgC. Of the 555 PgC released to the atmosphere from fossil fuel, cement, and land-use change, the ocean and terrestrial ecosystems absorbed 155 and 160 PgC, respectively. Remaining 240 PgC is accumulated in the atmosphere. As shown in Fig. 20.3a, annual amount of fossil fuel emissions is increasing. The fossil fuel emissions in 2011 are 9.5 PgC. The atmospheric CO_2 concentration in the Northern Hemisphere is larger than that in the Southern Hemisphere (Fig. 20.2a), strongly indicating the Northern Hemisphere source of CO_2 emissions.

20.3 Drivers of Past Climate Change

Earth's climate is governed by incoming solar and outgoing terrestrial radiation, and its partitioning into various components. Climate is not in equilibrium due to natural variability including solar variability, volcanic eruptions, and interactions among atmosphere, ocean, cryosphere, and biosphere, but can be thought as in quasi-equilibrium. Human intervention into the climate system is a new component.

Anthropogenic forcing since the industrial revolution is very large, changing the Earth's energy budget, and thus our climate. Radiative forcing is a metric to quantify the changes in net downward radiative flux at the tropopause. Positive (negative) value in radiative forcing warms (cools) the Earth's surface. Figure 20.4 shows radiative forcing estimates in 2011 relative to 1750. The total anthropogenic radiative forcing for 2011 is 2.29 W m^{-2}. The radiative forcing from emission of well-mixed greenhouse gases (CO_2, CH_4, N_2O, and Halocarbons) is 3.00 W m^{-2}, for which emissions of CO_2 alone contributes 1.68 W m^{-2}. The radiative forcing of the total aerosol effect is −0.9 W m^{-2}, though its estimate has large uncertainty. The radiative forcing by land use is negative because conversion from forest to grass or bare land increases surface albedo and acts to reflect more solar radiation. Overall, increasing atmospheric concentration of CO_2 is the largest contributor for the climate change. CO_2 is well mixed in the atmosphere and shows spatially homogeneous forcing, but aerosols and land use are highly heterogeneous. South Asia and East Asia are the regions of large negative forcing due to aerosols emissions.

20.4 Drivers of Future Climate Change

Future climate depends on future anthropogenic emissions of greenhouse gases. Cumulative emissions of CO_2 largely determine global mean surface warming by the late twenty-first century and beyond (IPCC 2014). Anthropogenic emissions of greenhouse gases are mainly driven by population size, economic activity, lifestyle, energy use, land use patterns, technology, and climate policy. For future climate

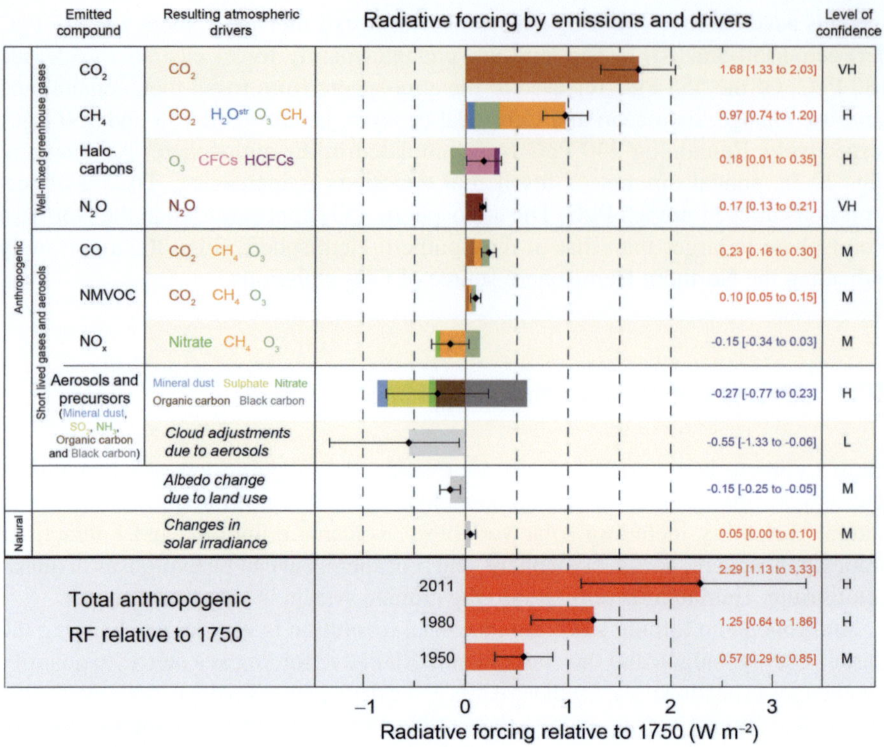

Fig. 20.4 Radiative forcing estimates in 2011 relative to 1750. Taken from IPCC WGI AR5 Figure SPM.5, IPCC (2013)

projections by climate models, scenarios of future time series of greenhouse gases emissions and atmospheric concentrations, air pollutant emissions and land use are needed. For that purpose, the Representative Concentration Pathways (RCPs) are created. The RCPs include a stringent mitigation scenario (RCP2.6), two intermediate scenarios (RCP4.5 and RCP6.0) and one scenario with very high greenhouse gases emissions (RCP8.5). RCP2.6 (RCP8.5) is a scenario with radiative forcing at 2100 is about 2.6 (8.5) W m^{-2}. Note that the total anthropogenic radiative forcing for 2011 is 2.29 W m^{-2}. RCP2.6 is representative of a scenario that aims to keep global warming likely below 2 °C above pre-industrial level. The atmospheric concentration of CO_2 at 2100 would be 421, 538, 670 and 936 ppm for RCP2.6, RCP45, RCP6.0, and RCP8.5, respectively.

20.5 Projections of Global and Regional Climate Change

Coordinated climate model experiments have been performed by research community under the Coupled Model Intercomparison Project Phase 5 (CMIP5) of the World Climate Research Programme (WCRP). World climate modeling centers/groups have participated in CMIP5 and provided future climate projections under the given future scenarios, i.e., the RCPs. Figure 20.5 shows time series of some projected variables. Global annual mean surface air temperature will rise 0.3–1.7 °C in RCP2.6 and 2.6–4.8 °C in RCP8.5 at the end of the twenty-first century (2081–2100) relative to 1986–2005. In other words, end-of-the-century temperature is projected to likely (probability > 66%) exceed 1.5 °C for RCP4.5, RCP6.0, and RCP8.5. Note that global temperature has already risen 0.61 °C from 1850–1900 to 1986–2005.

There are sources of uncertainty. Some models have a higher sensitivity to greenhouse gas forcing than others, and thus larger changes. Shadings indicate the 5–95%

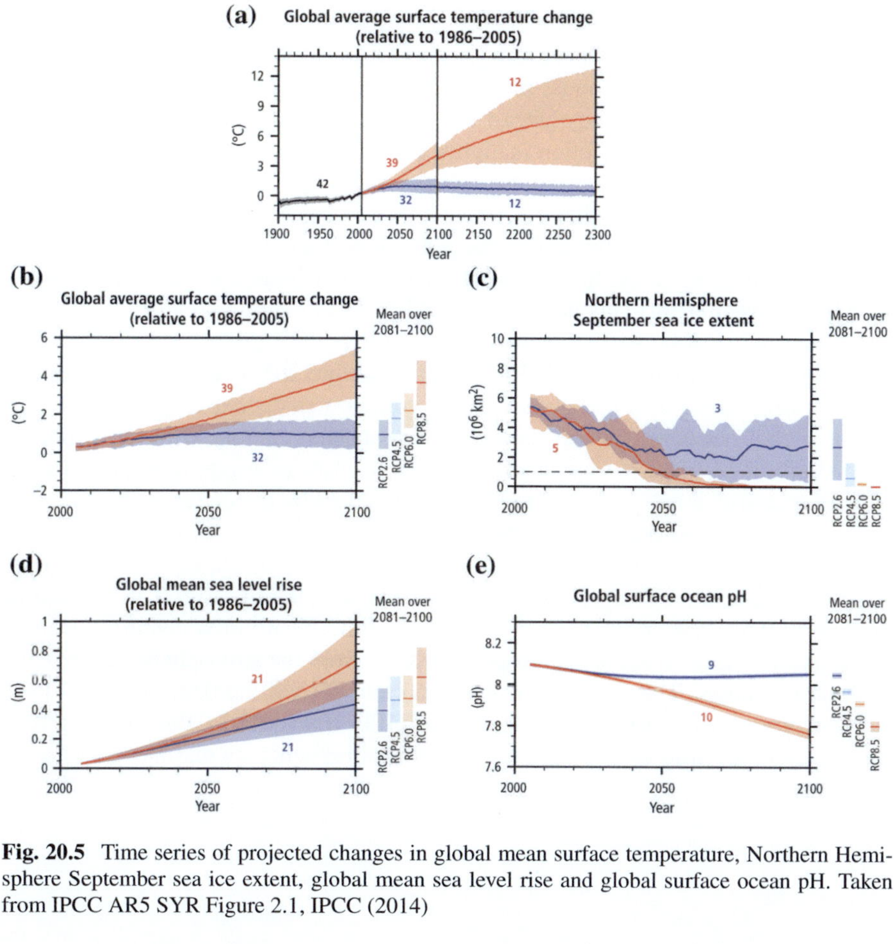

Fig. 20.5 Time series of projected changes in global mean surface temperature, Northern Hemisphere September sea ice extent, global mean sea level rise and global surface ocean pH. Taken from IPCC AR5 SYR Figure 2.1, IPCC (2014)

range across the climate models used (the number of CMIP5 models used is shown within the figure). For individual model projection, there is large interannual and decadal variability, but multi-model ensemble mean canceled out this variability. Another source of uncertainty is scenario. For the near-term projection (e.g., 2030s), model uncertainty is larger than scenario uncertainty, but at the end of the twenty-first century, scenario uncertainty dominates, implying that which emission scenario we actually adopt will determine the magnitude of future climate change.

It should be noted that temperature change is not globally uniform. The mean warming over land will be larger than over the ocean (the former is 1.4–1.7 times larger than the latter). The Arctic region will warm much more. Therefore, we will face a larger warming than the widely used/discussed value of global mean surface temperature change. As temperature rises, it is certain that there will be more frequent hot and fewer cold temperature extremes. Heat waves will occur with a higher frequency and longer duration.

In the long term, global mean precipitation will increase, but changes will not be spatially uniform. Precipitation will increase in high latitudes due to increased atmospheric moisture content by global warming. Many mid-latitude and subtropical arid and semi-arid regions where today's climate is dry will receive less precipitation. Thus the contrast of annual mean precipitation between dry and wet regions will increase. In addition, the contrast between wet and dry seasons will increase. However, precipitation projection is highly uncertain. Extreme precipitation events will become more intense and more frequent as already observed in many regions. The CMIP5 models project that the global average sensitivity of the 20-year return value of the annual maximum daily precipitation will increase by 5.3% per 1 °C of local temperature increase. In other words, a once in a 20-year event will become a once in a 14-year event. This value varies widely from region to region.

Reduction in Arctic sea ice is projected. Some models project a nearly ice-free Arctic Ocean in September for RCP8.5 scenario before 2050s. The Northern Hemisphere spring snow cover and global glacier volume will decrease.

The global ocean will warm during the twenty-first century with the strongest warming for the surface in tropical and Northern Hemisphere subtropical regions. In RCP4.5 scenario, a half (85%) of the energy absorbed by the ocean will be trapped in the upper 700 (2000) m. Because heat transport from surface to deep oceans takes long time, ocean warming will continue for centuries with resultant sea-level rise. Projected sea-level rise largely depends on scenarios ranging from 0.28–0.62 m in RCP2.6 to 0.52–0.98 m in RCP8.5 at the end of the twenty-first century. Because the oceans continue to uptake anthropogenic CO_2 from the atmosphere, ocean acidification continues. The decrease in surface ocean pH will be 0.06–0.07 (15–17% increase in acidity) for RCP2.6 and 0.30–0.32 (100–109%) for RCP8.5 by the end of the twenty-first century. This will have substantial influence on marine ecosystems.

20.6 Climate Stabilization

CO_2 emitted by anthropogenic activities since the Industrial Revolution is absorbed by about 30% each in the marine and terrestrial ecosystems, and the rest (>40%) has been accumulated in the atmosphere. This ratio will change both in the short term and in the long term. Absorption by terrestrial ecosystem is considered to become less. Absorption of CO_2 by the ocean will continue at least during the twenty-first century so that it is certain that ocean acidification becomes more.

IPCC (2013) showed that total amount of CO_2 emitted by human activities (cumulative carbon emissions) is a good indicator of global warming. The ratio of the change in global mean surface temperature to cumulative total CO_2 emissions is relatively constant irrespective of the time and does not depend on the scenario (Fig. 20.6). Due to uncertainty of climate model projections, temperature change per cumulative emissions 1000 PgC (1 trillion tons) is 0.8–2.5 °C (range of the lower shadow in Fig. 20.6). The upper shadow of the figure is the temperature change in the RCP scenario, which includes other greenhouse gases.

Lifetime of greenhouse gases and aerosols is different among them. Due to atmospheric adjustments, the effect of emissions of greenhouse gases and aerosols emissions decay quickly except for CO_2. As CO_2 remains in the atmosphere on time scales of centuries, CO_2 creates a long-lived warming effect. Warming due to anthropogenic CO_2 emissions is expected to remain for centuries even emissions are completely stopped. For this reason, as long as the net anthropogenic CO_2 emissions will not be negative over a long period of time, a large part of the climate change will be irreversible in the human time scale.

When we assume a normal distribution with temperature increase of 0.8–2.5 °C per 1000 PgC emissions, in order to limit temperature change since 1861–1880 less than 2 °C with probability >33%, >50% and >66%, it is necessary to suppress the cumulative CO_2 emissions since 1870 less than 1570, 1210 and 1000 PgC, respectively. A total 515 PgC has been emitted already in 1870–2011. This implies that the more CO_2 are emitted in near future, much fewer or even negative emissions is needed later. Considering greenhouse gases other than CO_2, the value reduces to about 900, 820, and 790 PgC, respectively.

From the above relationship, future greenhouse gas emissions and the stabilization target of global warming is linked. There will be a limit of cumulative total greenhouse gases emissions for a given target such as 2 and 3 °C of global average temperature rise for the end of the twenty-first century since the industrial revolution. There are multiple paths. For the same goal, late reduction of anthropogenic greenhouse gases emissions means a rapid reduction later. At the end of year 2015, countries worldwide submitted a plan called Intended Nationally Determined Contributions (INDC). Calculations based on that, global surface temperature will rise about 2.7 °C at the end of the twenty-first century.

Stabilization of the global average temperature does not mean that all the climate system stabilizes. Terrestrial and marine ecosystems, ice sheets and sea-level rise have their own (longer) time scales. As long as the oceans continue to absorb CO_2,

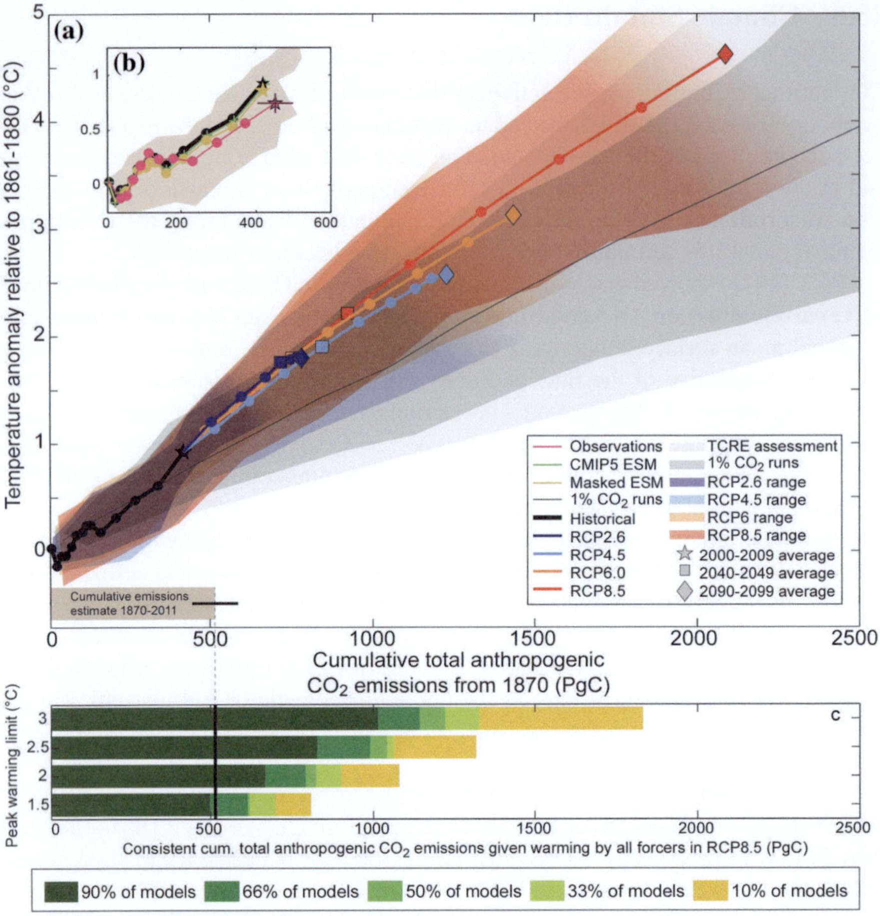

Fig. 20.6 **a** Global mean temperature increase since 1861–1880 as a function of cumulative total global CO_2 emissions. **b** Comparison of historical model results with observations. Taken from IPCC WGI AR5 TFE.8 Figure, IPCC (2013)

ocean acidification will also continue in the future. Sea level will rise beyond 2100 because thermal expansion of the ocean will continue for hundreds or thousands of years.

References

Future Earth (2014) Strategic research agenda 2014. ISBN: 978-0-930357-96-2

IPCC (2013) Climate Change 2013: The physical science basis. Contribution of Working Group I to the fifth assessment report of the Intergovernmental Panel on Climate Change [Stocker TF, Qin D, Plattner GK, Tignor M, Allen SK, Boschung J, Nauels A, Xia Y, Bex V, Midgley PM (eds)]. Cambridge University Press, Cambridge, United Kingdom and New York, NY, USA

IPCC (2014) Climate change 2014: synthesis report. Contribution of Working Group I, II, and III to the fifth assessment report of the Intergovernmental Panel on Climate Change [Core Writing Team, Pachauri RK, Meyr LA (eds)]. IPCC, Geneva, Switzerland

UNEP (2013) Research priorities on vulnerability, impacts and adaptation. Responding to the climate change challenge. ISBN: 978-92-807-3317-4

United Nations (2015) Transforming our world: the 2030 Agenda for Sustainable Development. Resolution adopted by the General Assemble on 25 September 2015, A/RES/70/1

WCRP (2015) IPCC AR5: workshop report: lessons learnt for climate change research and WCRP. International Space Science Institute, Bern, Switzerland, 8–10 September 2014. WCRP Report No. 5/2015

Chapter 21
Anthropogenic Lead Pollution in the Ocean

Toshitaka Gamo

Abstract Since the industrial revolution in the eighteenth to nineteenth centuries, humans have greatly increased the flux of lead (Pb) to the atmosphere and ocean, mainly through the combustion of coals and leaded gasoline. Although such anthropogenic Pb emissions gave little harmful consequences in the open ocean, there have been apparent changes in Pb concentration and its stable isotopic ratios in seawater. Time-series Pb records in the surface oceans have been and will be constructed by the measurement of not only seawaters but corals recording the past seawater Pb during their growth. The Pb concentrations in the North Atlantic surface water, which showed tenfold increase between 1880 and 1970 reflecting the anthropogenic Pb inputs, have decreased drastically since the middle of 1970s due to the prohibition of leaded gasoline in U.S. and European countries. On the other hand, surface waters of northern and tropical Indian Ocean as well as those in the western Pacific Ocean showed significantly higher Pb concentrations than those of the North Atlantic Ocean, implying recent rapid industrialization and a late phase-out of leaded gasoline in developing countries in Asia.

Keywords Leaded gasoline · Pb pollution · Seawater · Atlantic Ocean · Indian Ocean · Developing countries in Asia

21.1 Introduction

Human activities have increased the flux of lead (Pb) to the atmosphere, mainly through leaded gasoline combustion and high-temperature industries such as smelting, mining, and coal combustion, for the past two centuries since the industrial revolution. Lead is a volatile element at high temperature and emitted as atomic vapor when Pb-containing materials are sufficiently heated. Lead is also emitted as reactive PbBrCl compounds upon the leaded gasoline combustion.

T. Gamo (✉)
Atmosphere and Ocean Research Institute, The University of Tokyo, 5-1-5 Kashiwanoha, Kashiwa 277-8564, Japan
e-mail: gamo@aori.u-tokyo.ac.jp

Y. Himiyama et al. (eds.), *Human Geoscience*, Advances in Geological Science,
https://doi.org/10.1007/978-981-32-9224-6_21

Lead atoms emitted to the atmosphere are apt to attach to the surface of, or chemically react with, fine aerosol particles (Cziczo et al. 2009), since Pb is not stable as a gas at low temperatures. Such Pb-contaminated particles are dispersed far away with the global atmospheric transport system, and depositing on land or ocean surface (Reuer and Weiss 2002). Since Pb has relatively short residence time (within a few years) in the surface seawater as estimated from the distribution of the natural radioisotope ^{210}Pb (Bacon et al. 1976; Nozaki et al. 1976), the Pb concentration in the surface ocean tends to responsively follow the recent deposition of contaminated aerosols from the atmosphere.

The anthropogenic Pb emission could be regarded as one of the great "global geophysical experiments" whereby the human race alters the global environment and waits to observe the consequences (Revelle and Suess 1957), as similar as the cases of fossil-fuel CO_2, nuclear bomb fallout, and atmospheric trace gases such as CH_4, N_2O, freons, etc. Although Pb concentrations in the modern ocean are fortunately not so high as they cause harmful consequences to marine life or humans, the oceanic Pb distribution distinctly reminds us of the evolving extent of the human footprint upon the natural environment (Boyle et al. 2014).

21.2 Anthropogenic Emission of Leaded Gasoline

The temporal changes of the leaded gasoline consumption in recent decades clearly characterize the regional variation of anthropogenic activities in the world.

Figure 21.1 shows the history of gasoline Pb consumption in the USA and four European countries (France, Germany, Italy, and UK) between 1930 and 1993 (Wu and Boyle 1997). The gasoline consumption by the four European countries accounts

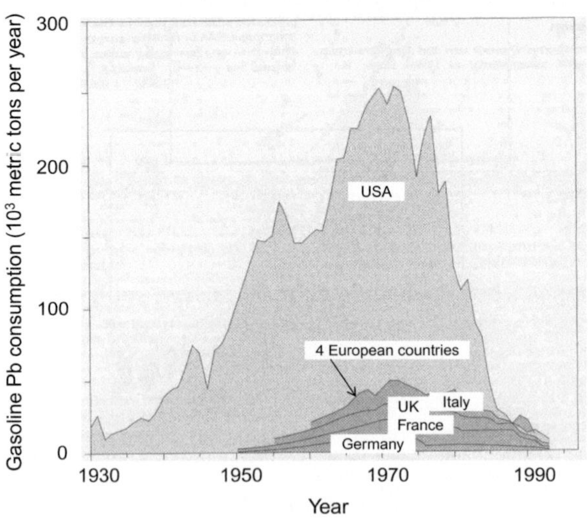

Fig. 21.1 Leaded gasoline consumption in the US and European countries (France, UK, Germany, and Italy) from 1930s to 1990s. Detailed data sources are shown in Wu and Boyle (1997)

for ~70% of the total consumption by all western European countries. It is apparent that the consumption in the USA was much higher than that in western Europe.

The phase-out of Pb from the US gasoline was started in 1970 when the US Environmental Protection Agency regulated the elimination of leaded gasoline (the US Clean Air Act of 1970). In the middle of 1980s, the Pb consumption by US gasoline decreased to less than 10% of its peak value in ~1970 (Fig. 21.1). Similar controls were also established in the western European counties. The decrease of the Pb emission since the 1970s is a notable result of such controls against the use of leaded gasoline.

On the other hand, such regulations were generally delayed in developing countries. Figure 21.2 shows the history of gasoline Pb emission from nine major countries around the Indian Ocean (Lee et al. 2014). If we compare Figs. 21.1 and 21.2, it is apparent that the phase-out of leaded gasoline in the developing countries was significantly late compared to that in the USA and western European countries. This gap or time lag is contrastively reflected in the Pb distribution in the Atlantic and Indian Oceans as described later.

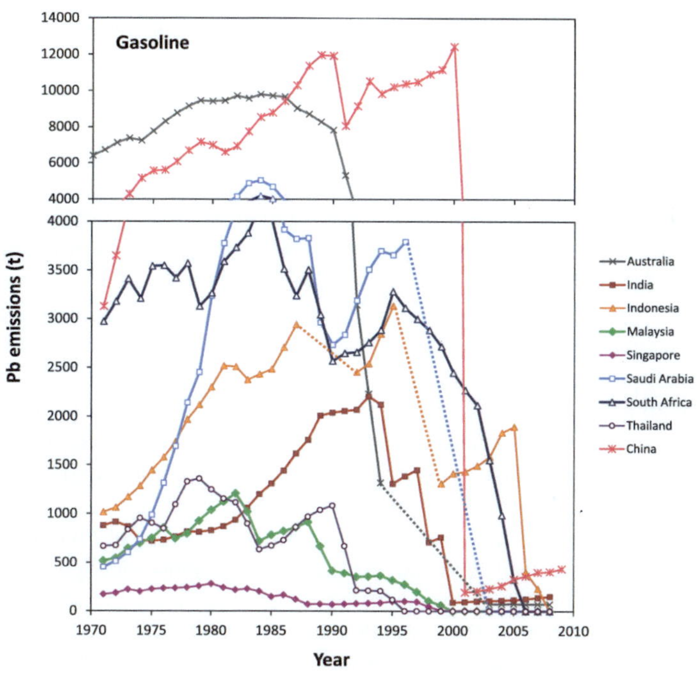

Fig. 21.2 Historical gasoline Pb emissions from the countries around the Indian Ocean, referred from Lee et al. (2014). Detailed data sources are available in Supplemental Material II of Lee et al. (2014)

21.3 Trace Pb Analysis in Seawater and Coral

The pervasive nature of Pb contamination during sampling and analysis, together with its very low concentrations (sub-nanomolar level) in seawater, had seriously hindered successful measurements of seawater Pb till the 1970s. It is only in the past 3–4 decades that we have succeeded to obtain precise records on the evolution of anthropogenic Pb in the ocean, which will be shown in the next section in detail.

As Pb-research pioneers, Clair Patterson and his co-workers in California Institute of Technology, U.S.A., first developed a reliable procedure for trace Pb analysis, calling attention to the overwhelming anthropogenic Pb fluxes into the global environment. They pointed out about half a century ago that humans had increased the flux of reactive Pb as alkyl leaded gasoline by more than a factor of 10 (e.g., Murozumi et al. 1969; Shirahata et al. 1980; Schaule and Patterson 1981; Flegal and Patterson 1983). Subsequently, there have been numerous publications showing oceanic Pb concentrations in a global scale (e.g., Boyle et al. 1986; Helmers and van der Loeff 1993; Alleman et al. 1999; Gallon et al. 2011; Echegoyen et al. 2014).

Lead has four stable isotopes: ^{204}Pb, ^{206}Pb, ^{207}Pb, and ^{208}Pb. Stable isotopic ratios, such as ^{206}Pb/^{207}Pb and ^{208}Pb/^{207}Pb, could give us useful information on the origins of Pb and ocean dynamics (circulation and mixing). In recent years, an international program GEOTRACES (collaborative studies of marine biogeochemical cycles of trace elements and their isotopes) has been playing a dominant role in advancing geochemical studies on trace elements in the ocean (e.g., SCOR Working Group 2007; Anderson et al. 2014; Mawji et al. 2015). From a scientific point of view, Pb and its stable isotopes in the oceanic environment are useful chemical tracers to elucidate not only in situ chemical processes but also the transport mechanism of trace metals in seawater, just similarly as harmless fluorocarbons in seawater are used to trace physical transport of oceanic waters. Studies on the temporal variation of the anthropogenic Pb distribution in the ocean could serve as an analogue for other anthropogenic pollutants.

Pb data obtained from chemical analysis of seawater samples are, however, restricted only the recent several decades, while the history of environmental Pb pollution ascends to the age of industrial revolution. We could expand the environmental Pb records to older years by including Pb data from coral core samples (e.g. Kelly et al. 2009; Lee et al. 2014). Corals take a slight amount of Pb in their $CaCO_3$ body as they grow in surface seawater, recording the temporal change of Pb concentration and its isotopic ratios along their growing annual bands. The Pb concentration of past seawaters could be estimated from the coral Pb/Ca data and an empirically determined partition coefficient of Pb between seawater and coral (Kelly et al. 2009).

21.4 Spatiotemporal Pb Distribution in the Ocean

21.4.1 Atlantic Ocean

The North Atlantic Ocean was the region where early industrializing economies impacted most heavily, and we have far more data on the Pb behavior in the North Atlantic Ocean than elsewhere. Beginning with the work of Schaule and Patterson (1983) on seawater samples collected in 1979, many following studies (Boyle et al. 1986; Shen and Boyle 1987, 1988; Helmers et al. 1990, 1991; Helmers and van der Loeff 1993; Veron et al. 1993, 1994; Hamelin et al. 1997; Wu and Boyle 1997; Alleman et al. 1999; Weiss et al. 2003; Desfenant et al. 2006; Noble et al. 2015) have given us useful information on the spatial and temporal variability of Pb in the North Atlantic Ocean.

In addition, Kelly et al. (2009) obtained annually banded data from three corals in the North Rock reef northeast of Bermuda Island (32°N, 65°W) in the North Atlantic Ocean, and combined them with annually averaged seawater data near Bermuda (~32°N, ~65°W). The temporal Pb variation in surface waters of the western North Atlantic near Bermuda with a timescale of century is summarized in Fig. 21.3a. Lead in the surface water began to rise from ~20 pmol kg^{-1} at the beginning of the nineteenth century, reflecting the development of the US Upper Mississippi Valley lead district in the mid-1800s (Heyl et al. 1959) as well as the US industrial revolution, reaching a plateau of ~80 pmol kg^{-1} in the 1920–1930s.

In the late 1940s after the end of World War II, the seawater Pb concentrations showed a marked increase, corresponding with the US and eventually European prosperity. The Bermuda surface ocean Pb levels peaked at >200 pmol kg^{-1} in the 1970s, when the US Clean Air Act of 1970 was established. The Act mandated catalytic converters to reduce atmospheric Pb pollution, which required removal of tetraethyl Pb from gasoline because it poisoned the catalysis (Nriagu 1989). Since the middle of 1970s, North Atlantic Pb concentrations have fallen sharply as shown in Fig. 21.3a. As a recent data, the value of 17 pmol kg^{-1} was observed during the US GEOTRACES North Atlantic Transect cruise in November 2011 (Noble et al. 2015).

It is noteworthy that, though not shown in Fig. 21.3a, there is an unusual peak in Pb concentration that occurred in the Atlantic coastal waters of southern Spain, where acid mine drainage from the Rio Tinto and Rio Odiel deliver water with Pb concentrations a million times more than that in Atlantic surface waters (e.g., van Geen et al. 1997). These high Pb values moved into the Mediterranean Sea through the northern branch of the Gibraltar inflow. Mediterranean surface water data from the 1980s shows very high values, up to 500 pmol kg^{-1}, and in 1982 the values from the deep Mediterranean Outflow at Gibraltar are ~250 pmol kg^{-1} (Moos and Boyle, unpublished data).

Surface seawaters spread into the ocean's interior by thermocline ventilation and deep-water formation. In the North Atlantic, thermocline ventilation occurs on time scales of years in the uppermost layer and about a century at 1000 m depth (Jenkins

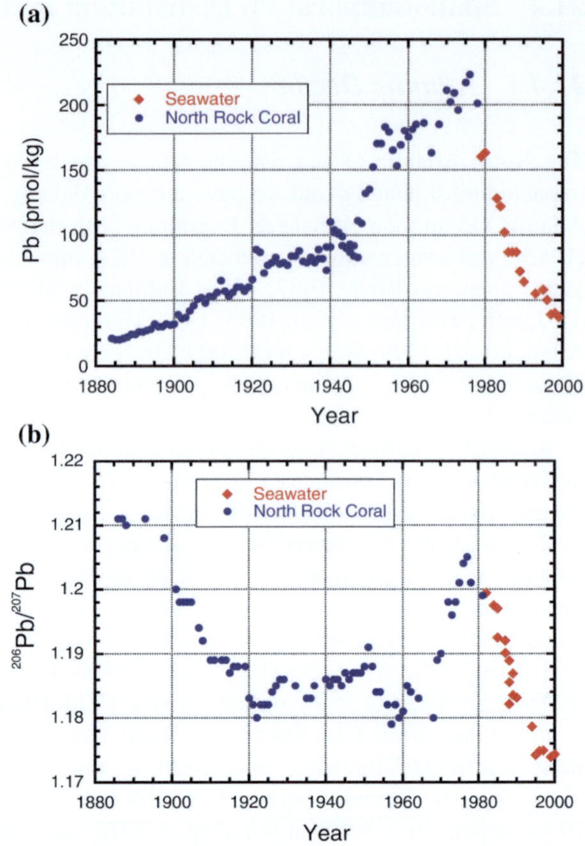

Fig. 21.3 **a** Past Pb concentrations and **b** past $^{206}Pb/^{207}Pb$ isotope ratios recorded in Bermuda corals together with annual averaged seawater data (red points). Data are from Kelly et al. (2009)

1998). Figure 21.4 shows time-series Pb profiles in the water column near Bermuda at roughly five-year intervals during the past 30 years. Seawaters that sank decades ago when surface Pb levels were order to magnitude higher, ~200 pmol kg^{-1}, are recognized in the profiles of 1979 and 1984 at mid-depths, with a maximum at ~400 m depth. This maximum deepened to 1–2 km depth in 2011 as a broader peak with less Pb concentration. It is likely that the decrease in water column Pb is driven not only by water transfer and mixing processes but also by scavenging onto settling particles. Figure 21.4 should, therefore, be regarded as a result of mixtures of these processes occurring in water columns.

Temporal changes in the isotope ratio ($^{206}Pb/^{207}Pb$) observed for surface waters (Fig. 21.3b) seem to be associated with the difference of Pb sources between US and European countries. US gasoline dominantly used tetraethyl Pb from the Ethyl Corporation with a $^{206}Pb/^{207}Pb$ ratio typically >1.17, whereas Europe chiefly used tetraethyl Pb from Associated Octel Ltd. with a lighter $^{206}Pb/^{207}Pb$ ratio typically <1.15 (Grousset et al. 1994). Figure 21.3b shows the time-series Pb isotope ratio of North Atlantic surface waters, implying variations in the relative contribution of US

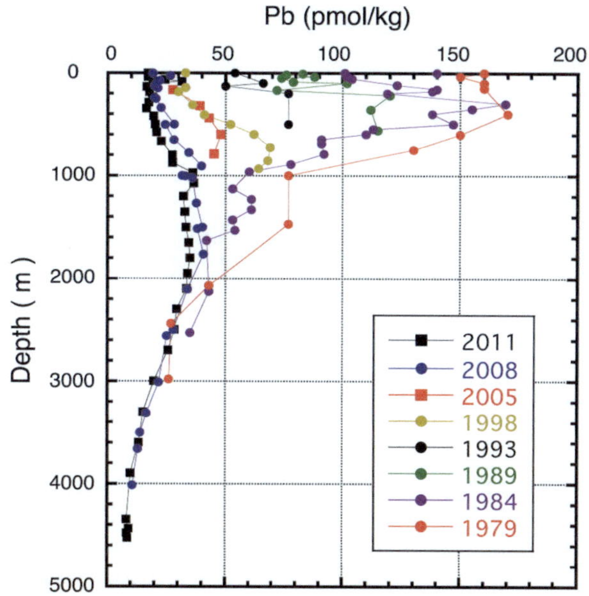

Fig. 21.4 Time-series Pb profiles in the water column near Bermuda between 1979 and 2011. Data are compiled from Schaule and Patterson (1983), Shen and Boyle (1987), Veron et al. (1993), Lee et al. (2011), and Noble et al. (2015)

and European sources during the twentieth century, together with some effects due to total emissions, atmospheric wind patterns, and proximity.

Since the United States used several times more Pb than Europe for a long time as shown in Fig. 21.1, the high US $^{206}Pb/^{207}Pb$ ratio dominated the North Atlantic. Figure 21.3b clearly shows the temporal decrease of $^{206}Pb/^{207}Pb$ ratio in the North Atlantic surface water since 1980, harmonized with the fact that the United States phased out leaded gasoline faster than Europe.

Figure 21.5 shows temporal variation of $^{206}Pb/^{207}Pb$ profiles of dissolved Pb in the water column near Bermuda between 1984 and 2011. It is clear that the earlier Pb reduction in the US also affected the intermediate water of the North Atlantic through thermocline ventilation.

21.4.2 *Pacific Ocean*

There are fewer data on Pb in the North Pacific Ocean than in the North Atlantic. Schaule and Patterson (1981) obtained the first valid oceanic Pb data of ~65 pmol kg^{-1} on surface water samples collected in 1976 near Hawaii. By the late 1990s, Pb concentrations near Hawaii had declined by a factor of two (Boyle et al. 2005).

It should be noted that the situation in the Pacific is not as simple as in the North Atlantic, where Pb emissions have been declining steadily for decades in almost all of the surrounding countries. There is a big problem in the Pacific Ocean that the Pb emission by China and Southern Asian countries has been increasing. The increase

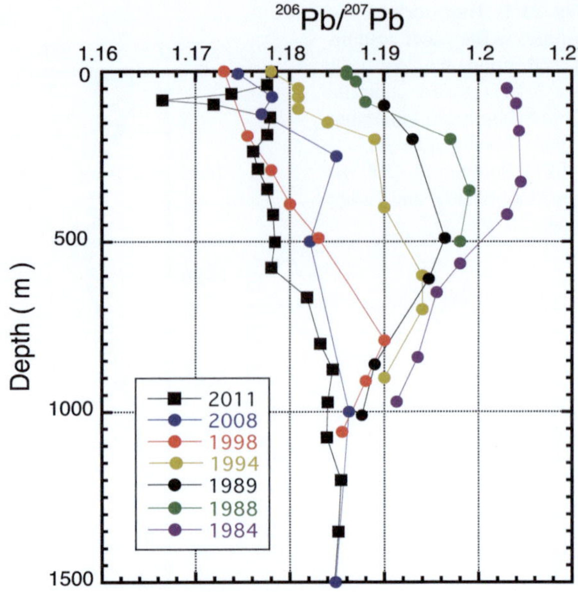

Fig. 21.5 Time-series $^{206}Pb/^{207}Pb$ profiles of dissolved Pb in the water column near Bermuda between 1984 and 2011. Data are from Shen and Boyle (1987), Sherrell et al. (1992), Veron et al. (1993), Reuer (2002), Reuer et al. (2003), Boyle et al. (2012), and Noble et al. (2015)

is dominantly derived from coal combustion, because leaded gasoline has not been used in China for the past 15 years (Flegal et al. 2013). Although there is scant older data for comparison, Pb concentrations in surface waters of the western North Pacific near Japanese Islands are quite high, ~100 pmol kg^{-1}, compared to those in the eastern area near Hawaii Islands of ~30 pmol kg^{-1}(Gallon et al. 2011). The Pb isotopic composition of these surface waters is within the values reported in Asian aerosols that are primary attributed to Chinese industrial Pb emissions (Gallon et al. 2011).

To the extent that data are available in deep profiles from the North Pacific, any change in Pb concentration in the central North Pacific appears confined to the upper few hundred meters (Boyle et al. 2005), which is also confirmed by observations from SAFe and GEOTRACES data in the eastern North Pacific (e.g. Wu et al. 2010).

Pb data from the South Pacific Ocean are still very sparse. Published data are limited to those from the equatorial zone of the western South Pacific (Flegal and Patterson 1983; Wu et al. 2010). These profiles show Pb concentrations of 13–23 pmol kg^{-1} in the surface waters, decreasing to ~4 pmol kg^{-1} in the deep waters.

Inoue et al. (2006), by using coral skeletons, determined temporal changes of Pb concentrations and its isotopic compositions in the surface seawater of the northwestern Pacific Ocean west of 160°E. The spatial distribution of Pb was shown to have a clear dilution pattern of Pb from Asian continent to the open ocean, suggesting the Pb emission mainly from industrial activities in Asian countries. Long-term record from corals at Ogasawara Island, Japan (~27°N, ~143°E) showed gradual increase of Pb content between 1900 and 2000, in clear contrast with the rapid decrease after ~1975 in the North Atlantic Ocean (Fig. 21.3a).

21.4.3 Indian Ocean

Vertical Pb profiles in the western Indian Ocean (between 40°E and 70°E) were measured by Vu and Sohrin (2013) between 17°N and 38°S and by Echegoyen et al. (2014) between 17°N and 62°S. Data were also obtained at two locations in the northern Indian Ocean: one in the Bay of Bengal (8.5°N, 86°E) and the other on the equator (80°E) (Vu and Sohrin 2013; Echegoyen et al. 2014). The seawater samples were collected during the Japanese GEOTRACES cruise in 2009–2010 (KH-09-5 cruise of R/V HakuhoMaru from the Bay of Bengal, Arabian Sea to the Southern Ocean). Surface Pb concentrations in the northern and central Indian Ocean north of 20°S were shown to have higher values (42–82 pmol kg^{-1}) than those in the North Atlantic near Bermuda and the North Pacific near Hawaii. This is probably due to the spurt of economic growth and industrialization in many countries surrounding the Indian Ocean, as well as a late phasing out of leaded gasoline in these countries as shown in Fig. 21.2. The surface Pb concentration decreased from north to south, 21 pmol kg^{-1} at 31°S to 7 pmol kg^{-1} at 62°S (Echegoyen et al. 2014), indicating limited regional anthropogenic Pb emissions in the Antarctic.

The deep Indian Ocean north of 20°S shows the lowest seawater Pb value ever observed, 1.5–3.3 pmol kg^{-1} (Echegoyen et al. 2014). This is probably the combined result of more than century-scale age of the deep water, plus scavenging of Pb along the northward flow path of the deep water from the Antarctic.

The Pb isotope ratios of the Indian Ocean water range 1.140–1.190 for $^{206}Pb/^{207}Pb$ and 2.417–2.468 for $^{208}Pb/^{207}Pb$, with lower ratios appearing in the shallow waters of North Indian Ocean and higher ratios appearing in the deep layers of the Southern Ocean (Lee et al. 2015). Comparison of these Pb isotope ratios with those of natural Pb ($^{206}Pb/^{207}Pb$: ~1.20, $^{208}Pb/^{207}Pb$: ~2.49) confirmed the results from the Pb concentrations (Echegoyen et al. 2014) that the Indian Ocean, particularly its northern part, is largely polluted by the input of anthropogenic Pb.

21.5 Concluding Remarks

Human activities, especially those after the industrial revolution, have emitted a great amount of Pb into the atmosphere to affect the Pb concentration and its isotope composition in the ocean. Oceanic Pb illustrates the evolving extent of the human footprint upon the natural environment. Time-series Pb record is thus regarded as an analogue of the human history of environmental destruction and protection, starting from unconscious and indiscreet emission of anthropogenic Pb to the environment to arrive at a modern stage of rational stoppage of the Pb emission in order to recover and preserve the environment.

Lead levels in surface ocean waters have been changing in line with regional emissions, with North Atlantic Pb levels increasing 10 times from the eighteenth century to ~1975 and decreasing since ~1975 down to nearly the original level. Pb concentrations in the Atlantic thermocline reflect the decreasing surface water concentrations due to water advection and scavenging by falling particles from surface to deeper ocean.

Although Pb has decreased in the Atlantic Ocean and in the east to central Pacific Ocean, a different story has arisen in the western Pacific and the Indian Ocean. Recent economic development and a late phase-out of leaded gasoline in China, India, Indonesia, and some other southern Asian countries have led to relatively high Pb concentrations in the northern and central Indian Ocean as well as in the western Pacific. Such a time lag for the environmental protection between advanced and developing countries is another problem to be solved in the future.

We must make continuous efforts for ocean observation to monitor spatiotemporal variation of anthropogenic Pb in sea water of the world oceans, as well as those for revealing past records from the analysis of coral cores. Comparative studies will also be desirable with other anthropogenic pollutants such as fossil fuel CO_2, nuclear bomb fallout, atmospheric trace gases such as CH_4, N_2O, freons, as well as various trace metals.

Acknowledgements A companionship for many years with the international GEOTRACES community is fairly helpful for the composition of this review. Financial supports are sincerely acknowledged: grants-in-aid for Scientific Research (A) (Nos. 23253001 and 16H02701) from the Ministry of Education, Culture, Sports, Science and Technology, Japan.

References

Alleman LY, Veron AJ, Flegal AR, Church TM, Hamelin B (1999) Invasion of the abyssal North Atlantic by modern anthropogenic lead. Geophys Res Lett 26:1477–1480

Anderson RF, Mawji E Cutter GA, Measures CI, Jeandel C (2014) GEOTRACES: changing the way we explore ocean chemistry. Oceanogr 27(1):50–61

Bacon MP, Spencer DW, Brewer PG (1976) $^{210}Pb/^{226}Ra$ and $^{210}Po/^{210}Pb$ disequilibria in seawater and suspended particulate matter. Earth Planet Sci Lett 32:277–296

Boyle EA, Shen GT, Chapnick SD, Bacon MP (1986) Temporal variability of lead in the western North Atlantic Ocean. J Geophys Res 91:8573–8593

Boyle EA, Bergquist BA, Kayser RA, Mahowald N (2005) Iron, manganese, and lead at Hawaii Ocean Time-series station ALOHA: temporal variability and an intermediate water hydrothermal plume. Geochim Cosmochim Acta 69:933–952

Boyle EA, John S, Abouchami W, Adkins JF, Echegoyen-Sanz Y, Ellwood M, Flegal AR, Fornace K, Gallon C, Galer S et al (2012) GEOTRACES IC1 (BATS) contamination-prone trace element isotopes Cd, Fe, Pb, Zn, Cu, and Mo intercalibration. Limnol Oceanogr Methods 10:653–665

Boyle EA, Lee J-M, Echegoyen Y, Noble A, Moos S, Carrasco G, Zhao N, Kayser R, Zhang J, Gamo T, Obata H, Norisuye K (2014) Anthropogenic lead emissions in the ocean. Oceanogr 27(1):69–75

Cziczo DJ, Stetzer O, Worringen A, Ebert M, Weinbruch S, Kamphus M, Gallavardin SJ, Curtius J, Borrmann S, Froyd KD et al (2009) Inadvertent climate modification due to anthropogenic lead. Nature Geosci 2:333–336

Desfenant F, Veron AJ, Camoin GF, Nyberg J (2006) Reconstruction of pollutant lead invasion into the tropical North Atlantic during the twentieth century. Coral Reefs 25:473–484

Echegoyen Y, Boyle EA, Lee J-M, Gamo T, Obata H, Norisuye K (2014) Recent distribution of lead in the Indian Ocean reflects the impact of regional emissions. PNAS 111:15328–15331

Flegal AR, Patterson CC (1983) Vertical concentration profiles of lead in the Central Pacific at 15°N and 20°S. Earth Planet Sci Lett 64:19–32

Flegal AR, Gallon C, Conaway CH, Ganguli P (2013) All the lead in China. Crit Rev Environ Sci Tech 43:1869–1944

Gallon C, Ranville MA, Conaway CH, Landing WM, Buck CS, Morton PL, Flegal AR (2011) Asian industrial lead inputs to the North Pacific evidenced by lead concentrations and isotopic compositions in surface waters and aerosols. Environ Sci Tech 45:9874–9882

Grousset FE, Quetel CR, Thomas B, Buat-Menard F, Donald OFX, Bucher A (1994) Transient Pb isotopic signatures in the Western European atmosphere. Environ Sci Tech 28:1605–1608

Hamelin B, Ferrand JL, Alleman L, Nicolas E (1997) Isotopic evidence of pollutant lead transport from North America to the subtropical North Atlantic gyre. GeochimCosmochimActa 61:4423–4428

Helmers E, Mart L, Schulz-Baldes M, Ernst W (1990) Temporal and spatial variations of lead concentrations in Atlantic surface waters. Mar Poll Bull 21:515–518

Helmers E, Mart L, Schrems O (1991) Lead in Atlantic surface waters as a tracer for atmospheric input. Fres J AnalChem 340:580–584

Helmers E, van der Loeff MMR (1993) Lead and aluminum in Atlantic surface waters (50°N to 50°S) reflecting anthropogenic and natural sources in the eolian transport. J Geophys Res 98:20261–20273

Heyl A Jr, Agnew A, Lyons E, Behre C Jr (1959) The geology of the Upper Mississippi Valley zinc-lead district. US GeolSurv Prof Paper, 309. http://pubs.er.usgs.gov/publication/pp309. Accessed 26 Nov 2013

Inoue M, Hata A, Suzuki A, Nohara M, Shikazono N, Yim WW-S, Hantoro WS, Donghuai S, Kawahata H (2006) Distribution and temporal changes of lead in the surface seawater in the western Pacific and adjacent seas derived from coral skeletons. Environ Poll 144:1045–1052

Jenkins WJ (1998) Studying subtropical thermocline ventilation and circulation using tritium and ^{3}He. J Geophys Res 103:15817–15831

Kelly AE, Reuer MK, Goodkin NF, Boyle EA (2009) Lead concentrations and isotopes in corals and water near Bermuda, 1780–2000. Earth Planet Sci Lett 28:93–100

Lee J-M, Boyle EA, Echegoyen-Sanz Y, Fitzsimmons JN, Zhang R, Kayser RA (2011) Analysis of trace metals (Cu, Cd, Pb, and Fe) in seawater using single batch nitrilotriacetate resin extraction and isotope dilution inductively coupled plasma mass spectrometry. Anal ChimActa 686:93–101

Lee J-M, Boyle EA, Nurhati IS, Pfeiffer M, Meltzner AJ, Suwargadi B (2014) Coral-based history of lead and lead isotopes of the surface Indian Ocean since the mid-20th century. Earth Planet Sci Lett 398:37–47

Lee J-M, Boyle EA, Gamo T, Obata H, Norisuye K, Echegoyen Y (2015) Impact of anthropogenic Pb and ocean circulation on the recent distribution of Pb isotopes in the Indian Ocean. Geochim-CosmochimActa 170:126–144

Mawji E, Schlitzer R, Dodas EM et al (2015) The GEOTRACES intermediate data product 2014. MarChem 177:1–8

Murozumi M, Chow TJ, Patterson C (1969) Chemical concentrations of pollutant lead aerosols, terrestrial dusts and sea salts in Greenland and Antarctic snow strata. GeochimCosmchimActa 33:1247–1294

Noble AE, Echegoyen-Sanz Y, Boyle EA, Ohnemus DC, Lam PJ, Kayser R, Reuer M, Wu J, Smethie W (2015) Dynamic variability of dissolved Pb and Pb isotope composition from the U.S. North Atlantic GEOTRACES transect. Deep-Sea Res II 116:208–225

Nozaki Y, Thomson J, Turekian KK (1976) The distribution of ^{210}Pb and ^{210}Po in the surface waters of the Pacific Ocean. Earth Planet Sci Lett 32:304–312
Nriagu JO (1989) The history of leaded gasoline. In: Vernet JP (ed) Heavy metals in the environment. Page Bros, Norwich, UK, pp 361–366
Reuer MK (2002) Centennial-scale elemental and isotopic variability in the tropical and subtropical North Atlantic Ocean. Dissertation, Massachusetts Institute of Technology and Woods Hole Oceanographic Institution
Reuer MK, Weiss DJ (2002) Anthropogenic lead dynamics in the terrestrial and marine environment. Philos Trans Royal Soc London A360:2889–2904
Reuer MK, Boyle EA, Grant B (2003) Lead isotope analysis of marine carbonates and seawater by multiple collector ICP-MS. ChemGeol 200:137–153
Revelle R, Suess HE (1957) Carbon dioxide exchange between atmosphere and ocean and the question of an increase of atmospheric CO_2 during the past decades. Tellus 9:18–27
Schaule B, Patterson CC (1981) Lead concentrations in the northeast Pacific: evidence for global anthropogenic perturbations. Earth Planet Sci Lett 54:97–116
Schaule B, Patterson CC (1983) Perturbations of the natural lead depth profile in the Sargasso Sea by industrial lead. In: Wong CS, Boyle EA, Bruland KW, Burton JD, Goldberg ED(eds) Trace metals in seawater. Plenum, NY, pp 487–504
SCOR Working Group (2007) GEOTRACES—an international study of the global marine biogeochemical cycles of trace elements and their isotopes. Chemie der Erde 67:85–131
Shen GT, Boyle EA (1987) Lead in corals: reconstruction of historical industrial fluxes to the surface ocean. Earth Planet Sci Lett 82:289–304
Shen GT, Boyle EA (1988) Thermocline ventilation of anthropogenic lead in the western North Atlantic. J Geophys Res 93:15715–15732
Sherrell RM, Boyle EA, Hamelin B (1992) Isotopic equilibrium between dissolved and suspended particulate lead in the Atlantic Ocean: evidence from ^{210}Pb and stable Pb isotopes. J Geophys Res 97:11257–11268
Shirahata H, Elias RW, Patterson CC (1980) Chronological variations in concentrations and isotopic compositions of anthropogenic atmospheric lead in sediments of a remote subalpine pond. GeochimCosmochimActa 44:149–162
van Geen A, Adkins JF, Boyle EA, Nelson CH, Palanques A (1997) A 120 year record of metal contamination on an unprecedented scale from mining of the Iberian Pyrite Belt. Geology 25:291–294
Veron AJ, Church TM, Flegal AR, Patterson CC, Erel Y (1993) Response of lead cycling in the surface Sargasso Sea to changes in tropospheric input. J Geophys Res 98:18269–18276
Veron AJ, Church TM, Patterson CC, Flegal AR (1994) Use of stable lead isotopes to characterize the sources of anthropogenic lead in North Atlantic surface waters. GeochimCosmochimActa 58:3199–3206
Vu HTD, Sohrin Y (2013) Diverse stoichiometry of dissolved trace metals in the Indian Ocean. Sci Rep 3:1745; https://doi.org/10.1038/srep01745
Weiss D, Boyle EA, Wu J, Chavagnac V, Michel A, Reuer M (2003) Spatial and temporal evolution of lead isotope ratios in the North Atlantic Ocean between 1981 and 1989. J Geophys Res 108(C10):3306. https://doi.org/10.1029/2000JC000762
Wu JF, Boyle EA (1997) Lead in the western North Atlantic Ocean: completed response to leaded gasoline phaseout. GeochimCosmochimActa 61:3279–3283
Wu J, Rember R, Jin M, Boyle EA, Flegal AR (2010) Isotopic evidence for the source of lead in the North Pacific abyssal water. GeochimCosmochimActa 74:4629–4638

Part VI
Global Sustainability Programmes and Human Geoscience

Chapter 22
Education for a Sustainable Society

Yoshiyasu Ida

Abstract The purpose of environmental education is to create solutions for issues related to the environmental issues and to promote the development of human beings. There are diverse views on environmental protection and development. The compromise is ESD; Education for Sustainable Development. ESD is one of keywords of environmental education. In other words, ESD exists within the framework of environmental education. Furthermore, Future Earth is proposed because global environmental issues have not been solved. It has been attempted to create solutions by integrating concepts and many points of view as Future Earth. Therefore, education for Future Earth would require the integration of fields and subjects, otherwise taking various disciplines into consideration.

Keywords Environmental education · ESD · Future Earth · School education

22.1 The Significance of Environmental Education

22.1.1 Environmental Education as Human Education

In recent years, damage such as the submerging of villages on islands in the South Pacific has been seen due to rises in sea level and recognized as stemming from global warming. The previously distinct dry and rainy seasons in the tropics are said to have become indistinct. The seasonal appeal that was suitable for tourism is no longer possible for travel agencies that previously explained, "Because it's the dry season, we are blessed with good weather for a great tourist season," as presently rainy days are occurring repeatedly even in the dry season, and clear skies are seen even during the rainy season. As for ocean, there is also the trend of shrinking ice in the Arctic Ocean. However, for many children, there are not many opportunities to directly see and feel for themselves the effects of climate change as global warming throughout the world. Nonetheless, we have become able to feel warming and abnormal weather,

---

Y. Ida (✉)
University of Tsukuba, Tsukuba, Japan
e-mail: ida@human.tsukuba.ac.jp

Y. Himiyama et al. (eds.), *Human Geoscience*, Advances in Geological Science,
https://doi.org/10.1007/978-981-32-9224-6_22

among other conditions, even in our immediate surroundings. We must consider how environmental and geography education should be conducted in this context.

Environmental education in Japan began with "nature study" and "nature conservation education" in the subject of science in the 1950s. In social studies, pollution education also began in the 1950s. Although lives were improved with the progress of industrialization, many individuals began to suffer from pollution. Consequently, through a revision of peoples' values, protection of nature, and clarification of who take responsibility for pollution, education that would attempt to reduce pollution came to be proposed. This was the starting point for environmental education in Japan.

On the other hand, regarding this pollution, companies attempting to raise profits through competition, consumers demanding cheap yet high-quality goods, and governments allowing this kind of society to exist, all became intertwined, and ultimately each ordinary citizen came to bear the responsibility of continuing to improve the environment. That is, environmental education is an education that causes individuals to think of their own responsibility and actions, and this education is necessary in society. Thus, at the 1972 United Nations Conference on the Human Environment in Stockholm, international "environmental education" was proposed. The goal was implementation of education for the purpose of planning forward-looking improvements of the human environment and conservation of nature, born out of the sense of crisis caused by the fact that humanity has been endangering its own survival.

Furthermore, the term "environment" is not limited only to the natural environment, but includes the social environment as well. Environmental education in social studies and geography education includes not only the relationship between humans and nature, but also the individual and group relationships between individuals and society. In other words, environmental education also includes elements such as peace and international understanding. This means that peace education and international understanding education overlap with environmental education. Namely, for each type of education, although there are unique aspects, there is also a common thread running through all of the subjects: human education.

22.1.2 Environmental Education Objectives and Significance

There are slight differences in nuance regarding the objective of environmental education arising from differences in educational stance. The goal of environmental education, as stated in the 1977 Tbilisi Declaration, is to nurture humans who have an interest, knowledge, skill, attitude, or intention of considering the environment and the various challenges tied to it, and who are able to contribute to solutions, not only for the present problems, but for new problems that may develop in the future as well (Miyazaki 2012). When organized based on Ida (2009), the objective of environmental education is structured into three points according to the child's developmental circumstances. The first point is to become familiar with natural and societal environments, **in** the environment. Whether it is a city or a rural area or a fishing village, familiarity with natural and societal environments will be promoted through the kind of nature the child is living, experiencing the kind of human relationships they are involved in, or other methods of creating awareness. This will give children the impetus to recognize that they are living on the land within those environments.

The second is to deepen knowledge **about** the environment, to understand environmental problems, to discuss the value of the environment, to deepen personal values, and to foster an attitude of formulating solutions to problems. The purpose of environmental education is not to provide knowledge about the environment. The emphasis is on a set of processes, based on knowledge and awareness of the environment, fostering value for the environment and the understanding of environmental problems, and attempting to reach solutions to these problems. Accordingly, in social studies as well, the goal is not only the acquisition of environmental knowledge, but also an understanding of the problems that arise from that knowledge, their own way of thinking about these problems and problem-solving.

On the other hand, for environmental education in social studies, there are differences in the importance placed on the contents and process of environmental education that seeks to teach geography and other so-called "proper topics" and environmental education requested by scholars of education, among others. Those requesting proper environmental education, such as geographers and geosciences researchers, place emphasis on mastering knowledge and deepening awareness of the environment, then making value judgments and considering what kind of actions should be taken. It aims to solve problems; however, problem-solving for intricately intertwined events is not easy and offering simple solutions is meaningless, therefore accurate awareness of the current situation is considered more important. In other words, logical learning based on knowledge is seen as important. On the other hand, for the environmental education spoken about by education scholars, although knowledge of the environment is necessary, importance is placed on learning the process of problem-solving and making value judgments utilizing that knowledge. Thus, a problem-solving process taking full advantage of the knowledge children possess should be considered currently. This knowledge cannot be learned sitting at a desk, but emphasis is instead given to knowledge gained by personal experience, as this is easier to utilize.

The third point is environmental education **for** the environment. In this step, the goal is to create and implement concrete strategies for the preservation and improvement of the environment. Although it might be implemented in the classroom, the opportunity for lifelong education is substantial. The scale includes all levels, from what can be done with the individuals around us, to a global scale on the level of international cooperation.

In addition to these ways of interacting with the environment—in, about, and for—environmental education with a place for character development **through** experiences of being sensitive to and promoting the environment have been proposed. Education is activities for promoting human formation. And environmental education which is for one's relationship with the environment as well, and be carrying out human development education has come to be emphasized. Namely, the significance of environmental education is to be character development through one's way of being involved with and promoting care of the environment, and the experience of feeling the environment.

22.2 Environmental Education and the Learning Process

22.2.1 The Learning Process

An appropriate learning process and method are necessary to achieve the goals of environmental education. The learning process can be organized as follows: ① understanding the theme, ② data collection, ③ data organization, ④ analysis and interpretation, ⑤ value judgment and decision-making, and ⑥ implementation and participation. For understanding the theme in a context of environmental education, discovery happens while becoming familiar with natural and societal environments. For elementary school, in particular, discovering problems while personally experiencing an event can increase the children's awareness of the problem. Meanwhile, for changes in the global environment such as climate, ocean, and water, it is also possible to set up challenges that mimic the changes that likely cannot be seen in our own environment. For the former, a division of roles is possible in which the issues are discovered during comprehensive studies and so, after which the investigation process is done in social studies. For the latter, first conceptual content is learned in social studies, then, using comprehensive studies to experience the concept firsthand, issues can be discovered while having concrete experiences in their own immediate area. As environmental studies are not conducted only in social studies, and the contents are connected to comprehensive studies, science, home economics and so, environmental education will be learned while utilizing the characteristics of those subjects.

Once the theme has been grasped, the next step is collecting data related to the issues. In addition to the textbook, data will be collected from sources such as supplementary materials, literature, and the Internet. "Flipped classrooms," in which data collection becomes an important component of home learning while discussion (analysis and interpretation) based on the collected data is conducted in class, have garnered attention. The step after data collection is organization of the data. With regard to the environment, organization expressed in maps is easier to analyze and interpret. There are many map exhibitions, including environmental map exhibitions, and although many elementary and middle school students are applying, the fact that there are many environment-related maps is evidence that environmental events are easy to arrange in this way.

The next step is to analyze and interpret the organized data. For analysis and interpretation, the point of view and method that are used depend on the respective academic subjects and lesson contents. For example, in social studies, a social studies perspective and way of thinking would be used, while a geographical perspective and way of thinking for learning geographical content would be used for analysis and interpretation. When the content is related to the environment for climate, land, ocean, water and so, analyses and interpretations should be conducted from the points of view of changes in the natural environment, the relationship between humans and nature and changes therein, and human attitudes when facing nature.

In many cases, the results derived from this series of learning processes are presented, some comments and impressions are given, and the lesson is complete. When the lesson is completed at this point, the students finish with an understanding up to the present situation, and at a level of knowledge mastery.

In environmental education, for the abovementioned learning process, it is desired that learning by experience outdoors, in an area influenced by nature, can be integrated. Learning in field work is possible for the steps of understanding the issues and data collection, and analysis and interpretation can be done while observing the real thing in the field. In any event, rather than awareness via in-class learning, awareness from experience will continue to be emphasized. Based on this kind of learning process, while questioning personal values, the values of others, and societal values, students continue to make decisions and move on to the value judgments and decision-making step. This step is important in environmental education and based on knowledge and awareness of the environment, as well as values and decision-making for the environment. In other words, questions like "How should things be?" and "What should be done?" are asked. This kind of value judging and decision-making exercise connects environmental education to society participation as disaster preparedness. In other words, students reach the level of investigation and utilization.

22.2.2 Coordination in Environmental Education from Elementary to High School

Environmental education does not end with elementary school, but continues into middle and high school. In elementary school, investigation activities and textbooks, understanding of issues, collection of data, organization of data, and, though student viewpoints are not likely to deepen, analysis and interpretation are also conducted. In elementary school, because the learning method is the emphasis of the course content, viewpoints informing analysis and interpretation will broaden. However, as students possess insufficient systematic knowledge, deepening these points is difficult. Nonetheless, there are also cases in which, as investigative learning progresses, sufficient knowledge is gained, and analysis and interpretation are deepened. In middle school, knowledge is conceptually and systematically arranged through textbooks, and viewpoints on analysis and interpretation are clarified. Based on educational content during these school stages (as well as discussion of analyses and interpretations in high school) from sources of knowledge and multilateral, multifaceted viewpoints, value judgments and decisions are made and students continue to progress towards public participation and a future-focused perspective grounded in the present reality.

Throughout elementary school learning, there are value judgments and decision-making learning that leads to society participation. Nonetheless, in many cases, the conclusion is already decided, and it cannot be said that students reach decision-making and value judgments based on analyses and interpretations. For example, in a lesson on the issue of solid waste, the value judgment and decision that waste should be separated and put out for collection is already decided before the lesson, and the real content of the lesson becomes understanding of the reason for physical separation. Of course, through the contents of that lesson, improvements in a proactive attitude towards separating waste will likely be seen, but it is difficult to say if they have created new values amidst abundant knowledge and the clash of disparate opinion. Students do find solutions to environmental issues; however, they are often unrealistic. Nonetheless, gaining environmental awareness by learning through experience with investigative activities in the field, becoming capable of critical thinking and debating about the future, and possessing their own opinions, are important aspects of students' elementary school environmental education.

In Japanese middle schools, geography and history are studied in the first and second years, and civics is taught in the third year. Students are offered an opportunity to rethink an environmental solution, which they thought of themselves in elementary school, in their geography and history studies. Therefore, they conduct deeper analysis and interpretation, and better understand the regional features and complexity of yet more environmental issues. Discussion of solutions is executed after learning these contents by students. The structure of a lesson might be as follows. First, regional characteristics of environmental problems are understood from a global perspective, after which analysis and interpretation are conducted at the national (e.g., Japan) level. The children debate from the environment-related perspectives of the various people of the region, and it is in this context that they come to consider the difficulty of a solution and suggest their own alternative. Furthermore, in civics, perspectives such as economics and politics are incorporated, and environmental issues can be considered more multilaterally.

In high school geography, students revisit the environmental issues studied in middle school and how each country or region is responding. Ways to create resolutions that transcend national borders are identified through peer discussion, and students can then develop their own views regarding what should be done by individuals and countries. This is ideally a learning process in which students advance towards future, forward-looking public participation.

Through the development of this kind of spiraling environmental learning from elementary school through to high school, children's environmental knowledge and proactive behavior for building a sustainable society will continue to be nurtured. Accordingly, environmental education is established from elementary to high school as a continuous curriculum, and thus, the school faculty at each level must keep this in mind.

22.2.3 *Improvement of Middle School Textbooks Dealing with Environmental Issues*

There is currently no specified textbook that goes along with the aforementioned environmental educational learning process in Japan, however, Ida (2014) proposed the introduction of a textbook conforming to this learning process for the geography topic of "resources, energy, and industry," as a middle school case study. The details are shown in Table 22.1.

Based on the textbook from Table 22.1, an actual trial curriculum at a middle school was conducted. Although it was a geography education, it was also regarded as a class related to environmental education, and learning was conducted from the perspective of how energy that is the foundation of people's lives should be handled while relating to the natural environment according to region. As the topic was considered on a national and global scale, the awareness by personal experience aspect was minimal, but even so, because it caused students to consider what natural energy might be appropriate for their region, it is possible that their thoughts reflected their own life experiences. Additionally, through the utilization of role-playing and

Table 22.1 Middle school textbook draft based on the learning process (geography: "resources, energy, and industry")

Learning process	Contents	Assigned pages
Understanding the theme, data collection, and data organization (acquisition)	Present a diagram of global resource distribution, indicating that underground resources are unevenly distributed. On the topic of consumption, as is relevant to resource-less Japan, try to foster understanding of the import of resources from the figures	2
Analysis and interpretation (application)	Analyze the features of Japan's energy generation from diagrams and pictures, and discuss the reasons behind it. First, while comparing countries of the world with Japan on a graph, discover the features of Japan's energy generation and analyze its main sources. Read and understand that, from the standpoint of Japan's thermal and nuclear power plants, Japan is dependent on foreign raw materials, and try to discover the issues and advantages of thermal power generation. Analyze why people widely oppose nuclear power plant sites in Japan from a geographical point of view	2

(continued)

Table 22.1 (continued)

Learning process	Contents	Assigned pages
Value judgments, decision-making, public participation (inquiry)	Prepare to make value judgments and decisions about energy suited to one's own region. Understand that natural energy sources entails variety of problems, and, from photos and figures, understand that there is an uneven regional distribution of resources and regional situations even in natural energy production. Use this understanding as to the material for value judgments and decision-making. Through real debate, foster the realization that there are divergent opinions on geothermal energy in accordance with various positions	2

Ida (2014)

other activities, awareness of the social environment through personal experience was also possible. In the textbook draft, the time given to explore two pages was four hours, however, the children's interest was high, and the debate was profound.

22.3 ESD and Environmental Education

Looking for a resolution to global environmental issues, in 1980 the concept of "sustainable development" was proposed by three groups, including the International Union for the Conservation of Nature and Natural Resources (IUCN) (Kawashima et al. 2002). The following is a commentary based on Ida (2011, 2013).

As the Earth's environment has deteriorated over the course of industrial development by advanced countries, some argue for suppressing development and for conservation and protection of the environment, while others argue that development through industrialization and the like must progress in order for developing countries—where development has not progressed as much—to rise above poverty. "Sustainable development" has been proposed as a compromise for these arguments. "Sustainable development" has become a keyword in environmental and geography education, and has become a popular concept throughout the world. With the tendency for social studies in Japan to keep up with modern understandings, Japanese education has been impacted in no small way by forward-looking educational viewpoints such as those represented by environmental education. Still, the need for further future-oriented education is suggested, such as Education for Sustainable Development (ESD). ESD, aiming to overcome global challenges and build a sustainable society, was decided upon through Japan's proposal at the United Nations

General Assembly in 2002 with UNESCO as the promoting organization. The years from 2005 to 2014 were "The 10 Years of ESD." Beginning in Japan, it has come to be espoused by many countries.

In South Korea and other Asian countries like Singapore, ESD has been treated as an evolution in environmental education and has come to strongly reflect a future-orientated approach to experiential learning in environmental education. The objective of ESD includes understanding the background of phenomena, developing systematic thinking, and promoting thinking skills with an emphasis on critical thinking ability, fostering the ability to analyze data and information, enhancing communication skills, and cultivating values for a sustainable society. Regarding curriculum contents, eradication of poverty, environmental conservation, climate change as global warming, energy reduction, disaster preparedness, and population shifts are included and instructed through the participatory learning method. In Japan, it is common for ESD not to be regarded merely as an evolution in environmental education, but to be understood as something incorporating a wider educational values, one that includes environmental education. Consequently, even if only environmental education was undertaken, with the influence of ESD, it would become a more strongly future-oriented curriculum.

In 2008 and 2009, even in the reported national curriculum standards in Japan, ESD was intentionally described as, "creation of a sustainable society," and implementation in each subject was encouraged. When ESD is implemented in social studies, learning that addresses the following is important: exploratory learning, development of thinking ability, analysis of materials, development of skills, exchange of opinions, and development of communication skills. Through these kinds of learning, a cultivation of values such as what students themselves think, and what they think should be done, is accomplished. These values can become values for the foundation of a sustainable society. As shown in Fig. 22.1, the goals and methods of ESD are combined with environment-related content to become social studies reflecting ESD, and while relating to other subjects, can lead students to heightened expectations for a future society and public participation. Even within the structure of social studies, environmental education intertwined with ESD, can be positioned as a major lesson of this school subject. Education that includes the concept of ESD, with the environment atits core (relating to poverty, human rights and peace, while nurturing skills in thinking and judgment, emphasizing experiential learning, participation in the future world, the raising of children that will bear the responsibility of the Earth's future, and therefore defying the current boundaries of social studies) will become what can be called environmental education. Environmental education is the framework through which we can fundamentally solve the global-scale environmental problems that are worsening on a daily basis (Mizota 2014). Given that there was awareness of sustainability before the dissemination of the concept of ESD, it can be said that environmental education fundamentally includes ESD. Moreover, rather than being conflicting concepts, environmental education and ESD have much in common, and are concepts that can provide mutual encouragement in the right direction.

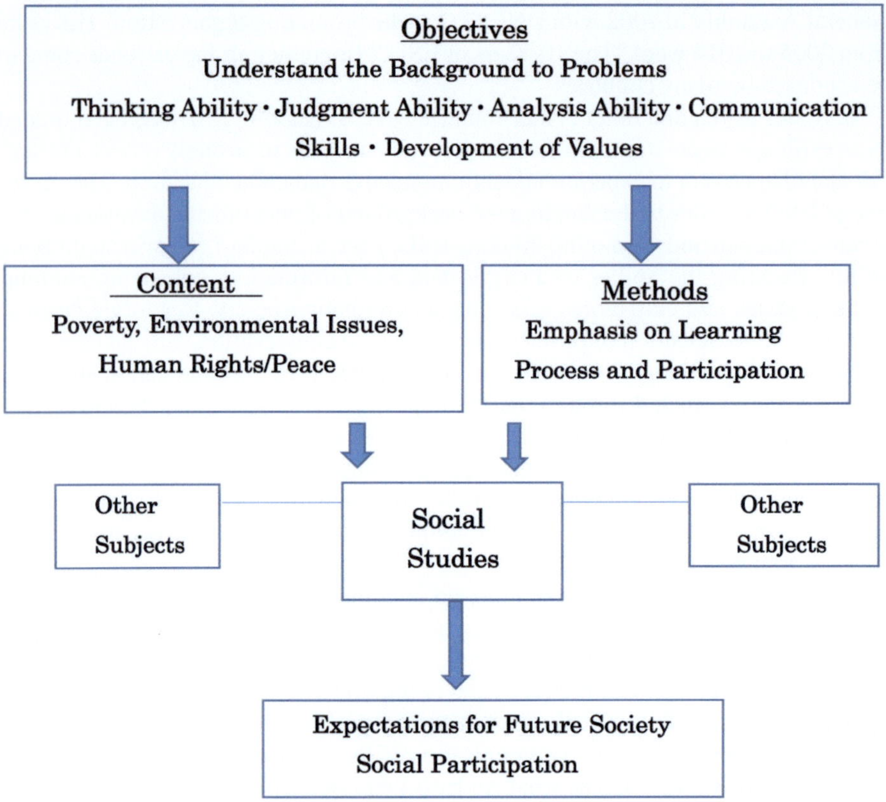

Fig. 22.1 Social studies framework for the creation of a sustainable society (Ida 2011)

22.4 Education in Future Earth

Global environmental change, extinction of species, ruined land and bringing the global environment closer to its limits have created an even greater sense of crisis for the human race and the Earth. Growth limits and sustainable development continue to be mentioned, but a solution has not been reached. Accordingly, integration of research from further spheres was extolled and the "Future Earth" concept proposed. In Future Earth, research has advanced with the three themes of dynamic understanding of the Earth, development on a global scale, and transformation into a sustainable society. From these three topics, through interdisciplinary research spanning natural and social science, geoscience, engineering, agriculture, medicine and more, not only can knowledge and understanding related to the abovementioned themes be gained, but also through cooperation with research groups and various stakeholders (administration/policymakers, research donors, industry, educators, media, and civic groups), we can aim for a transformation into a sustainable society. In other words, it is the integration of knowledge from stakeholders with research groups at the center and things learned from the experience of various people, the sharing of a common point of view, and planning to act together, which can lead to the creation of a sustainable society. The concept of Future Earth has been promoted by the Science Council of Japan (Science Council of Japan Future Earth Promotional Committee 2014).

To implement Future Earth, education is essential. For the sake of the environment, environmental education toward its promotion, along with the concept of Future Earth, must be heavily involved. Environmental education at this point includes an even wider range of academic disciplines, cooperation from many individuals, and hopes to nurture the ability to act for a sustainable society from a global perspective. Through ESD and Future Earth, environmental education can gain an even greater expansion. On the other hand, through enlarging the scope of environmental education in this way, what is used as the foundation, what kind of knowledge and abilities are learned and nurtured, and how environmental education is further prompted, all become serious challenges faced in each school such as elementary, middle, and high schools. Furthermore, regarding education for ESD and Future Earth, whether they are included as environmental education, or whether they are thought of as having educational goals, content, and methods differing from environmental education may require discussion. Nevertheless, whether it is environmental education, ESD, or Future Earth, the heart of the issue remains the same: that Earth's current environmental issues must be solved as quickly as possible for the sake of the future of the human race. The time is coming in which the respective environmental education pieces must be organized in an integrated way into lifelong education for elementary, middle, high school, and university students, as well as working adults. Otherwise, even if concepts were proposed for new solutions to global environmental problems, our children who will be responsible for the future, would not able to accept the concepts and not to utilize the ability to solve global environmental problems.

In this way, environmental education is a cross-curricular topic; however, it begins with geography education. These achievements in various subjects are mutually intertwined, and they must be effectively advanced while utilizing "Period for integrate study". Moreover, while tying together new concepts such as ESD and Future Earth, further advancement of environmental education will be accomplished. Along with this development of concept regulation and technological advances, children's interest in environmental issues can be nurtured. Regarding what is necessary for children, those who will find themselves responsible for the future, not only the teacher's understanding of environmental education, but also the teacher's skill in observing the children and their ability to judge what should be taught, have become even more important.

In addition, though the concept of Future Earth has been proposed, practically implementing education under Future Earth is difficult. Then, this paper advocates that the concept should be expanded from a growth-oriented perspective of environmental education. It is said that perhaps. With the development of curricula and concepts in environmental education, ESD and Future Earth, it is likely that concrete educational objectives, content, and methods can be presented.

References

Ida Y (2009) Environmental education as social studies. In: Kato Y, Akashi Y (eds) Reading and understanding elementary school social studies. Nihon Bunkyo Shuppan, pp 58–63 (in Japanese)
Ida Y (2011) "Social Studies" and "Geography and History Subjects" for sustainable development: a proposal for integrated geography and history subjects. J Soc Stud 113:1–8 (in Japanese)
Ida Y (2013) Outline survey of the various regions of the world. In: Ida Y (ed) Ultimate middle school social studies—geography edition. Nihon Bunkyo Shuppan, pp 94–95 (in Japanese)
Ida Y (2014) Resources, energy, and industry. In: Yatabe R (ed) Survey research for the improvement of social studies textbooks—to bring textbooks closer to real society. Textbook Research Center, pp 85–93 (in Japanese)
Kawashima M, Ichikawa S, Imamura M (eds) (2002) Invitation to environmental education. Minerva Shobo, (in Japanese)
Mizota K (2014) Education for sustainable development and environmental education. In: Miyagi University of Education Research Center for Education in International Understanding (ed) What is ESD? vol 2. pp 17–18 (in Japanese)
Miyazaki S (2012) Environmental education in social studies. In: Ida Y, Ushiro Y, Tsukada Y (eds) Theory and teaching of academic education humanities. ed. Kyodo Shuppan, pp 187–204 (in Japanese)
Science Council of Japan Future Earth Promotional Committee (2014) Future Earth. Research Institute for Humanity and Nature (in Japanese)

Chapter 23
Geo-information Sharing for Sustainable Development of Human Society

Eikichi Tsukuda, Shinji Takarada, Koki Iwao and Joel Bandibas

Abstract We believe that data sharing of visualized geological information of invisible earth subsurface is extremely important for the sustainable development of human society. Space and temporal coverage data for climate changes, geo-hazard mitigation, and evaluation of energy and natural resources are essential for future society. In this section, we would like to introduce some recent activities of Geoinformation sharing, mostly efforts by the Geological Survey of Japan in the Asia-Pacific region, that is, general information on Global Earth Observation (GEO), One Geology, G-EVER project, and CCOP Geoinformation Sharing Infrastructure.

Keywords Global earth observation · Interoperability · Data sharing · Geo-hazards · Disaster mitigation

23.1 Introduction

The 2011 off the Pacific coast of Magnitude 9.0 Tohoku Earthquake clearly showed that the Asia-Pacific region is an area with high risk of catastrophic natural disasters such as earthquakes and volcanic eruptions. Once a disaster occurs, in current highly globalized economy, it can create unpredictable turmoil all over the world, not only in the affected area. These global-scale disasters are crucial for the sustainable development of the global economy to ensure human security. It is not a fire along the other side of river bank side. Japanese islands are located along plate convergence boundaries among the Pacific plate, North American plate, Philippine Sea plate and Eurasia plates, therefore historically had been affected by gigantic earthquakes and volcanic eruptions. After the 1995 Hyogoken-Nanbu (Kobe) earthquake, extensive studies and deployment of geophysical observation networks have been done in Japan, such as nation-wide active fault surveys, development high-density seismograph networks and GNSS crustal deformation monitoring system. Our knowledge on earthquakes is improved dramatically and observation and study results are systematically trans-

E. Tsukuda (✉) · S. Takarada · K. Iwao · J. Bandibas
Geological Survey of Japan, National Institute of Advanced Industrial Science and Technology, Higashi 1-1-1, Tsukuba 305-8567, Japan
e-mail: e-tsukuda@aist.go.jp

Y. Himiyama et al. (eds.), *Human Geoscience*, Advances in Geological Science,
https://doi.org/10.1007/978-981-32-9224-6_23

ferred to the Japanese society. However, it is fair to say that our understanding of earthquakes is still limited and farther studies should be done for the mitigation of seismic disasters.

One of impediments to the development of geologic hazard studies is the infrequency of occurrence in a limited area. This makes it difficult to promote hypothesis and test processes. This may be solved by the international correlational studies because geological hazards are not less frequent phenomena any more. We strongly believe that it is very urgent issue to establish an effective international framework where we collaborate and develop a network system gathering information on disaster mitigation in Asia-Pacific region.

Data sharing and interoperability in the area of earth science which handles huge numbers of data set. We are facing a lot of difficulties in scientific area as well as social communications. We still need tremendous continuous efforts to make it visible and understandable to wide range of decision-makers from government officials and ordinary citizens. These activities accord with GEO, Group on Earth Observations (GEO Website 2019).

23.2 Global Earth Observation

Several international bodies related to earth science have been developing global and regional earth observation data and consolidating the system. For example, under the principle of "unrestricted disclosure" of non-military scientific data and research information, ICSU, the International Council for Science envisioned the WDC, World Data Center, during the Cold War era, and has been taken over by the WDS, World Data System, since 2008.

They envision to establish long-term data archive of valuable scientific data as human beings assets, provide free data under quality controlled, and ensure the interoperability.

GEO is a voluntary partnership of governments and organizations that envisions "a future wherein decisions and actions for the benefit of humankind are informed by coordinated, comprehensive and sustained Earth observations and information (GEO Website 2019)".

At the first Earth Observation Summit in 2003, ad hoc Earth Observation Working Group (ad hoc GEO) was established, in which the 10-year implementation plan has been started. At the 3rd Earth Observation Summit in 2005, 10 year implementation plan for GEOSS and GEO has been established. 9 Societal benefit areas, namely, disasters, health, energy, climate, water, weather, ecosystems, agriculture, and biodiversity, are the main target areas and global, regional data has been established for the areas and common platform to share the data and knowledge was constructed. GEO is now entering the second phase.

In the United Nations, at its substantive session in July 2011, ECOSOC considered the report of the Secretary General (E/2011/89) and adopted a resolution to create the United Nations Committee of Experts on Global Geospatial Information Management (UNCEGGIM Terms of Reference 2019). It is a first mechanism to discuss on Global Geospatial Information Management under the framework of UN. It could be a successor of International Steering Committee for Global Mapping (ISCGM), in cooperation with relevant organizations of respective countries and regions around the world (Global Map Data Archives 2019). They have been working for the Global Mapping Project, an effort to develop, provide and promote the use of geospatial information necessary for addressing global challenges, including environmental problems and disaster prevention, and advancing sustainable development of the world.

These three efforts (WDS, GEOSS, and UNGGIM) are taking advantage of each of the feats, which are complementary.

Each body has been consolidating a system to maintain and utilize the data which ensure the interoperability. For example, GEO established GCI GEOSS Common Infrastructure. Users are enabled to access the GEOSS data and services through GCI. In addition, GEO accomplished the GEOSS Data Sharing Principle as one of the early achievement (GEO Data Sharing Principle Website 2019).

It is composed of three key principles as follows:

- There will be full and open exchange of data, metadata, and products shared within GEOSS, recognizing relevant international instruments and national policies and legislation;
- All shared data, metadata, and products will be made available with minimum time delay and at minimum cost;
- All shared data, metadata and products being free of charge or no more than cost of reproduction will be encouraged for research and education.

Under the leadership of ICSU, Committee on Data for Science and Technology, CODATA, team of experts create the draft of white paper on an overview of the data policy, the implementation of the data sharing principles, and summarizing the guidelines.

This document is not legal documents but expected to be a common basis to discuss how to share data and information provided in GEOSS. Followed by the rule, LANDSAT data becomes free in October 2008. ASTER Global Digital Elevation Model also becomes free for the purpose of global environmental issues.

23.3 OneGeology

The mission of the OneGeology project is to make web-accessible the best available geological and other geoscience data worldwide at the best possible scales, starting with at least 1:1 million scale (OneGeology Website 2019). OneGeology is a Geological Survey initiative launched in the International Year of Planet Earth in 2007, which will make public and internet-accessible the best available digital geological map data and all forms of relevant geoscience digital data worldwide to better address the needs of society. Currently, more than 170 Geological Surveys in the world are involved in this project. The Geological Survey of Japan is the representative of the Asian region which includes the CCOP members countries. GSJ provides the server that hosts the Web Map Services of the geological maps of many countries in East and Southeast Asia which are registered to OneGeology. The WMSs of the 1:1M scale geological maps of Indonesia, Malaysia, Vietnam, Myanmar, Philippines, and Papua New Guinea are hosted by GSJ servers. Only the 1:1M geological maps of Laos, Thailand and South Korea are hosted by their own countries' servers. The OneGeology covering East Asia website is also available (OneGeology Asia Portal Site 2018).

23.4 G-EVER Earthquake and Volcanic Hazards Information Mapping and Assessment

23.4.1 The Asia-Pacific Region Earthquake and Volcanic Hazards Mapping Project

The Asia-Pacific Region Earthquake and Volcanic Eruption Risk Management (G-EVER) Consortium, consisting of geohazard research institutes in the Asia-Pacific region, was established in 2012 with the main objective of reducing the risk caused by earthquakes, tsunamis and volcanic eruptions worldwide (Takarada et al. 2014; G-EVER Website 2017). The G-EVER consortium published the Eastern Asia Earthquake and Volcanic Hazards Information Map in 2016 (Takarada et al. 2016; Fig. 23.1a, b). The online Asia-Pacific Region Earthquake and Volcanic Hazard Information System (G-EVER Asia-Pacific Region Earthquake and Volcanic Hazard Information System 2017) were made accessible to the public in 2015. Detailed data of earthquake source regions, active faults, tephra falls, caldera, ignimbrite (pyroclastic flow), and fatalities caused by earthquake and volcanic events are available in formats compatible with major GIS software.

The database of the Hazard Information System (Fig. 23.2) contains geospatial data about geological hazards which were used for the publication of the Map (Takarada ct al. 2016). Geological maps, active faults, earthquakes hypocenters, and source areas, fatalities of major earthquakes, tsunami hazards, distribution of volcanoes, calderas, pyroclastic falls and ignimbrites (large-scale pyroclastic flows) and fatalities of major volcanic events are available on the system. The geospatial contents of the system's database was recently updated using more detailed data. This includes datasets containing 267 earthquake source regions worldwide (>Magnitude 6). Active fault distribution in Eastern Asia was also compiled and updated using recent data from the Philippines and Thailand. Three tsunami-affected areas of the 1707 Hoei, 1868 Meiji-Sanriku, 1993 Hokkaido-Nanseioki earthquakes in Japan were also added. Around 236 earthquake fatality data covering East and Southeast Asia since 1850AD, classifying the cause of fatalities into building collapse, fire, tsunami, landslide, and other reasons, were compiled. The database also contains

(a)

(b)

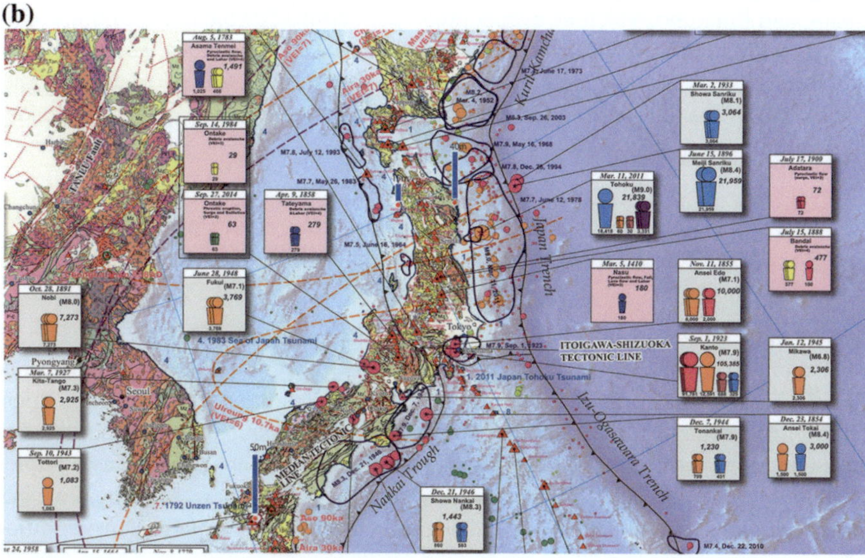

Fig. 23.1 **a** Eastern Asia earthquake and volcanic hazards information map (Geological Survey of Japan, AIST, Eastern Asia earthquake and volcanic hazards information map. http://creativecommons.org/licenses/by-nd/2.1/jp/). **b** A close-up view of the hazard information map showing Japan and its surrounding areas (Geological Survey of Japan, AIST, Eastern Asia earthquake, and volcanic hazards information map. http://creativecommons.org/licenses/by-nd/2.1/jp/)

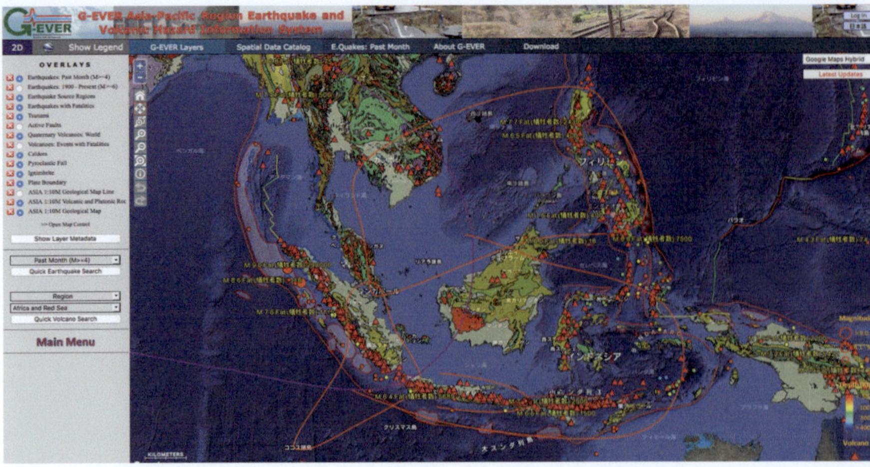

Fig. 23.2 G-EVER Asia-pacific region earthquake and volcanic hazard information system

22 calderas, 18 large-scale eruption ignimbrites and 24 tephra fall distributions with isopach maps covering East and Southeast Asia. The ignimbrites were the results of the major caldera-forming eruptions such as Toba, Tambora, Aso-4, Aira, Krakatau, Rabaul, and Pinatubo. Volcanic events involving about 150 volcanic eruptions since 1400AD with their fatality data are included. The causes of fatalities are identified as pyroclastic flow, debris avalanche, tephra fall, lahar, tsunami, volcanic gas, and other reasons. All of the hazard data can be viewed and most of them can be downloaded using the information system.

23.4.2 G-EVER Volcanic Hazard Assessment Support System

The Volcanic Hazards Assessment Support System (G-EVER Volcanic Hazard Assessment Support System 2017) was developed by G-EVER consortium since 2012 (Takarada 2017). It aims to provide a user-friendly, WebGIS-based, open-access online system for potential volcanic hazard assessment and risk mitigation of the Quaternary volcanoes in the world. The system does not require any complex installation procedure and Unix command operations. Web services using the Open Geospatial Consortium (OGC 2019) standards are used for spatial data rendition and processing were. The OGC-based web services used are the Web Map Service (WMS), Web Coverage Service (WCS) and Web Processing Service (WPS). The use of the system makes it easy to compare the simulation results with other maps. It has interface to easily search for volcanoes and integrate digital elevation model covering more than 2700 Quaternary volcanoes in the world. The system provides three simulations algorithms which are the Energy Cone (Fig. 23.3a), Titan2D (Fig. 23.3b) and Tephra2. These algorithms can estimate affected areas caused by pyroclastic flows,

(a)

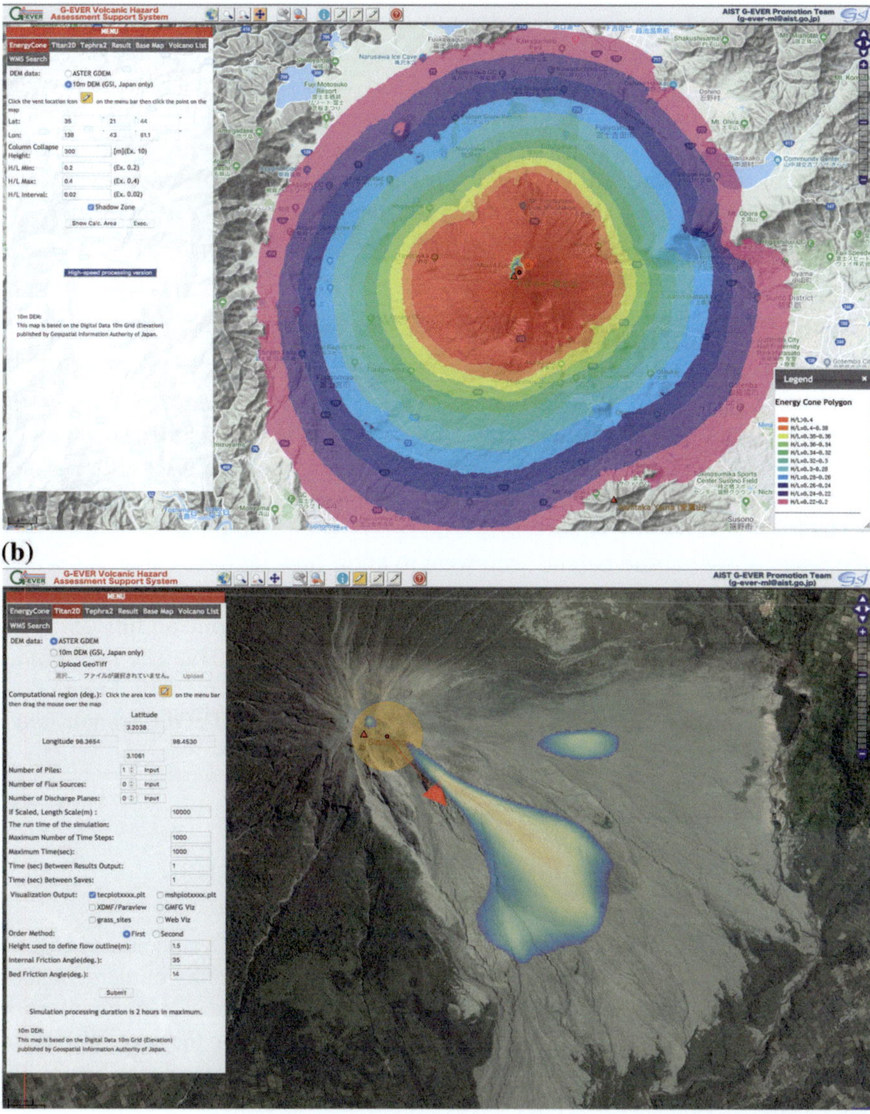

Fig. 23.3 **a** An energy-cone simulation result at Fuji Volcano, Japan. **b** A Titan2D simulation result at Sinabung Volcano, Indonesia

debris avalanche, and tephra falls. The system has an easy to understand graphical input interface and provides a data download capability (e.g., kml and shapefiles). The system was developed using eruption chronology, volcanic eruption databases, and numerical simulations, which are useful for volcanic hazard assessment. It can provide the users map estimating the area that would be affected by volcanic eruptions at any locations near the volcano using numerical simulations. The system can estimate volcanic hazards by overlaying the generated map of volcanic gravity flows or tephra falls on major roads, houses and evacuation areas using WebGIS.

23.4.3 PHIVOLCS FaultFinder Mobile Application

The PHIVOLCS FaultFinder mobile application was developed to determine the distance between the user's location or any selected location and the nearest active fault in the Philippines. The application is the product of the joint collaboration between the Philippine Institute of Volcanology and Seismology (PHIVOLCS) and the Geological Survey of Japan (GSJ). The app is developed to be easily used by a wide range of users from parents who want to know how far their children's school from the nearest active fault is to people who wanted to know if the house they want to buy is safe. PHIVOLCS officially introduced the app to the public on July 25, 2016. The App was published in Google Play Store on April 13, 2017. The FaultFinder App site is accessed by around 780,000 users as of January 16, 2019. Figure 23.4 shows one of the user interfaces of the mobile application (PHIVOLCS FaultFinder 2017).

23.5 CCOP Geoinformation Sharing Infrastructure

The CCOP Geoinformation Sharing Infrastructure for East and Southeast Asia (GSi) project (Fig. 23.5a) is one of the most important activities currently implemented by CCOP. The scope of the GSi project is to (1) compile various geoscientific information in the CCOP member countries and construct a database on the open Web using world standard formats and GIS, (2) promote digitization of geoscience data in the CCOP member countries at a high-quality level, and (3) establish a comprehensive Geoinformation database and infrastructure in Asia. The project aims to make various geoscientific information on the GSi system such as geological maps, geohazards, geophysical, mineral resources, geo-environment, groundwater, topographic maps, and remote sensing data internet-accessible.

The major goals of the GSi project are classified into four: (1) Geoinformation sharing, (2) delivery of geoscience knowledge to society, (3) international standardization, and (4) capacity building. Geoinformation sharing aims to enhance the collaboration and communication among the CCOP member countries, to establish the comprehensive database in East and Southeast Asia, to make data archive, to promote

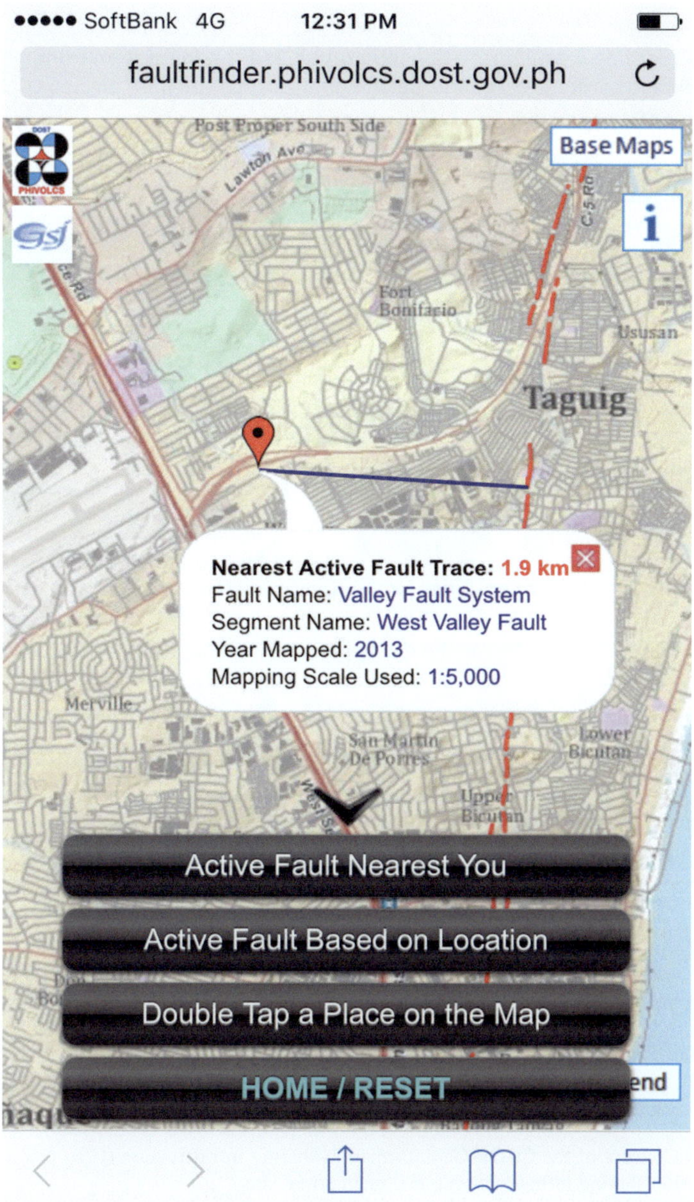

Fig. 23.4 The PHIVOLCS Faultfinder mobile application

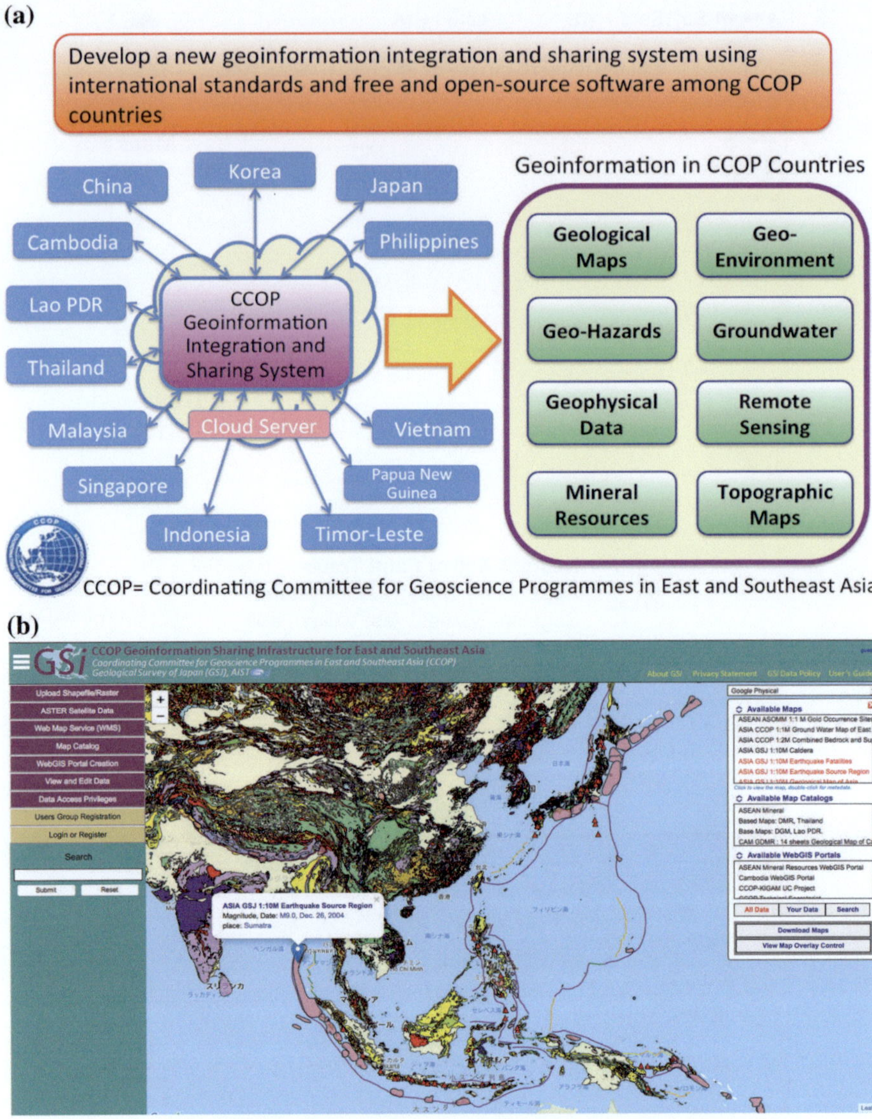

Fig. 23.5 **a** Concept of the CCOP Geoinformation sharing infrastructure (GSi) project. **b** CCOP Geoinformation sharing infrastructure (GSi) main portal site

digitization, and to provide data analysis tools on the web. Delivery of geoscience knowledge to society aims to make geoscience information relevant and useful for society, to increase user accessibility, to provide the information on hazard mitigation and geo-environment, to make geoscience data freely available and understandable to users, and to provide visualization tools and data for outreach programs. International standardization aims to promote inter-operability using OGC [Open Geospatial Consortium)-based standards and web services (i.e., Web Map Service (WMS), Web Processing Service (WPS), Web Feature Service (WFS), Web Coverage Service (WCS), and Web Map Tile Service (WMTS)], to use Free and Open Source Software (FOSS), to increase the transparency of geological survey activities, and to collaborate with other projects such as Onegeology. Capacity building aims to hasten the transfer of WebGIS and database technology to the countries that need them through training courses, workshops, and the publication of cookbooks and online training materials.

The GSi main portal site (Fig. 23.5b; CCOP Geoinformation Sharing Infrastructure System 2018) provides web-based functions for spatial data rendering and analysis using WMS and WPS, respectively. It can also be used to download data in several formats (shapefile, KML, PNG, PDF, and excel). The information system follows the standard model of the Spatial Data Infrastructure (SDI). It also provides the interface for the creation of a customized WebGIS portal for spatial data viewing and processing. There are more than 15 GSi-generated portal sites which include member country's sites, CCOP Groundwater, ASEAN Mineral Resources and OneGeology covering East Asia (Figs. 23.4 and 23.5). More than 570 maps are stored on the GSi system which includes the following: 1:10M-1:50k geological maps, 1:1M-1:200k seamless geological maps, 1:10k-1:50k geological map of volcanoes, 1:10k-1:50k hazard zoning maps (earthquake, liquefaction, tsunami, volcano, flood and landslide), 1:1M seismotectonic map, 1:50k coastal erosion map, 1:250k Quaternary geology map, 1:1M geochemical map, 1:1M magnetic anomaly map, 1:750k-1:1M groundwater map, 1:50k hot spring distribution map, 1:250k-1:1M mineral resources map, ASTER satellite data, 1:100k road map, and 1:50k city map. GSi's geospatial data can be rendered using GIS software and other WMS clients. Mobile version of GSi system is also available.

Currently, the GSi project has four cloud servers, which are the CCOP GSi, GSJ Geoinfo, Geological Agency of Indonesia, and MGB servers. The GSi system will also host and provide data for some applications and data analysis tools. Several mobile applications related to seismic and geological hazard mapping, field data capture and spatial data analysis are being developed.

Collaborating with other projects such as OneGeology, ASEAN Mineral Resources projects, and those implemented by CCOP such as Groundwater, CCS-M, 1:1M Harmonized Geological Map and KIGAM Unconventional Oil and Gas Resources Project is in progress.

The 3rd GSi International Workshop was held in Langkawi, Malaysia, from September 18–20, 2018. It was co-funded by the Department of Mineral and geoscience (JMG), Malaysia. Forty-three (43) participants from the CCOP member countries (Cambodia, Indonesia, Japan, Republic of Korea, Lao PDR, Malaysia, Myanmar, Papua New Guinea, Philippines, Thailand, and Vietnam) including the staff of JMG and CCOP TS attended the workshop. The GSi main portal and customized portal sites are officially opened to the public on the first day of the workshop.

References

CCOP Geoinformation Sharing Infrastructure System (2018) https://ccop-gsi.org/main/. Last accessed 20 Jan 2019

GEO Data Sharing Principles Implementation Website https://www.earthobservations.org/geoss_dsp.shtml. Last accessed 20 Jan 2019

GEO Website (2019) https://www.earthobservations.org/wigeo.php. Last accessed 20 Jan 2019

G-EVER Asia-Pacific Region Earthquake and Volcanic Hazards Information System (2017) http://ccop-geoinfo.org/G-EVER/. Last accessed 20 Jan 2019

G-EVER Volcanic Hazards Assessment Support System (2017) http://volcano.g-ever1.org/. Last accessed 20 Jan 2019

G-EVER Website (2017) http://g-ever.org/. Last accessed 20 Jan 2019

Global Map Data Archives (2019) https://globalmaps.github.io. Last accessed 20 Jan 2019

OneGeology Asian Region Portal Site (2018) https://ccop-gsi.org/gsi/onegeologyasia/. Last accessed 20 Jan 2019

OneGeology Website (2019) http://www.onegeology.org/. Last accessed 20 Jan 2019

Open Geospatial Consortium (2019) http://www.opengeospatial.org/. Last accessed 20 Jan 2019

PHIVOLCS FaultFinder (2017) http://faultfinder.phivolcs.dost.gov.ph/. Last accessed 20 Jan 2019

Takarada S (2017) The volcanic hazards assessment support system for the online hazard assessment and risk mitigation of quaternary volcanoes in the world. Front Earth Sci 5:102. https://doi.org/10.3389/feart.2017.00102

Takarada S, Bandibas JC, Ishikawa Y, G-EVER Promotion Team (2014) Global earthquake and volcanic eruption risk management activities, volcanic hazard assessment support system and Asia-Pacific region hazard mapping project in G-EVER. Episodes 37:321–328

Takarada S, Ishikawa Y, Maruyama T, Yoshimi M, Matsumoto D, Furukawa R, Teraoka Y, Bandibas JC, Kuwahara Y, Azuma T, Takada A, Okumura K, Koizumi N, Tsukuda E, Solidum RU, Daag AS, Cahulogan M, Hidayati S, Andreastuti S, Li X, Nguyen HP, Lin C-H (2016) Eastern Asia earthquake and volcanic hazards information map. Geological Survey of Japan, AIST, Japan

United Nations Committee of Experts on Global Geographic Information Management (UNCEGGIM) Terms of Reference (2019) http://ggim.un.org/meetings/2010-NY/documents/papers/UNCEGGIM-TOR-final.pdf. Last accessed 20 Jan 2019

Chapter 24
Global Sustainability Programmes and Human Geoscience

Yukio Himiyama

Abstract Human geoscience is deeply involved in global sustainability programmes, such as Future Earth, SDGs and ESD, as a discipline and as an academic community. This chapter discusses why and how human geoscience can contribute to these programmes, and argues that their mutual collaborations, as well as the enhancement of each, is the key for their success in achieving their common goal of the world sustainability, and that human geoscience can assist and promote such mutual collaborations as well as the progress of each programme.

Keywords Sustainability · Future earth · SDGs · ESD · Earth system

24.1 Introduction

Human geoscience is the geoscience of human sphere. In other words, it is the science about the natural phenomena that occur on the surface of the earth and their relations with human activities. There are many pressing issues on sustainability which *human geoscience* is concerned about, such as the use and management of water, land and other resources, global environmental problems including those related with climate change, land degradation, environmental pollution and population increase, and geo-disasters and their reduction including those related with earthquakes, tsunamis, volcanic activities, floodings, typhoons and land slides. They are all issues to be addressed in the three main global sustainability programmes, namely FE (Future Earth, 2015–present), SDGs (Sustainable Development Goals, 2015–present), and ESD (Education for Sustainable Development, 2005–present). Figure 24.1 shows the three sustainability programmes and their relations with the United Nations (UN), UNESCO and International Science Council (ISC). The consolidation of the links between the three programmes, i.e. FE-SDGs, ESD-SDGs and FE-ESD, is no less important for the world sustainability than the enhancement of each programme. It is noted that the strong relation expressed by double lines does not necessarily mean a strong cooperation.

Y. Himiyama (✉)
Hokkaido University of Education, Sapporo, Japan
e-mail: himiyamay@kkd.biglobe.ne.jp

Y. Himiyama et al. (eds.), *Human Geoscience*, Advances in Geological Science,
https://doi.org/10.1007/978-981-32-9224-6_24

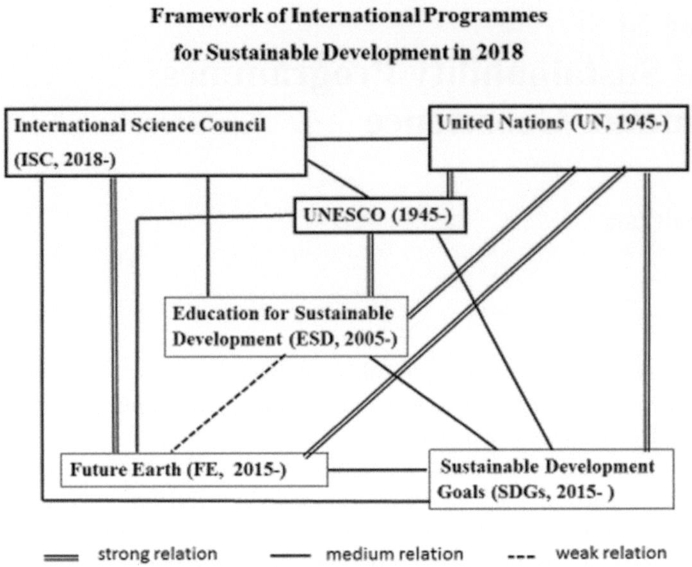

Fig. 24.1 Framework of international programmes for sustainable development in 2019

Human geoscience, both as a discipline and as an academic community, has strength in its wide coverage of spatial scales (human to global) and temporal scales (hourly to geological), interdisciplinarity (human behaviour to earth system dynamics), field observations (field surveys, mapping, monitoring), data analyses (spatial, local, regional, global) and in education of geosciences and sustainability. Therefore, it is related not only with a wide range of themes of the global sustainability programmes, but also with the linkages among them which may otherwise be overlooked or poorly addressed. This Chapter discusses why and how *human geoscience* contributes to them.

24.2 Future Earth

Future Earth is a 10-year international research programme launched in June 2012 at the UN Conference on Sustainable Development (Rio+20). It is a new comprehensive framework and integrative research platform for sustainability, and was established on the achievements and experiences of previous global environmental change programmes, such as IGBP (International Geosphere-Biosphere Programme, 1987–2015), IHDP (International Human Dimensions Programme, 1990–2014), DIVERSITAS (1991–2014) and WCRP (World Climate Research Programme, 1980–). *Future Earth Initial Design: report of the transition team* (Future Earth 2013) says that Future Earth would provide critical knowledge required for societies to face the challenges posed by global environmental change and to identify opportunities for a transition to global sustainability.

Future Earth made its full start in 2015, and as of late 2018 there are over twenty Global Research Projects (GRP) joining Future Earth mainly from the former IGBP, IHDP and DIVERSITAS in order to continue and develop their missions in more interdisciplinary and transdisciplinary contexts, and eleven Knowledge-Action-Networks (KAN) on the following topics: Water-Energy-Food Nexus; Ocean; Transformations; Natural Assets; Sustainable Development Goals; Urban; Health; Finance and Economics; Systems of Sustainable Consumption and Production; Decarbonisation; Emergent Risks and Extreme Events. There is high expectation on the role of the KANs in implementing the SDGs. The SDGs-KAN, in particular, is intended to facilitate synthesis and integral aspects of the SDGs framework (Schmalzbauer and Visbeck 2016). The inclusion of *Emergent Risks and Extreme Events* KAN is noted as an important step forward for Future Earth, as disasters used to be somewhat put aside. *Human geoscience* certainly welcomes it.

Future Earth held a global summit meeting in August 2018, and the report of the meeting, namely *Future Earth Global Summit—meeting summary report* (Future Earth 2018), indicates that the idea of *Global Systemic Challenges* has been put forward as an effort to consolidate the Future Earth community around a few flagship initiatives on which GRPs, KANs, national and regional entities of Future Earth and partners can work together. J. Rockström, Co-chair of the Future Earth Advisory Committee, says that the first of them to be developed is not only about quantifying the scientific targets for sustainable earth systems, but also about operationalizing and understanding transformations and social tipping point, and that Future Earth community needs to go further and identify two to three *Global Systemic Challenges* with bottom-up initiatives highlighting the idea that there are some challenges that cannot be accomplished on their own.

It is reminded that *Future Earth Initial Design* (Future Earth 2013) devoted one chapter to the education and capacity building strategy, saying that education and capacity building are core capabilities that earth needs to nurture through partnerships. Regrettably, education has been somewhat left out from the mainstream of Future Earth, while the ICSU (International Council for Science) and its successor ISC (International Science Council) and many other international academic organizations are increasingly committed to education. As mentioned in Chap. 1 of the present book, the *human geoscience* roadmap issued from Science Council of Japan in 2014 says that the main mission of the discipline is to grasp the situation and problems of human sphere and find ways to solve the problems, and to enhance research and education of *human geoscience*, ensuring its contribution to the world sustainability by working closely with Future Earth. So, how about making *education for sustainable development* the core of *Global Systemic Challenge*?

24.3 SDGs

The United Nations (UN) Sustainable Development Summit held in September 2015 in New York adopted the *Sustainable Development Goals* (SDGs) as the core of the *2030 Agenda for Sustainable Development*. The SDGs, which set 17 thought-out Goals, are becoming a major driving device to achieve the sustainable world. They have their roots in *Agenda 21*, a comprehensive plan for sustainable development adopted at the Earth Summit in Rio de Janeiro in 1992, and in Millennium Development Goals (MDGs) to reduce extreme poverty by 2015 built on the Millennium Declaration adopted at the Millennium Summit in September 2000.

The SDGs are a product of long-time work of countries, the UN, scientists and others (Griggs et al. 2013). The annual High-level Political Forum on Sustainable Development serves as the central UN platform for the follow-on and review of the SDGs, and the Division for Sustainable Development Goals (DSDG) in the UN Department of Economic and Social Affairs (UNDESA) provides substantive support and capacity-building for the SDGs and their related thematic issues (UN 2018).

Many world scientific communities responded positively to the launch of the SDGs. ICSU, which was then the largest and most influential international academic body, published *A Guide to SDG Interactions: from Science to Implementation* (ICSU 2017) in order to promote and guide research on the interactions between the 17 Goals of SDGs by showing some preliminary findings of the pilot studies on Food/Agriculture (SDG2), Health (SDG3), Energy (SDG7) and Oceans (SDG14).

Implementation of the SDGs requires relevant regional/spatial data and information as well as scientific and administrative supports from all over the world. United Nations Environment Programme (UNEP) publishes *Global Environment Outlook* (GEO) every five or so years. As discussed in Chaps. 1 and 6 of the present book, it is one of the most credited periodic reports of the environmental state of the whole world, and the issues covered by the book largely coincide those of *human geoscience*.

What is notable with the latest GEO, i.e. *GEO-6: Global Environment Outlook: All Regional Assessments* (UNEP 2016) in SDGs context is that for the first time in the GEO series it is mostly made up with the regional assessments of the six UNEP regions, namely Africa, Asia and the Pacific, Europe, Latin America and the Caribbean, North America, and West Asia, whereas the previous GEOs put stronger emphasis on global, rather than regional, assessments on cross-cutting subjects, such as atmosphere, land, water, and biodiversity. This shift of structure of GEO is significant in view of the SDGs, as countries trying to achieve their SDGs goals might wish to know more about the regional detail and priority than the global.

The first priority issue of each UNEP region shown in the world map attached to GEO-6 is as follows:

- Africa: land management and use
- Asia Pacific: increasing vulnerability
- Europe: low-carbon resource-efficient societies, smart cities, improved health and adaptation to climate change

- Latin America and the Caribbean: sustainable management of biological resources
- North America: reducing GHG emissions, energy transition, and city-level innovation
- West Asia: peace, security, and the environment.

This summary clearly indicates that the first priority issue of each region varies greatly from region to region. For example, disasters are the primary concern in Asia Pacific region, while in the other regions, particularly in Europe and North America, they are not. It may explain the relatively weak interest in disasters and land use among the people in Europe or North America, including the specialists of the sustainability issues. It is also noted that disasters are in fact often related with some global, regional, or local environmental problems directly and/or indirectly, while the specialists of disasters and those of environmental issues tend to work separately from each other, particularly in the regions other than Asia Pacific.

Human geoscience can be a platform where these two groups of specialists, together with the others concerned, gather and work together. One negative aspect of highlighting the first priority issue of each region in *GEO-6* is that it may obscure the other important issues in each region. Another drawback of *GEO-6* is lack of global analyses and insights, which have given way to the regional ones. Future Earth, which intends to contribute to implementation of the SDGs, is desired to promote research to fill such gaps, and *human geoscience* with its wide coverage of geo disasters and global/regional/local environmental issues can be a reliable partner.

24.4 ESD

The UN General Assembly at its 57th Session held in December 2002 adopted resolution on *UN Decade of Education for Sustainable Development*, and proclaimed the years 2005–2014 to be the *UN Decade of Education for Sustainable Development* (DESD) (UN 2002). It can be seen as an effort to implement Chap. 36 of *Agenda 21* on *promotion of education, public awareness and training* adopted at the UN Conference on Environment and Development held in Rio de Janeiro in 1992. It confirmed the importance of *education for sustainable development*, and designated UNESCO as the lead agency for the promotion of the Decade. UNESCO is in charge of management, coordination and implementation of the Global Action Programme (GAP) on the ESD.

The 2014 UNESCO World Conference on ESD held in Nagoya, Japan, marked the end of the DESD, facilitating the launch of the Aichi-Nagoya Declaration on ESD and the roadmap for the implementation of the GAP on ESD. UNESCO Associated Schools are a global network of more than 10,000 educational institutions in 181 countries as of late 2018, and they are playing important roles in the promotion of ESD (UNESCO 2018). UNESCO considers the ESD as a key instrument to achieve the SDGs, and provides guidelines and ample information at its websites, such as *Education for Sustainable Development Goals: Learning Objectives* on how ESD can support each of the 17 SDGs (UNESCO 2017). On the other hand, SDG's Goal4 on education Target 4.7 refers to ESD as follows:

> By 2030, ensure that all learners acquire the knowledge and skills needed to promote sustainable development, including among others, through *education for sustainable development* and sustainable lifestyles, human rights, gender equality, promotion of a culture of peace and non-violence, global citizenship and appreciation of cultural diversity and culture's contribution to sustainable development.

ESD can be enriched and consolidated more by strengthening some of its key subjects such as geoscience in general and *human geoscience* in particular. It is very important in grasping, avoiding and overcoming various geo disasters and environmental problems, as seen in the previous chapters of the present book. However, geoscience education, in general, hasn't been as successful as desired worldwide. Multi-scale views of *human geoscience* are important in dealing with global, regional and local problems or activities, as they are vital for the improvement of environmental education, which has been under persistent challenge of the division of the globalists and localists (Himiyama 2010).

Future Earth, which stressed the importance of *education for sustainable development* in its initial design can enhance cooperation with the ESD by mobilizing the many academic communities under its influence, including geoscience communities. An encouraging sign for geoscience education in the recent years is that in the geoscience conferences, poster presentations by high school students are becoming a common scene internationally. *Human geoscience* can contribute to the ESD not only by improving and enhancing geoscience education, but also by linking it with the large geoscinece community. For more detailed discussion on cooperation of Future Earth and ESD, see Chap. 22.

24.5 Human Geoscience for the World Sustainability

Global sustainability programmes represented by Future Earth, SDGs and ESD need to be promoted vigorously in order to achieve the world sustainability without delay. They need strong support and participation from the wide range of academic/educational communities, national and local governments and the societies. *Human geoscience* will increase its contribution to each of these sustainability programmes both as a discipline and as an academic community.

The close collaborations of the three sustainability programmes, as well as the enhancement of each programme, will be the key for their success because they are expected to have substantial qualitative as well as quantitative synergetic effects. *Human geoscience* can assist such collaborations as well as the progress of each programme. It can also serve as a platform where individuals cultivate their skills pertaining to it.

Future Earth and ESD are both cooperating with SDGs, and FE-SDGs and ESD-SDGs linkages are to be developed further. However, the cooperation of Future Earth and ESD remains to be seen. The relation of Future Earth and ESD is similar to that of science and education. By working closely, science and education can enhance their understanding, learn, help, and encourage each other (Himiyama 2002). It is particularly important in the case of sustainability science and education.

Future Earth will be able to activate not only itself, but also the whole triangular system of the three sustainability programmes shown in Fig. 24.1, if it consolidates its relation with the ISC and its member unions and organizations. One of such unions, namely International Geographical Union (IGU), which proclaims geography to be the science for sustainability, established a new commission for Future Earth in 2016 in order to support Future Earth from the geographer's perspective. Many others will follow, if Future Earth shows its leadership and commitment to science and education for sustainability. *Human geoscience* will be a useful and reliable supporter of such initiative.

References

Future Earth (2013) Future Earth initial design: report of the transition team. International Council for Science (ICSU), Paris. http://futureearth.org/sites/default/files/Future-Earth-Design-Report_web.pdf

Future Earth (2018) Future earth global summit—meeting summary report. http://futureearth.org/sites/default/files/files/FE_Summit_report_V8%20(2).pdf

Griggs D, Stafford-Smith M, Gaffney O, Rockström J, Öhman MC, Shyamsundar P, Steffen W, Glaser G, Kanie N, Noble I (2013) Sustainable development goals for people and planet. Nature 495(7441):305–307

Himiyama Y (2002) Regional sustainable development review: Japan. In: Knowledge for sustainable development—an insight into the encyclopedia of life support systems, vol 1. UNESCO/EOLSS Publishers, Paris and Oxford, pp 1067–1086

Himiyama Y (2010) Glocal environmental education: a prescription for tomorrow. In: Himiyama Y, Singh RB, Kanda F, Hindson J (eds) Glocal environmental education. Rawat Publications, Jaipur, India, pp 29–45

International Council for Science (ICSU) (2017) A guide to SDG interactions: from science to implementation, In: Griggs DJ, Nilsson M, Stevance A, McCollum D (eds). International Council for Science, Paris/Kiel

Schmalzbauer B, Visbeck M (eds) (2016) The contribution of science in implementing the sustainable development goals. German Committee Future Earth, Stuttgart/Kiel

UNEP (2016) GEO-6: Global environment outlook: all regional assessments. UNEP, Nairobi, Kenya. https://europa.eu/capacity4dev/unep/blog/geo-6-global-environment-outlook-all-regional-assessments

UNESCO (2017) Education for sustainable development goals—learning objectives. http://unesdoc.unesco.org/images/0024/002474/247444e.pdf

UNESCO (2018) Education for sustainable development. https://en.unesco.org/themes/education-sustainable-development

United Nations (1992) Agenda 21. https://sustainabledevelopment.un.org/outcomedocuments/agenda21

United Nations (2002) United Nations decade of education for sustainable development. http://www.un-documents.net/a57r254.htm

United Nations (2018) Sustainable development goals knowledge platform. https://sustainabledevelopment.un.org/sdgs

MIX
Papier aus verantwortungsvollen Quellen
Paper from responsible sources
FSC® C105338

If you have any concerns about our products,
you can contact us on
ProductSafety@springernature.com

In case Publisher is established outside the EU,
the EU authorized representative is:
Springer Nature Customer Service Center GmbH
Europaplatz 3, 69115 Heidelberg, Germany

Printed by Libri Plureos GmbH
in Hamburg, Germany